Molecular Biology of Free Radical Scavenging Systems

SERIES EDITORS
John Inglis and Jan A. Witkowski
Cold Spring Harbor Laboratory

CURRENT COMMUNICATIONS
In Cell & Molecular Biology

---

1   *Electrophoresis of Large DNA Molecules: Theory and Applications*
2   *Cellular and Molecular Aspects of Fiber Carcinogenesis*
3   *Apoptosis: The Molecular Basis of Cell Death*
4   *Animal Applications of Research in Mammalian Development*
5   *Molecular Biology of Free Radical Scavenging Systems*

Forthcoming
6   *Molecular Immunobiology of Lyme Disease*

CURRENT COMMUNICATIONS 5
In Cell & Molecular Biology

Molecular Biology of Free Radical Scavenging Systems

Edited by
John G. Scandalios
North Carolina State University

Cold Spring Harbor Laboratory Press 1992

CURRENT COMMUNICATIONS 5
In Cell & Molecular Biology
Molecular Biology of Free Radical Scavenging Systems

Copyright 1992 Cold Spring Harbor Laboratory Press
All rights reserved
Printed in the United States of America
Cover design by Leon Bolognese & Associates Inc.

Front Cover: The three-dimensional structure of manganese superoxide dismutase from the thermophilic bacterium, *Thermus thermophilus*, was determined by Bill Stallings, Kitty Pattridge, and Martha Ludwig in collaboration with Jim Fee at the University of Michigan in the Biophysics Research Division. The model has been refined at 1.8 Å resolution by crystallographic techniques. This enzyme is a tetramer of identical subunits with a manganese cofactor (filled spheres) at the four active sites. The polypeptide chain fold of each subunit also resembles that of iron superoxide dismutases. The subunit chains of iron- and manganese-dependent superoxide dismutases from other species aggregate as dimers or tetramers, conserving only the top-bottom subunit interfaces, which create channels leading to the catalytic centers. Anita Metzger (University of Michigan) generated this image.

Back cover: Cat1 genomic map showing introns, exons, and 5' and 3' non-coding regions. The 2.5-kb promoter region is to the left (see Scandalios, this volume).

Library of Congress Cataloging-in-Publication Data

Molecular biology of free radical scavenging systems / edited by John G. Scandalios.
    p. cm. -- (Current communications in cell & molecular biology ; 5)
  Includes bibliographical references and index.
  ISBN 0-87969-409-2
  1. Active oxygen--Pathophysiology. 2. Antioxidants. 3. Free radicals (Chemistry) I. Scandalios, John G. II. Series.
RB170.M65 1992
574.87'65--dc20   92-2731
                                                       CIP

The articles published in this book have not been peer-reviewed. They express their authors' views, which are not necessarily endorsed by Cold Spring Harbor Laboratory.

Authorization to photocopy items for internal or personal use, or the internal or personal use of specific clients, is granted by Cold Spring Harbor Laboratory Press for libraries and other users registered with the Copyright Clearance Center (CCC) Transactional Reporting Service, provided that the base fee of $3.00 per article is paid directly to CCC, 27 Congress St., Salem, MA 01970. [0-87969-409-2/92 $3.00 + .00]. This consent does not extend to other kinds of copying, such as copying for general distribution, for advertising or promotional purposes, for creating new collective works, or for resale.

All Cold Spring Harbor Laboratory Press publications may be ordered directly from Cold Spring Harbor Laboratory Press, 10 Skyline Drive, Plainview, New York 11803. Phone: 1-800-843-4388. In New York (516) 349-1930. FAX: (516) 349-1946.

Contents

Preface, vii

DNA Damage by Endogenous Oxidants and Mitogenesis as Causes of Aging and Cancer   1
B.N. Ames and M.K. Shigenaga

Small Molecule Antioxidant Defenses in Human Extracellular Fluids   23
B. Frei, R. Stocker, and B.N. Ames

DNA Damage by Oxygen-derived Species: Its Mechanism, and Measurement Using Chromatographic Methods   47
B. Halliwell and O.I. Aruoma

Possible Protective Mechanisms of Tumor Necrosis Factors against Oxidative Stress   69
G.H.W. Wong, A. Kamb, L.A. Tartaglia, and D.V. Goeddel

Regulation of Bacterial Catalase Synthesis   97
P.C. Loewen

Regulation of the Antioxidant Defense Genes Cat and Sod of Maize   117
J.G. Scandalios

Regulation of Yeast Catalase Genes   153
H. Ruis and B. Hamilton

Production and Scavenging of Active Oxygen in Chloroplasts   173
K. Asada

Iron and Manganese Superoxide Dismutases: Catalytic Inferences from the Structures   193
W.C. Stallings, C. Bull, J.A. Fee, M.S. Lah, and M.L. Ludwig

Superoxide Radical in Escherichia coli   213
S.I. Liochev and I. Fridovich

Regulation and Protective Role of the Microbial Superoxide Dismutases 231
D. Touati

The Human CuZn Superoxide Dismutase Gene and Down's Syndrome 263
Y. Groner, O. Elroy-Stein, K.B. Avraham, M. Schickler, H. Knobler, D. Minc-Golomb, O. Bar-Peled, R. Yarom, and S. Rotshenker

Index 281

Preface

Molecular oxygen presents a paradox to aerobic organisms: it is both essential and potentially toxic. Although unreactive in its ground state, oxygen is reduced to water under normal metabolic conditions, via a stepwise pathway during which partially reduced, very reactive intermediates are produced. These reactive oxygen species include the superoxide radical ($\cdot O_2^-$), hydrogen peroxide (H_2O_2), and the most potent oxidant, the hydroxyl radical ($\cdot OH$). Such active oxygen intermediates are also generated in living cells exposed to a great variety of environmental insults including radiation; air pollutants such as ozone, SO_2, and acid rain; herbicides; and various other redox active compounds and environmental stressors.

Increased levels of active oxygen species or free radicals create a situation known as oxidative stress, which leads to a variety of biochemical and physiological lesions often resulting in metabolic impairment and cell death. These highly reactive oxygen intermediates can readily react with various biological macromolecules such as DNA, proteins, and lipids to cause mutations, peroxidation of membrane lipids, and protein destruction. These lesions in turn lead to various diseases and degenerative processes such as aging, carcinogenesis, and immunodeficiencies in animals, and membrane leakage, senescence, chlorophyll destruction, and decreased photosynthesis in plants.

To cope with oxidative stress, aerobic organisms evolved protective antioxidant defenses, both enzymatic and nonenzymatic. However, aside from numerous correlative responses (i.e., increases in oxidative stress that lead to increased levels of some antioxidant defenses), little is known of the underlying molecular mechanisms by which the genome perceives oxidative insult and mobilizes a response to it. Such information is interesting in and of itself but is also essential in any future attempts to raise tolerance to environmental oxidative stress in organisms and reduce cellular damage by active oxygen. During the past several years, interest in the role of free radicals and other active oxygen intermediates in biological systems has grown, producing numerous international conferences on the chemical and clinical aspects of active oxygen species.

Lately, the techniques of molecular biology have been applied to the study of antioxidant defenses in both prokaryotes and eukaryotes. Many of the genes for antioxidant enzymes have been isolated, cloned, and characterized from diverse organisms. Currently, many laboratories are studying the regulation and expression of such genes and attempting to engineer organisms for increased tolerance to oxidative stress.

Aspects of oxygen free radical research were discussed in November 1990 at one of the prestigious Banbury Center Conferences at Cold Spring Harbor Laboratory. This book is not a summary of its proceedings, but many of the authors attended the meeting and benefited from stimulating discussions among a small group of active investigators primarily concerned with the molecular mechanisms by which cells cope with oxidative stress. The great success of the meeting encouraged me to accept the Series Editors' invitation to compile this book. The invited chapters clearly reflect the current state of the field and indicate future research in this very important and exciting area of modern biology.

It is with great pleasure that I thank Jan Witkowski, Director, and other staff of the Banbury Center, whose enthusiasm and efforts made the conference possible, and the many corporate sponsors who supported it. The excellent work and editorial expertise of John Inglis, Director, and other staff of the Cold Spring Harbor Laboratory Press in preparing this volume, especially Patricia Barker and Joan Ebert, are gratefully acknowledged. Finally, I wish to express deep appreciation to Katya Davey, hostess of Robertson House, for her most gracious and warm hospitality. For those of us who started our careers at Cold Spring Harbor, Katya provides continuity, warmth, and inspiration to keep returning.

J.G.S.

Special Support

The meeting at the Banbury Center on which this book is based was supported by funding from:

- Abbott Laboratories
- CNS Research
- Metropolitan Life Foundation
- The Ohio State University
- Ross Laboratories

Corporate Sponsors

The meetings' program at Cold Spring Harbor Laboratory is supported by:

- Alafi Capital Company
- American Cyanamid Company
- AMGen Inc.
- Applied Biosystems, Inc.
- Becton Dickinson and Company
- Boehringer Mannheim Corporation
- Bristol-Myers Squibb Company
- Ciba-Geigy Corporation/Ciba-Geigy Limited
- Diagnostic Products Corporation
- E.I. du Pont de Nemours & Company
- Eastman Kodak Company
- Genentech, Inc.
- Genetics Institute
- Hoffmann-La Roche Inc.
- Johnson & Johnson
- Kyowa Hakko Kogyo Co., Ltd.
- Life Technologies, Inc.
- Eli Lilly and Company
- Millipore Corporation
- Monsanto Company
- Pall Corporation
- Perkin-Elmer Cetus Instruments
- Pfizer Inc.
- Pharmacia Inc.
- Schering-Plough Corporation
- SmithKline Beecham Pharmaceuticals
- The Wellcome Research Laboratories, Burroughs Wellcome Co.
- Wyeth-Ayerst Research

ducts, and a repair system for apurinic sites that are produced by spontaneous depurination (Lindahl 1982). The measurement of DNA adducts by new methods shows that DNA damage produced by oxidation (see below) could be the most significant endogenous damage.

OXIDATIVE DNA DAMAGE

Oxidants are produced as by-products of normal metabolism and of lipid peroxidation (Figs. 1 and 2). Nonspecific DNA repair enzymes excise DNA adducts to release deoxynucleotides, and specific DNA repair glycosylases release free bases. Deoxynucleotides are enzymatically hydrolyzed to deoxynucleosides that are not usually further metabolized, and both these and the free bases may be recovered in the urine. Two products of oxidative damage to DNA are thymine glycol and 5-hydroxymethyluracil. A specific DNA repair enzyme, a DNA glycosylase in mouse cells, repairs 5-hydroxymethyluracil and differs from the specific DNA glycosylase repair enzyme for thymine glycol in mouse cells (Hollstein et al. 1984). The existence of these *specific* repair enzymes points to the importance of this type of DNA damage in vivo.

The postulated importance of endogenously produced oxidative damage to DNA in aging and age-related degenerative

$$O_2 \xrightarrow{e^-} O_2^- \xrightarrow{e^-} H_2O_2 \xrightarrow{e^-} \cdot OH \xrightarrow{e^-} H_2O$$

$$O_2 \xrightarrow[\text{LIGHT}]{\text{DYE}} {}^1O_2$$

FIGURE 1 Oxidants from normal metabolism. The formation of superoxide, hydrogen peroxide, and hydroxyl radicals by successive additions of electrons to oxygen. Cytochrome oxidase adds four electrons fairly efficiently during energy generation in mitochondria, but some of these toxic intermediates are inevitable by-products. The same oxidants are produced in copious quantities from phagocytic cells. Singlet oxygen is generated from oxygen by the absorption of energy from a dye activated by light.

Lipid Peroxidation

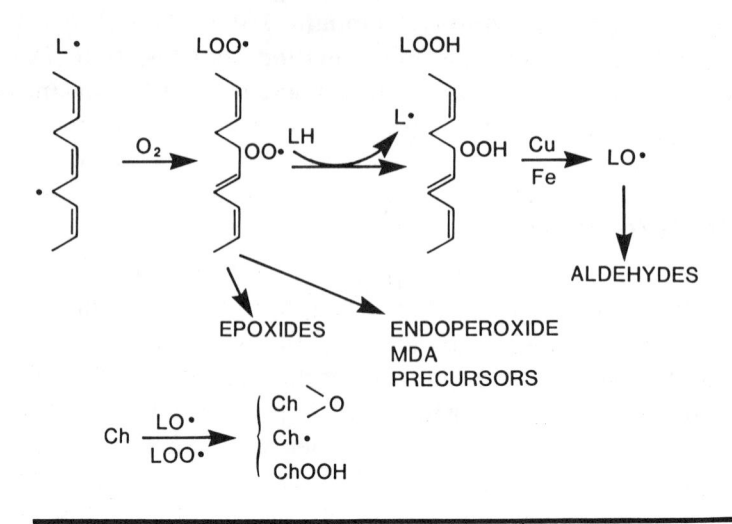

Mutagens/Carcinogens

FIGURE 2 Lipid peroxidation (Ames 1988). (L) Lipid radical; (LΣ) alkoxy lipid radical; (LOOΣ) hydroperoxy lipid radical; (LOOH) lipid hydroperoxide; (Ch) cholesterol; (Ch>O) cholesterol epoxide; (L>O) lipid epoxide; (MDA) malondialdehyde.

pathologies such as cancer has prompted efforts to develop rapid methods that measure this damage (Cathcart et al. 1984; Adelman et al. 1988; Richter et al. 1988; Ames 1989; Fraga et al. 1990). Endogenously produced oxidative damage to DNA has been assayed by measuring the urinary levels of the known radiation damage products thymine glycol, thymidine glycol, hydroxymethyluracil, and hydroxymethyl-deoxyuridine by high performance liquid chromatography (HPLC) with UV detection (Cathcart et al. 1984; Saul et al. 1987; Adelman et al. 1988). Our results indicate that normal humans excrete a total of about 100 nmole per day of the first three compounds. We have considerable evidence that most of this total is derived from repair of oxidized DNA, rather than from alternative sources, such as diet or bacterial flora (Cathcart et al. 1984; Saul et al. 1987; Ames and Saul 1988). This 100 nmole may therefore represent an average of about 10^3 oxidized thymine residues per day for each of the body's 6×10^{13} cells. Because these products are only 3 of more than 20 products of oxidative damage of DNA (Cadet and Berger 1985; von Sonntag 1987), the total number of all types of oxidative hits to DNA per cell per day may be about 10^4 in man and about 10^5 in the rat.

A more easily assayed product of oxidative DNA damage is 8-hydroxydeoxyguanosine (oh^8dG), which can be measured with high sensitivity by HPLC with electrochemical detection (EC) (Floyd et al. 1986). oh^8dG is a mutagen (Wood et al. 1990; Shibutani et al. 1991) formed in DNA by gamma irradiation (Dizdaroglu 1985) and various carcinogens (Kasai et al. 1987; Fiala et al. 1989; Rosier and VanPeteghem 1989). Polyclonal and monoclonal antibodies that recognize oh^8dG have been produced, and their binding properties have been characterized (Degan et al. 1991; Park et al. 1992). An immunoaffinity column facilitates the isolation of oh^8dG, oh^8Gua, and oh^8G from urine (Degan et al. 1991; Park et al. 1992) (Fig. 3). We have now produced monoclonal antibodies to oh^8dG that are even more effective. The steady-state level of oh^8dG lesions in liver DNA from year-old Fischer 344 rats is 64,000 residues per cell or about 1 per 224,000 bases; about 4,000 molecules of oh^8dG and oh^8Gua are excreted per cell per day (Fraga et al. 1990). Taking into account that this lesion is only one of more than 20 known gamma-irradiation-induced DNA damage products (von Sonntag 1987), we estimate that each rat cell contains approximately 10^6 oxidatively damaged bases in its DNA (Richter et al. 1988; Fraga et al. 1990) and that about 10^5 oxidative hits to the DNA occur per rat cell per day (Cathcart et al. 1984; Fraga et al. 1990). This estimate based on oh^8dG is in agreement with the earlier estimate based on thymidine glycol. The repair rate almost equals the damage rate; however, we estimate in DNA isolated from kidney of young and old rats that 80 oh^8dG residues accumulate per cell per day (Fraga et al. 1990).

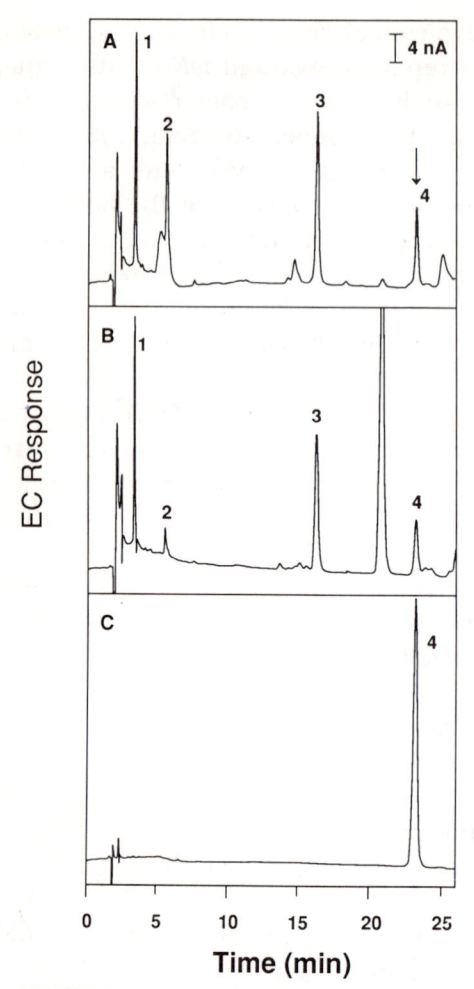

FIGURE 3 Gradient HPLC-EC chromatograms of urine samples processed by C18/OH solid phase extraction and the anti-oh[8]dG immunoaffinity columns (Degan et al. 1991). (A) Rat urine; (B) human urine; (C) oh[8]dG standard (5 pmole). The amounts of urine analyzed were equivalent to 0.26 and 0.34 ml of urine for the rat and human samples, respectively. The arrow denotes the retention time of oh[8]dG (peak 4). Peaks 1, 2, and 3 correspond to the retention times for uric acid, oh[8]Gua, and oh[8]G, respectively.

The oxidative DNA damage rate as measured by thymidine glycol excretion in urine for mouse, rat, monkey, and man is related to the metabolic rate and is inversely related to the age-specific cancer rate and life span (Adelman et al. 1988). We have found a similar relationship with oh^{8}dG excretion (Shigenaga et al. 1989).

DEFENSES AGAINST OXIDANTS

Many defense mechanisms within the organism have evolved to limit the levels of reactive oxidants and the damage they induce. Among the defenses are enzymes such as SOD, catalase, and glutathione peroxidase, as well as the dietary antioxidants β-carotene, α-tocopherols, and vitamin C. We have been particularly interested in antioxidant micronutrients because their levels in humans can be altered, and we have discussed five previously unappreciated antioxidants (Frei and Ames, this volume).

A study of the effect of seminal fluid ascorbic acid on the levels of oh^{8}dG in sperm DNA (Fraga et al. 1991) showed the following: (1) In individuals whose dietary intake of ascorbate was decreased from 250 to 5 mg per day, the levels of seminal fluid ascorbate levels were halved, and the levels of oh^{8}dG in sperm DNA were doubled. (2) Repletion of dietary ascorbate led to a decrease in the steady-state levels of oh^{8}dG. In another group of 25 subjects, the steady-state levels of oh^{8}dG correlate inversely with the content of seminal plasma ascorbate. These results indicate both that human sperm DNA is subjected to endogenous oxidative damage and that ascorbate appears to protect against such oxidative damage. The high level of oxidative damage to sperm DNA reported here could be related to a higher risk of birth defects, particularly in populations with low ascorbate levels, such as smokers.

Because of the time interval between the generation of oxidants and their destruction by various defense mechanisms, low levels of oxidants can persist for sufficient time to produce damage to cellular macromolecules (Chance et al. 1979). For nuclear DNA, however, the mammalian cell has three more levels of defense. First, nuclear DNA is compartmentalized away from mitochondria and peroxisomes, where most oxidants are probably generated. Second, most nonreplicating nuclear DNA is surrounded by histones and polyamines that may protect against oxidants. Finally, most types of DNA damage can be repaired by efficient enzyme systems. The net result of this multilevel defense is that nuclear DNA is very well protected, but not completely protected from oxidants.

MUTAGENESIS, MITOGENESIS, AND CARCINOGENESIS

Geneticists have long known that cell division is critical for mutagenesis. If one accepts that mutagenesis is important for carcinogenesis, it follows that mitogenesis rates must be important. The inactivation of tumor suppressor genes is also known to be important in carcinogenesis, and recent evidence suggests that one of the functions of tumor suppressor genes is to inhibit mitogenesis (Stanbridge 1990). Once the first copy of a tumor suppressor gene is mutated, the inactivation of the second copy (loss of heterozygosity) is more likely to be caused by processes whose frequency is dependent on cell division (mitotic recombination, gene conversion, and nondisjunction) than by an independent second mutation (Ames and Gold 1990a,b). Therefore, loss of heterozygosity will be stimulated by increased mitogenesis. Thus, while the stimulation of mitogenesis increases the chance of every mutational step, it is a much more important factor for tumor induction after the first mutation has occurred. This explains why mutagenesis and mitogenesis are synergistic (Ames and Gold 1990a,b) and why mitogenesis after the first mutation is more effective than before (Fig. 4).

Thinking of chemicals as "initiators" or "promoters" confuses mechanistic issues (Iversen 1988). The idea that "promoters" are not in themselves carcinogens is not credible on mechanistic grounds and is not correct on experimental grounds (Iversen 1988; Ames and Gold 1990a,b). Every classic promoter that has been tested adequately, e.g, phenobarbital, catechol, TPA, is a carcinogen. The very word promoter confuses the issue, since mitogenesis may be caused by one dose of a chemical and not by a lower dose. Dominant oncogenes and their clonal expansion by mitogenesis can clearly be involved in carcinogenesis, adding complexity; however, these

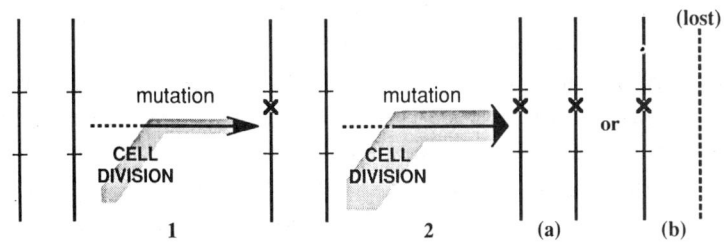

FIGURE 4 Mitogenesis (induced cell division) is a major multiplier of endogenous (or exogenous) DNA damage leading to mutation (Ames and Gold 1990a). The pathway to inactivating (X) both copies of a recessive tumor-suppressor gene is shown (two vertical lines represent the pair of chromosomes carrying the genes). Cell division increases mutagenesis due to the following: DNA adducts converted to mutations before they are repaired (1 and 2a); mutations due to DNA replication (1 and 2a); vulnerability of replicating DNA to damage (1 and 2a). Mitotic recombination (2a), gene conversion (2a), and nondisjunction (2b) are more frequent, and the first two give rise to the same mutation on both chromosomes. This diagram does not attempt to deal with the complex mutational pathway to tumors (Fearon et al. 1990; Kinzler et al. 1991).

mechanisms are still consistent with the view that mitogenesis is an important factor in carcinogenesis. Nongenotoxic agents, e.g., saccharin, can be carcinogens at high doses just by causing cell killing with chronic mitogenesis and inflammation, and the dose response would be expected to show a threshold (Ames and Gold 1990a; Cohen and Ellwein 1990; Butterworth et al. 1991). Epigenetic factors are also involved in carcinogenesis. However, both mitogenesis (e.g., through mitotic recombination) and DNA damage can cause loss of 5-methylcytosine or other epigenetic modification, as we have discussed previously (Ames and Gold 1990a). Chronic mitogenesis by itself can be a risk factor for cancer: Theory predicts it and a large literature supports it (Ames and Gold 1990a; Preston-Martin et al. 1990). The 40% of rodent carcinogens that are not detectable mutagens should be investigated to see if their carcinogenic effects at high dose result from induction of mitogenesis; if so, then such rodent carcinogens would be unlikely to be a risk at low doses.

Genotoxic chemicals, because they hit DNA, are even more effective than nongenotoxic chemicals at causing cell killing and cell replacement at high doses. Since genotoxic chemicals also act as mutagens, they can produce a multiplicative interaction not found at low doses, leading to an upward curving dose response for carcinogenicity (Ames and Gold 1990a; Cohen and Ellwein 1990; Butterworth et al. 1991). Mitogenesis can often be the dominant factor in chemical carcinogenesis at the high, nearly toxic doses used in rodent bioassays, even for mutagens. Mitogenesis can be caused by toxicity of chemicals at high dose (cell killing and subsequent replacement), by interference with cell-cell communication at high doses (Trosko et al. 1983, 1990a,b; Trosko 1989), by substances such as hormones binding to receptors that control cell division (Preston-Martin et al. 1990), by oxidants (the wound healing response), by viruses, etc. (Ames and Gold 1990a). The important factor is not toxicity, but increased mitogenesis in those cells that are not discarded. Work on radiation has also supported the idea of both mutagenesis and mitogenesis being important in tumor induction (Jones et al. 1983; Jones 1984; Little et al. 1985; Ootsuyama and Tanooka 1991).

Epigenetic changes in DNA such as 5-methylcytosine appear important in turning off genes in differentiation and could play a role in both cancer (Holliday 1987a; Vorce and Goodman 1989a,b; Fearon et al. 1990) and aging (Holliday 1987a,b). It has been observed that the 5-methylcytosine level decreases with age (Wilson et al. 1987), and it is known that cells are dedifferentiating with age (Hartman 1983; Hartman and Morgan 1985). Folate deficiency (Bhave et al. 1988; Cravo et al. 1991), mitogenesis, and DNA damage (Cannon et al. 1988) would all be expected to increase the rate of loss of 5-methylcytosine.

CAUSES OF HUMAN CANCER

The high endogenous level of oxidative adducts reinforces evidence from epidemiology that both deficiency of antioxidants (National Research Council 1989; Bendich and Butterworth 1991) and mitogenesis (Ames and Gold 1990a; Preston-Martin et al. 1990) are likely to be important risk factors for cancer.

Mitogenesis and Cancer

Henderson and co-workers (Henderson et al. 1988; Preston-Martin et al. 1990) and others (Ames and Gold 1990a) have discussed the importance of chronic mitogenesis for many, if not most, of the known causes of human cancer, e.g., hormones in breast cancer; hepatitis B (Dunsford et al. 1990) or C viruses or alcohol in liver cancer; high salt or *Helicobacter* (*Campylobacter*) infection in stomach cancer; papillomavirus in cervical cancer; asbestos or tobacco smoke in lung cancer; and excess animal fat and low calcium in colon cancer. For chemical carcinogens associated with occupational cancer, worker exposure has been primarily at high, near-toxic, doses that might be expected to induce mitogenesis. Permitted worker exposure levels for some rodent carcinogens are too close to the doses that induce tumors in test animals (Gold et al. 1987). For high occupational exposures, little extrapolation is required from the doses used in rodent bioassays, and therefore, assumptions about extrapolation are of less importance.

Chronic inflammation is a risk factor for cancer (Templeton 1980; Lewis and Adams 1987; Madsen 1989; Weitzman and Gordon 1990). Oxidants induce both mutagenesis and mitogenesis, a major risk factor for carcinogenesis (see below), and antioxidants are important for decreasing this stimulus. The oxidants produced by phagocytic cells and activated neutrophils during inflammation are signals for mitogenesis (to promote wound healing) (Chan et al. 1986; Craven et al. 1987; Crawford and Cerutti 1988; Sieweke et al. 1989; Burdon et al. 1990). The expression of both *fos* and *jun* proto-oncogenes is stimulated by oxidants (Abate et al. 1990). Inflammatory responses induce oxidative damage and mutagenesis in addition to their stimulation of mitogenesis. Activated neutrophils damage DNA in neighboring cells (Shacter et al. 1988). Macrophages induce in mouse mammary tumor cells the formation of drug-resistant variants (Yamashina et al. 1986). Ascorbate effectively neutralizes oxidants derived from activated neutrophils in vitro (Anderson and Lukey 1987) and may play an important protective role against inflammation-induced mitogenesis and mutagenesis in vivo.

Dietary Imbalances

Epidemiologists have been accumulating evidence that unbalanced diets are major contributors to heart disease and cancer and are likely to be as important as smoking. The main dietary imbalances are too few fruits and vegetables and too much fat. Particular micronutrients in fruits and vegetables that appear to be important in disease prevention are antioxidants (carotenoids, tocopherols, ascorbate) and folic acid, but many more vitamins and essential minerals may also be of interest (Reddy and Cohen 1986; National Research Council 1989; Bendich and Butterworth 1991). Micronutrients are components of the defenses against oxidants and other endogenous mutagens contributing to the degenerative diseases associated with aging: cancer, heart disease, cataracts, etc. Since endogenous oxidative DNA damage is enormous, there are good theoretical reasons for thinking that antioxidants should be as important as they are being found to be. Surveys have indicated that 91% of the U.S. population is not eating sufficient fruit and vegetables; almost half of the population had eaten neither fruits nor vegetables on the day of the survey (Patterson and Block 1991).

Work on folate deficiency in mice showing that it results in chromosome breakage (MacGregor et al. 1990) reinforces the extensive literature (National Research Council 1989; Bendich and Butterworth 1991) indicating that folate deficiency is an important cause of chromosome breaks, cancer, and birth defects. Again, a sizable proportion of the population (30% or more) may not be ingesting sufficient folate. Hypomethylation of DNA that is associated with folate deficiency could also contribute to epigenetic effects such as dedifferentiation of cells that occurs during tumorigenesis.

In the quest to delay aging and prevent cancer and heart disease, it is important to understand what level of each micronutrient is optimal for long-term effects. The U.S. recommended daily allowance (RDA) is based on the level necessary to prevent an immediate pathological effect, but long-term optimal levels may be higher. The great genetic variability of the human species makes it likely that many people require a higher than average optimal RDA for particular micronutrients. This will require the development of in vivo assays of short-term damage that can be measured in humans. The measurement of DNA damage in humans is clearly relevant (Degan et al. 1991; Park et al. 1992).

CALORIE RESTRICTION: EFFECTS ON MITOGENESIS

In rodents, calories may be the most interesting carcinogen (Boutwell and Pariza 1987; Roe 1989; Roe et al. 1991). A calorie-restricted diet, compared to an ad libitum diet, significantly increases the life span of rats and mice and dramatically decreases the cancer rate. Markedly lowered mitogenesis rates are observed in a variety of tissues (Heller et al. 1990; Lok et al. 1990), an effect that is consistent with suppression of growth-related hormones (Armario et al. 1987) and proto-oncogene expression (Nakamura et al. 1989) of rats fed such a diet. Decreased mitogenesis in calorie-restricted rats is likely to account for much of the decrease in the cancer rate.

Calorie restriction that results in decreased growth rates and a significant delay in maturation of reproductive function (Holehan and Merry 1985) is associated with depressed levels of circulating mitogenic hormones such as insulin and growth hormone (Armario et al. 1987). The lowered incidence of mammary tumors observed in calorie-restricted rats has been attributed to reduced circulating levels of the mammotropic hormones estrogen and prolactin (Sylvester et al. 1982). In the absence of these mitogenic stimuli that are necessary for reproductive functions and growth, it has been postulated that cellular energy is diverted to maintenance activities that lead to more efficient DNA repair (Weraarchakul et al. 1989), better coupled mitochondrial respiration (Weindruch et al. 1980), and increased antioxidant defenses (Koizumi et al. 1987). The higher levels of antioxidant defenses could in principle account for delay in the loss of the immune response observed in calorie-restricted rodents (Kubo et al. 1984).

POSSIBLE ROLE OF OXIDANTS AND ANTIOXIDANTS IN METASTASES

Cancer cells assume increasingly aggressive phenotypes as tumor growth progresses to metastases (Nowell 1986). This phenotype is frequently associated with point mutations in the Ha-*ras* proto-oncogene and increased collagenase secretion that is necessary for basement membrane breakdown and metastasis (Garbisa et al. 1987). Collagenase is released from the cells in a latent form that is inhibited by tissue inhibitors of metalloproteinase (TIMPs) (Carmichael et al. 1986). Activation of collagenase or inactivation of TIMPs would be a critical step in progression of tumors to metastases (Liotta et. al. 1991). The HOCl produced by polymorphonuclear leukocytes (PMNs) has been shown to autoactivate collagenase that leads to breakdown of basement membranes (Weiss et al. 1985). A similar effect of oxidants in tumors could facilitate metastases. Of possible relevance in this regard are the oxidants that have been shown to be elaborated from tumor cells or tumoricidal macrophages. Cells isolated from various tumors, for example, have recently been shown to produce a copious flux of H_2O_2 comparable to that of phorbol-ester-triggered PMNs (Szatrowski and Nathan 1991). Oxidative activation of collagenases either directly by oxidation of the critical metal complexed thiol or oxidative inactivation of TIMP would lead to the conversion of collagenase from its inactive to active form. Direct evidence for the elaboration of oxidants from tumor cells in vivo has not been obtained. However, the increase in antioxidant activities that have been detected in aggressive tumor cell populations (Borunov et al. 1989) suggests the induction of an adaptive response that allows the tumor cell to survive the high flux of self-generated oxidants. Human small-cell lung cancer cells isolated at different stages of tumor progression exhibit increasing resistance to the oxidant-inducing chemotherapeutic drug adriamycin (de Vries et al. 1989). Such adaptive responses may also play an important role in radioresistance that is observed in many malignant cancers.

CONCLUSIONS AND FUTURE DIRECTIONS

In this review, we present the argument that oxidant-induced DNA damage and mitogenesis are two important factors in the causation of cancer. Oxidants that are produced by normal aerobic metabolism contribute to the extensive oxidative DNA damage that has been measured in rodents and humans. The oxidative damage rate in mammalian species with a high metabolic rate, short life span, and high age-specific cancer rate is much higher than the rate in humans, long-lived mammals with a lower metabolic rate and a lower age-specific cancer rate. Ascorbate protects human sperm DNA from oxidative damage (Fraga et al. 1991). Similar effects of dietary antioxidants in somatic tissues would be desirable, as would studies designed to understand the relationship between risk of certain cancers and the protective effect of antioxidants on oxidative DNA damage (Patterson and Block 1991).

Mitogenesis increases the probability of producing the mutations that are required for the development of malignant tumors. Risk factors for increased mitogenesis include chronic inflammatory diseases (Preston-Martin et al. 1990) and ingestion of excessive calories that result in increases in circulating mitogenic hormones, e.g., insulin, growth hormone, estrogens. It remains to be seen whether mitogenesis that accompanies chronic inflammatory diseases can be suppressed in a human population by dietary antioxidants.

The elaboration of oxidants by tumor cell lines deserves further attention. The production of large amounts of oxidants by tumor cells could facilitate the activation of latent collagenases to an active form needed for metastases. Such cells could also be viewed as possessing a mutator phenotype, i.e., a cell with a higher mutation rate. Further studies will need to be performed to delineate the role of oxidant-producing tumor cells in tumor progression and metastases and to determine the possible beneficial effects of dietary antioxidants in delaying this progression.

ACKNOWLEDGMENTS

This work was supported by National Cancer Institute Outstanding Investigator grant CA-39910 and by National Institute of Environmental Health Sciences Center grant ESO-1896 to B.N.A. M.K.S. was supported by a National Institute of Aging postdoctoral fellowship AG-05489. This paper was adapted in part from B.N. Ames and L.S. Gold (1990a,b) and

B.N. Ames and L.S. Gold, *J. Natl. Cancer Inst.* (in press) and B.N. Ames and L.S. Gold, *Mutat. Res.* (1991).

REFERENCES

Abate, C., L. Patel, F.J. Rauscher III, and T. Curran. 1990. Redox regulation of *fos* and *jun* DNA-binding activity *in vitro*. *Science* **249:** 1157.

Adelman, R., R.L. Saul, and B.N. Ames. 1988. Oxidative damage to DNA: Relation to species metabolic rate and life span, *Proc. Natl. Acad. Sci.* **85:** 2706.

Ames, B.N. 1983. Dietary carcinogens and anticarcinogens. Oxygen radicals and degenerative diseases. *Science* **221:** 1256.

―――. 1988. Measuring oxidative damage in humans: Relation to cancer and aging. *IARC Publ.* **89:** 407.

―――. 1989. Endogenous oxidative DNA damage, aging and cancer. *Free Radical Res. Commun.* **7:** 121.

Ames, B.N. and L.S. Gold. 1990a. Chemical carcinogenesis: Too many rodent carcinogens. *Proc. Natl. Acad. Sci.* **87:** 7772.

―――. 1990b. Too many rodent carcinogens: Mitogenesis increases mutagenesis. *Science* **249:** 970.

―――. 1991. Endogenous mutagens and the causes of aging and cancer. *Mutat. Res.* **250:** 3.

Ames, B.N. and R.L. Saul. 1988. Cancer, aging, and oxidative DNA damage. In *Theories of carcinogenesis* (ed. O.H. Iversen), p. 203. Hemisphere Publishing, Washington, D.C.

Anderson, R. and P.T. Lukey. 1987. A biological role for ascorbate in the selective neutralization of extracellular phagocyte-derived oxidants. *Ann. N. Y. Acad. Sci.* **498:** 229.

Armario, A., J.L. Montero, and T. Jolin. 1987. Chronic food restriction and circadian rhythms of pituitary-adrenal hormones, growth hormone and thyroid-stimulating hormone. *Ann. Nutr. Metab.* **31:** 81.

Bendich, A. and C.E. Butterworth, Jr., eds. 1991. *Micronutrients in health and in disease prevention*. Marcel Dekker, New York.

Bhave, M.R., M.J. Wilson, and L.A. Poirier. 1988. c-H-*ras* and c-K-*ras* gene hypomethylation in the livers and hepatomas of rats fed methyl-deficient, amino acid-defined diets. *Carcinogenesis* **9:** 343.

Borunov, E.V., L.P. Smirnova, I.A. Shchepetkin, V.Z. Lankin, and N.V. Vasil'ev. 1989. High activity of antioxidant enzymes in a tumor as a factor of "avoidance of control" in the immune system. *Byull. Eksp. Biol. Med.* **107:** 467.

Boutwell, R.K. and M.W. Pariza. 1987. Historical perspective: Calories and energy expenditure in carcinogenesis. *Am. J. Clin. Nutr.* **45:** 151.

Burdon, R.H., V. Gill, and C. Rice-Evans. 1990. Oxidative stress and tumour cell proliferation. *Free Radical Res. Commun.* **11:** 65.

Butterworth, B.E., T.J. Slaga, W. Farland, and M. McClain, eds. 1991. *Chemically induced cell proliferation: Implications for risk assessment.* Wiley-Liss, New York.

Cadet, J. and M. Berger. 1985. Radiation-induced decomposition of the purine bases within DNA and related model compounds. *Int. J. Radiat. Biol. Relat. Stud. Phys. Chem. Med.* **47:** 127.

Cannon, S.V., A. Cummings, and G.W. Teebor. 1988. 5-Hydroxymethylcytosine DNA glycolases activity in mammalian tissue. *Biochem. Biophys. Res. Commun.* **151:** 1173.

Carmichael, D.F., A. Sommer, R.C. Thompson, D.C. Anderson, C.G. Smith, H.G. Welgus, and G.P. Stricklin. 1986. Primary structure and cDNA cloning of human fibroblast collagenase inhibitor. *Proc. Natl. Acad. Sci.* **83:** 2407.

Cathcart, R., E. Schwiers, R.L. Saul, and B.N. Ames. 1984. Thymine glycol and thymidine glycol in human and rat urine: A possible assay for oxidative DNA damage. *Proc. Natl. Acad. Sci.* **81:** 5633.

Chan, T.M., E. Chen, A. Tatoyan, N.S. Shargill, M. Pleta, and P. Hochstein. 1986. Stimulation of tyrosine-specific protein phosphorylation in the rat liver plasma membrane by oxygen radicals. *Biochem. Biophys. Res. Commun.* **139:** 439.

Chance, B., H. Sies, and A. Boveris. 1979. Hydroperoxide metabolism in mammalian organs. *Physiol. Rev.* **59:** 527.

Cohen, S.M. and L. Ellwein. 1990. Cell proliferation in carcinogenesis. *Science* **249:** 1007.

Craven, P.A., J. Pfanstiel, and F.R. DeRubertis. 1987. Role of activation of protein kinase C in the stimulation of colonic epithelial proliferation and reactive oxygen formation by bile acids. *J. Clin. Invest.* **79:** 532.

Cravo, M., J. Mason, R.N. Salomon, J. Ordovas, J. Osada, J. Selhub, and J.H. Rosenberg. 1991. Folate deficiency in rats causes hypomethylation of DNA. *FASEB J.* **5:** A914.

Crawford, D. and P. Cerutti. 1988. Expression of oxidant stress-related genes in tumor promotion of mouse epidermal cells JB6. In *Anticarcinogenesis and radiation protection* (ed. O. Nygaard and M.G. Simic), p. 183. Plenum Press, New York.

Cutler, R.G. 1984. Antioxidants, aging, and longevity. In *Free radicals in biology* (ed. W.A. Pryor), p. 371. Academic Press, New York.

de Vries, E.G.E, C. Meijer, H. Timmer-Bosscha, H.H. Berendsen, L. de Leij, R.J. Scheper, and N.H. Mulder. 1989. Resistance mechanisms in three human small cell lung cancer cell lines established from one patient during clinical follow-up. *Cancer Res.* **49:** 4175.

Degan, P., M.K. Shigenaga, E.-M. Park, P.E. Alperin, and B.N. Ames. 1991. Immunoaffinity isolation of urinary 8-hydroxy-2′-deoxyguanosine and 8-hydroxyguanine and quantitation of 8-hydroxy-2′-deoxyguanosine in DNA by polyclonal antibodies. *Carcinogenesis* **5:** 865.

Dizdaroglu, M. 1985. Formation of 8-hydroxyguanine moiety in deoxyribonucleic acid on gamma-irradiation in aqueous solution. *Biochemistry* **24:** 4476.

Doll, R. 1971. The age distribution of cancer: Implications for models of carcinogenesis. *J. R. Stat. Soc.* **A134:** 133.

Dunsford, H.A., S. Sell, and F.V. Chisari. 1990. Hepatocarcinogenesis due to chronic liver cell injury in hepatitis B virus transgenic mice. *Cancer Res.* **50:** 3400.

Fearon, E.F., K.R. Cho, J.M. Nigro, S.E. Kern, J.W. Simons, J.M. Ruppert, S.R. Hamilton, A.C. Preisinger, G. Thomas, K.W. Kinzler, and B. Vogelstein. 1990. Identification of a chromosome 18q gene that is altered in colorectal cancer. *Science* **247:** 49.

Fiala, E.S., C.C. Conaway, and J.E. Mathis. 1989. Oxidative DNA and RNA damage in the livers of Sprague-Dawley rats treated with the hepatocarcinogen 2-nitropropane. *Cancer Res.* **49:** 5518.

Floyd, R.A., J.J. Watson, P.K. Wong, D.H. Altmiller, and R.C. Richard. 1986. Hydroxyl free radical adduct of deoxyguanosine: Sensitive detection and mechanisms of formation. *Free Radical Res. Commun.* **1:** 163.

Fraga, C.G., M.K. Shigenaga, J.-W. Park, P. Degan, and B.N. Ames. 1990. Oxidative damage to DNA during aging: 8-Hydroxy-2' deoxyguanosine in rat organ DNA and urine. *Proc. Natl. Acad. Sci.* **87:** 4533.

Fraga, C.G., P.A. Motchnik, M.K. Shigenaga, H.J. Helbock, R. Jacob, and B.N. Ames. 1991. Ascorbic acid protects against endogenous oxidative DNA damage in human sperm. *Proc. Natl. Acad. Sci.* **88:** 11003.

Garbisa, S., R. Pozzatti, R.J. Muschel, U. Saffiotti, M. Ballin, R.H. Goldfarb, G. Khoury, and L.A. Liotta. 1987. Secretion of type IV collagenolytic protease and metastatic phenotype: Induction by transfection with c-Ha-*ras* but not c-Ha-*ras* plus Ad2-E1a. *Cancer Res.* **47:** 1523.

Gold, L.S., G.M. Backman, N.K. Hooper, and R. Peto. 1987. Ranking the potential carcinogenic hazards to workers from exposures to chemicals that are tumorigenic in rodents. *Environ. Health Perspect.* **76:** 211.

Harman, P. 1981. The aging process. *Proc. Natl. Acad. Sci.* **78:** 7124.

———.1991. The aging process: Major risk factor for disease and death. *Proc. Natl. Acad. Sci.* **88:** 5360.

Hartman, P.E. 1983. Mutagens: Some possible health impacts beyond carcinogenesis. *Environ. Mutagen.* **5:** 139.

Hartman, P.E. and R.W. Morgan. 1985. Mutagen-induced focal lesions as key factors in aging: A review. In *Molecular biology of aging: Gene stability and gene expression, aging series* (ed. R.S. Sohal et al.), vol. 29, p. 93. Raven Press, New York.

Heller, T.D., P.R. Holt, and A. Richardson. 1990. Food restriction retards age-related histological changes in rat small intestine. *Gastroenterology* **98:** 387.

Henderson, B.E., R. Ross, and L. Bernstein. 1988. Estrogens as a cause of human cancer: The Richard and Hinda Rosenthal Foundation Award Lecture. *Cancer Res.* **48:** 246.

Holehan, A.M. and B.J. Merry. 1985. Modification of the oestrous cycle hormonal profile by dietary restriction. *Mech. Ageing Dev.* **32:** 63.

Holliday, R. 1987a. DNA methylation and epigenetic defects in carcinogenesis. *Mutat. Res.* **181:** 215.

———. 1987b. The inheritance of epigenetic defects. *Science* **238:** 163.

Hollstein, M.C., P. Brooks, S. Linn, and B.N. Ames. 1984. Hydroxymethyluracil DNA glycosylase in mammalian cells. *Proc. Natl. Acad. Sci.* **81:** 4003.

Iversen, O.H., ed. 1988. Initiation, promotion: critical remarks on the two-stage theory. In *Theories of carcinogenesis*, p. 119. Hemisphere Publishing, Washington, D.C.

Jones, T.D. 1984. A unifying concept for carcinogenic risk assessments: Comparison with radiation-induced leukemia in mice and men. *Health Phys.* **4:** 533.

Jones, T.D., G.D. Griffin, and P.J. Walsh 1983. A unifying concept for carcinogenic risk assessments. *J. Theor. Biol.* **105:** 35.

Kasai, H., S. Nishimura, Y. Kurokawa, and Y. Hayashi. 1987. Oral administration of the renal carcinogen, potassium bromate, specifically produces 8-hydroxydeoxyguanosine in rat target organ DNA. *Carcinogenesis* **8:** 1959.

Kinzler, K.W., M.C. Nilbert, B. Vogelstein, T.M. Bryan, D.B. Levy, K.J. Smith, A.C. Preisinger, S.R. Hamilton, P. Hedge, A. Markham, M. Carlson, G. Joslyn, J. Groden, R. White, Y. Miki, Y. Miyoshi, I. Nishiso, and Y. Nakamura. 1991. Identification of a gene located at chromosome 5q21 that is mutated in colorectal cancers. *Science* **251:** 1366.

Koizumi, A., R. Weindruch, and R.L. Walford. 1987. Influences of dietary restriction and age on liver enzyme activities and lipid peroxidation in mice. *J. Nutr.* **117:** 361.

Kubo, C., B.C. Johnson, N.K. Day, and R.A. Good. 1984. Calorie source, calorie restriction, immunity and aging of (NZB/NZW)F1 mice. *J. Nutr.* **114:** 1884.

Lewis, J.G. and D.O. Adams. 1987. Inflammation, oxidative DNA damage, and carcinogenesis. *Environ. Health Perspect.* **76:** 19.

Lindahl, T. 1982. DNA repair enzymes. *Annu. Rev. Biochem.* **51:** 61.

Liotta, L.A., P.S. Steeg, and W.G. Stetler-Stevenson. 1991. Cancer metastasis and angiogenesis: An imbalance of positive and negative regulation. *Cell* **64:** 327.

Little, J.B., A.R. Kennedy, and R.B. McGandy. 1985. Effect of dose rate on the induction of experimental lung cancer in hamsters by radiation. *Radiat. Res.* **103:** 293.

Lok, E., F.W. Scott, R. Mongeau, E.A. Nera, S. Malcolm, and D.B.

Clayson. 1990. Calorie restriction and cellular proliferation in various tissues of the female Swiss Webster mouse. *Cancer Lett.* **51:** 67.

Madsen, C. 1989. Squamous-cell carcinoma and oral, pharyngeal and nasal lesions caused by foreign bodies found in food. Cases from a long-term study in rats. *Lab. Anim.* **23:** 241.

MacGregor, J.T., R. Schlegel, C.M. Wehr, P. Alperin, and B.N. Ames. 1990. Cytogenetic damage induced by folate deficiency in mice is enhanced by caffeine. *Proc. Natl. Acad. Sci.* **87:** 9962.

Nakamura, K.D., P.H. Duffy, M.H. Lu, A. Turturro, and R.W. Hart. 1989. The effect of dietary restriction on *myc* protooncogene expression in mice: A preliminary study. *Mech. Ageing Dev.* **48:** 199.

National Research Council. 1989. *Diet and health, implications for reducing chronic disease risk.* National Academy Press, Washington, D.C.

Nowell, P.C. 1986. Mechanisms of tumor progression. *Cancer Res.* **46:** 2203.

Ootsuyama, A. and H. Tanooka. 1991. Threshold-like dose of local β-irradiation repeated throughout the life span of mice for induction of skin and bone tumors. *Radiat. Res.* **125:** 98.

Park, E.-M., M.K. Shigenaga, P. Degan, T.S. Korn, J.W. Kitzler, C.M. Wehr, P. Kolachana, and B.N. Ames. 1992. The assay of excised oxidative DNA lesions: Isolation of 8-oxoguanine and its nucleoside derivatives from biological fluids with a monoclonal antibody column. *Proc. Natl. Acad. Sci.* **89:** (in press).

Patterson, B.H. and G. Block. 1991. Fruit and vegetable consumption national survey data. In *Micronutrients in health and in disease prevention* (ed. A. Bendich and C.E. Butterworth, Jr.), p. 409. Marcel Dekker, New York.

Portier, C.J., J.C. Hedges, and D.G. Hoel. 1986. Age-specific models of mortality and tumor onset for historical control animals in the National Toxicology Program's carcinogenicity experiments. *Cancer Res.* **46:** 4372.

Preston-Martin, S., M.C. Pike, R.K. Ross, and P.A. Jones. 1990. Increased cell division as a cause of human cancer. *Cancer Res.* **50:** 7415.

Reddy, B.S. and L.A. Cohen, ed. 1986. *Diet, nutrition, and cancer: A critical evaluation,* vols. I and II. CRC Press, Boca Raton, Florida.

Richter, C., J.-W. Park, and B.N. Ames. 1988. Normal oxidative damage to mitochondrial and nuclear DNA is extensive. *Proc. Natl. Acad. Sci.* **85:** 6465.

Roe, F.J.C. 1989. Non-genotoxic carcinogenesis: Implications for testing and extrapolation to man. *Mutagenesis* **4:** 407.

Roe, F.J.C., P.N. Lee, G. Conybeare, G. Tobin, D. Kelly, D. Prentice, and B. Matter. 1991. Risks of premature death and cancer predicted by body weight in early adult life. *Hum. Exp. Toxicol.* **10:** 285.

Rosier, J.A. and C.H. VanPeteghem. 1989. Peroxidative *in vitro* metabolism of diethylstilbestrol induces formation of 8-hydroxy-2′ deoxyguanosine. *Carcinogenesis* **10:** 405.

Saul, R.L. and B.N. Ames. 1986. Background levels of DNA damage in the population. In *Mechanisms of DNA damage and repair: Implications for carcinogenesis and risk assessment* (ed. M.G. Simic et al.), p 529. Plenum Press, New York.

Saul, R.L., P. Gee, and B.N. Ames. 1987. Free radicals, DNA damage, and aging. In *Modern biological theories of aging* (ed. H.R. Warner et al.), p. 113. Raven Press, New York.

Shacter, E., E.J. Beecham, J.M. Covey, K.W. Kohn, and M. Potter. 1988. Activated neutrophils induce prolonged DNA damage in neighboring cells. *Carcinogenesis* **9:** 2297.

Shibutani, S., M. Takeshita, and A.P. Grollman. 1991. Insertion of specific bases during DNA synthesis past the oxidation-damaged base 8-oxodG. *Nature* **349:** 431.

Shigenaga, M.K., C.J. Gimeno, and B.N. Ames. 1989. Urinary 8-hydroxy-2′-deoxyguanosine as a biomarker of *in vivo* oxidative DNA damage. *Proc. Natl. Acad. Sci.* **86:** 9697.

Sieweke, M.H., A.W. Stoker, and M.J. Bissell. 1989. Evaluation of the cocarcinogenic effect of wounding in Rous sarcoma virus tumorigenesis. *Cancer Res.* **49:** 6419.

Stanbridge, E.J. 1990. Human tumor suppressor genes. *Annu. Rev. Genet.* **24:** 615.

Sylvester, P.W., C.F. Aylsworth, D.A. Van Vugt, and J. Meites. 1982. Influence of underfeeding during the "critical period" or thereafter on carcinogen-induced mammary tumors in rats. *Cancer Res.* **42:** 4943.

Szatrowski, T.P. and C.F. Nathan. 1991. Production of large amounts of hydrogen peroxide by human tumor cells. *Cancer Res.* **51:** 794.

Templeton, A. 1980. Pre-existing, non-malignant disorders associated with increased cancer risk. *J. Environ. Pathol. Toxicol.* **3:** 387.

Totter, J.R. 1980. Spontaneous cancer and its possible relationship to oxygen metabolism. *Proc. Natl. Acad. Sci.* **77:** 1763.

Trosko, J.E. 1989. Towards understanding carcinogenic hazards: A crisis in paradigms. *J. Am. Coll. Toxicol.* **8:** 1121.

Trosko, J.E., C.C. Chang, and B.V. Madhukar. 1990a. *In vitro* analysis of modulators on intercellular communications: Implications for biologically based risk assessment models for chemical exposure. *Toxicol. In Vitro.* **4:** 635.

———. 1990b. Chemical, oncogene, and growth factor inhibition of gap junctional communication: An integrative hypothesis of carcinogenesis. *Pathobiology* **58:** 265.

Trosko, J.E., C.C. Chang, and A. Medcalf. 1983. Mechanisms of tumor promotion: Potential role of intercellular communication. *Cancer Invest.* **1:** 511.

von Sonntag, C. 1987. *The chemical basis of radiation biology.* Taylor and Francis, London.

Vorce, R.L. and J.I. Goodman. 1989a. Hypomethylation of *ras* oncogenes in chemically induced and spontaneous B6C3F1 mouse liver tumors. *Mol. Toxicol.* **2:** 99.

———. 1989b. Altered methylation of *ras* oncogenes in benzidine-induced B6C3F1 mouse liver tumors. *Toxicol. Appl. Pharmacol.* **100:** 398.

Weindruch, R.H., M.K. Cheung, M.A. Verity, and R.L. Walford. 1980. Modification of mitochondrial respiration by aging and dietary restriction. *Mech. Ageing Dev.* **12:** 375.

Weiss, S.J., G. Peppin, X. Ortiz, C. Ragsdale, and S.T. Test. 1985. Oxidative autoactivation of latent collagenase by human neutrophils. *Science* **227:** 747.

Weitzman, S.A. and L.J. Gordon. 1990. Inflammation and cancer: Role of phagocyte-generated oxidants in carcinogenesis. *Blood* **76:** 655.

Weraarchakul, N., R. Strong, W.G. Wood, and A. Richardson. 1989. The effect of aging and dietary restriction on DNA repair. *Exp. Cell Res.* **181:** 197.

Williams, G.C. and R.M. Nesse. 1991. The dawn of Darwinian medicine. *Q. Rev. Biol.* **66:** 1.

Wilson, V.L., R.A. Smith, S. Ma, and R.G. Cutler. 1987. Genomic 5-methyldeoxycytidine decreases with age. *J. Biol. Chem.* **262:** 9948.

Wood, M.L., M. Dizdaroglu, E. Gajewski, and J.M. Essigman. 1990. Mechanistic studies of ionizing radiation and oxidative mutagenesis: Genetic effects of a single 8-hydroxyguanine (7-hydro-8-oxoguanine) residue inserted at a unique site in a viral genome. *Biochemistry* **29:** 7024.

Yamashina, K., B.E. Miller, and G.H. Heppner. 1986. Macrophage-mediated induction of drug-resistant variants in a mouse mammary tumor cell line. *Cancer Res.* **46:** 2396.

Small Molecule Antioxidant Defenses in Human Extracellular Fluids

B. Frei,[1] R. Stocker,[2] and B.N. Ames[3]
[1]Department of Nutrition, Harvard School of Public Health
Boston, Massachusetts 02115
[2]The Heart Research Institute
Sydney, N.S.W. 2050, Australia
[3]Division of Biochemistry and Molecular Biology
University of California, Berkeley, California 94720

Throughout our lifetimes, we are constantly exposed to oxidants such as superoxide radicals ($\cdot O_2^-$), hydrogen peroxide (H_2O_2), hydroxyl radicals, and singlet oxygen. These oxidants are generated as by-products of normal metabolism, e.g., by spontaneous autoxidation of electron transport carriers in mitochondria (Chance et al. 1979), and are also of exogenous origin such as cigarette smoke, polluted air, and natural radioactive gases (Ames 1983; Cerutti 1991). One consequence of this life-long exposure to oxidants is oxidation of lipids (called lipid peroxidation) in cell membranes and lipoproteins. It is becoming increasingly evident that peroxidative damage to low-density lipoprotein (LDL) plays a key role in the etiology of atherosclerosis (Steinberg et al. 1989; Schwartz et al. 1991) and that lipid peroxidation may also contribute to other human diseases, such as certain types of cancer, rheumatoid arthritis, and myocardial reoxygenation injury, as well as to the degenerative processes associated with aging (Halliwell and Gutteridge 1989).

Physiological defenses against oxidative stress include small molecule antioxidants and antioxidant proteins. The antioxidant proteins can act in several ways: They destroy oxidants catalytically (e.g., superoxide dismutase, catalase, glutathione peroxidase), or they scavenge oxidants in a sacrificial manner (e.g., albumin) (Halliwell 1988), or they sequester transition metals in a way that prevents these metal ions from catalyzing free radical generation (e.g., transferrin and ceruloplasmin) (Gutteridge 1983; Aruoma and Halliwell 1987). The small molecule antioxidants either act in a sacrificial manner by scavenging oxidants, or they chelate transition metal ions; they can be separated into water-soluble antioxidants (e.g., ascorbic acid, glutathione, uric acid) and lipid-soluble antioxidants (e.g., α-tocopherol, β-carotene, ubiquinols), the latter associated with membranes and lipoproteins (Stocker and Frei 1991).

PROTEINACEOUS EXTRACELLULAR ANTIOXIDANT DEFENSES

Although intracellular antioxidant defenses are provided mainly by CuZn- and Mn-dependent superoxide dismutases (SODs), catalase, glutathione peroxidase, glutathione reductase, glutathione S-transferase, phospholipid hydroperoxide glutathione peroxidase, and the hexose monophosphate shunt (Halliwell and Gutteridge 1989), extracellular fluids are very poor in those antioxidant enzymes (Halliwell and Gutteridge 1990; Stocker and Frei 1991). For example, human blood plasma, the most readily available extracellular fluid, contains no significant SOD, catalase, or peroxidase activities (see Table 1). Although there is an extracellular form of a CuZn-containing SOD distinct from the intracellular, cytosolic CuZn-SOD, the former appears to be bound to the extracellular surface of endothelial cells, providing these cells with a protective coat against $\cdot O_2^-$ attack (Karlsson and Marklund 1987). The levels of free, circulating extracellular SOD in plasma are very low compared to intraerythrocytic levels (Stocker and Frei 1991). Likewise, there is only little if any catalase activity in plasma. Although the presence in serum of a posttranslationally modified form of erythrocyte catalase was reported by Goth (1991), it is possible that the catalase is released from erythrocytes by lysis during blood collection and thus represents an artifact. The only other extracellular antioxidant enzyme that has been identified in plasma is a selenium-dependent glutathione peroxidase. This enzyme, like extracellular SOD, is glycosylated (Takahashi et al. 1987) but is

TABLE 1 ANTIOXIDANT DEFENSES IN HUMAN BLOOD PLASMA

Antioxidant	Normal plasma or serum concentrations
Small molecule antioxidants[a]	
water-soluble	
uric acid	160–450 μM
ascorbic acid	30–150 μM
bilirubin	5–20 μM
glutathione	<2 μM
lipid-soluble (lipoprotein-associated)	
α-tocopherol	15–40 μM
ubiquinol-10	0.4–1.0 μM
lycopene	0.5–1.0 μM
β-carotene	0.3–0.6 μM
lutein	0.1–0.3 μM
Antioxidant proteins[b]	
nonenzymatic	
protein thiols (?)	350–500 μM
albumin	38–52 g/l
transferrin	1.5–3.4 g/l
lactoferrin	0.03–0.28 mg/l
(ferritin)	(0.02–0.44 mg/l)
ceruloplasmin (Cu binding; $\cdot O_2^-$ scavenging)	0.18–0.40 g/l
haptoglobin	0.5–3.6 g/l
hemopexin	0.5–1.2 g/l
enzymatic	
ceruloplasmin (ferroxidase activity)	0.18–0.40 g/l
CuZnSOD	5–20 units/ml
catalase	?
(glutathione) peroxidase	0.4 units/ml

See text for further details.

[a]Concentrations of small molecule antioxidants are as listed (with permission) in Table 3 of Stocker and Frei (1991).

[b]Concentrations for antioxidant proteins are from Lentner (1984), except for protein thiols, which are from Wayner et al. (1987), and CuZnSOD, which are from Karlsson and Marklund (1987).

secreted from hepatic rather than endothelial cells (Avissar et al. 1989). Interestingly, the extracellular selenium-dependent glutathione peroxidase has a low affinity for its reducing substrate glutathione (apparent K_m between 4.3 and 5.3 mM) (Maddipati and Marnert 1987; Takahashi et al. 1987) and thus cannot function effectively at the low micromolar levels of glutathione found in plasma (Cantin et al. 1987).

Because catalytic antioxidant defenses are so low in extracellular fluids, the nonenzymatic antioxidants are of major importance. These consist of proteins that bind metals or biological iron complexes, the protein thiol groups, and the small molecule antioxidants (Table 1). Since metal ions and biological iron complexes can catalyze hydroxyl radical production and initiate lipid peroxidation (Kappus 1985), binding of these metal ions and iron complexes in inactive forms is an important antioxidant defense mechanism (Halliwell and Gutteridge 1990). Thus, binding of iron to transferrin or lactoferrin, of copper to ceruloplasmin or albumin, of heme to hemopexin, and of hemoglobin to haptoglobin greatly diminishes or even completely eliminates the effectiveness of these metal ions and iron complexes to stimulate lipid peroxidation (Halliwell and Gutteridge 1990; Stocker and Frei 1991). Ceruloplasmin also displays ferroxidase activity, i.e., it can catalyze oxidation of ferrous ions (Fe^{++}) to ferric ions (Fe^{+++}) (Gutteridge 1983), thereby inhibiting both iron-stimulated lipid peroxidation and the Fenton reaction (Kappus 1985). Uric acid, besides scavenging oxidants (Ames et al. 1981), has also been shown to complex iron tightly (Davies et al. 1986). On the other hand, iron bound to ferritin, a tissue iron storage protein present in small concentrations in human extracellular fluids (Jacobs et al. 1975; Lentner 1984), can be readily mobilized to catalyze free radical generation (Biedmond et al. 1984). Whereas plasma is well endowed with iron- and copper-binding proteins and thus does not contain detectable amounts of "catalytic" iron or copper (Gutteridge et al. 1981; Gutteridge 1984), other human extracellular fluids such as cerebrospinal fluid and synovial joint fluid contain less transferrin, ceruloplasmin, and albumin (Halliwell and Gutteridge 1985; Gutteridge 1986). Thus, these extracellular fluids are more sensitive to metal-ion-dependent free radical damage. Catalytic iron or copper has been detected in cerebrospinal fluid, synovial knee-joint fluid from rheumatoid arthritis patients, and plasma from patients with idiopathic hemochromatosis (Gutteridge et al. 1981; Gutteridge 1984; Grootveld et al. 1989).

The protein thiols also belong to the category of proteinaceous extracellular antioxidants. In plasma, most of the protein-associated thiol groups are found on albumin (Radi et al. 1991). Since albumin is present at very high concentrations in plasma, and has a half-life of only about 20 days, oxidation of its thiol groups may have no significant biological consequences. Therefore, albumin has been suggested to act as a sacrificial antioxidant (also because it confines copper-mediated oxidative damage to its own binding site; see above) (Halliwell 1988). However, oxidation of critical thiol groups of proteins other than albumin, e.g., enzymes or antibodies, might lead to partial or total loss of the physiological activity of these proteins and thus should be considered oxidative damage rather than antioxidant protection. In addition, the thiyl radical formed upon protein thiol oxidation is itself a potential source of oxidative damage and is able to initiate lipid peroxidation (Schöneich et al. 1989), which is another reason that thiol oxidation might not afford much antioxidant protection.

SMALL MOLECULE EXTRACELLULAR ANTIOXIDANTS

The small molecule antioxidants in plasma and other human extracellular fluids are the water-soluble antioxidants, present homogeneously distributed in the aqueous phase (except for the albumin-bound fractions of bilirubin and uric acid), and lipoprotein-associated, lipid-soluble antioxidants. These extracellular small molecule antioxidants include ascorbic acid and α-tocopherol, other dietary components such as carotenoids and ubiquinols, and end products of metabolic pathways, i.e., bilirubin and uric acid (Table 1). Glutathione, an important intracellular antioxidant, is only present in small concentrations in extracellular fluids, with the only possible exception of alveolar epithelial lining fluid (Cantin et al. 1987). The concentration of the various small molecule antioxidants varies widely from one extracellular fluid to another. For example, the normal concentrations of ascorbic acid are about 30 μM to 150 μM in human plasma (Table 1), two- to fivefold higher in cerebrospinal fluid (Specter 1977), and up to 1.5 mM in the aqueous humor of the eye, i.e., 20–30 times the concentration in plasma (Lentner 1981). Uric acid concentrations are between 160 µM and 450 µM in plasma (Table 1), about the same in synovial joint fluid (Lentner 1981) and saliva (Ames et al. 1981), but only 4 µM to 37 µM in cerebrospinal fluid (Lentner 1981).

The small molecule antioxidants mostly act by directly scavenging oxidants and, therefore, are consumed during the course of their antioxidant action (noncatalytic, sacrificial antioxidant activity). This is in contrast to antioxidant enzymes, or metal- and iron complex-binding proteins, which act catalytically or "anticatalytically," respectively. Thus, a steady supply or regeneration of the small molecule antioxidants is necessary in order for them to maintain antioxidant protection under continued oxidative stress conditions. Certain antioxidants can regenerate other antioxidants. For example, ascorbic acid can regenerate uric acid from the uric acid anion radical (Maples and Mason 1988), as well as protein thiols from thiyl free radicals (D'Aquino et al. 1989). Ascorbic acid and α-tocopherol are thought to act synergistically against lipid peroxidation, with ascorbic acid in the aqueous phase reducing the α-tocopheroxyl radical in LDL (Sato et al. 1990). Although the synergistic action of ascorbic acid and α-tocopherol is well documented in vitro, the in vivo evidence is still tenuous. Finally, it has been suggested that ubiquinol-10 regenerates α-tocopherol within lipids (Maguire et al. 1989; Frei et al. 1990a).

RELATIVE EFFECTIVENESS OF SMALL MOLECULE ANTIOXIDANTS AND PROTEIN THIOLS IN PLASMA AGAINST LIPID PEROXIDATION INDUCED BY AQUEOUS AND LIPID-SOLUBLE PEROXYL RADICALS

The relative importance of an individual extracellular antioxidant against lipid peroxidation (anti*per*oxidative activity) depends on the nature of the oxidants to be scavenged and the site of production of oxidants, i.e., aqueous phase or lipids. In addition, one has to distinguish between *quantitative* contribution of an individual antioxidant to the total antiperoxidative capacity of an extracellular fluid and the *qualitative* importance of an antioxidant. Whereas both quantitative and qualitative importance depend on the concentration of the antioxidant, the former also depends on the number of oxidant molecules scavenged by each molecule of the antioxidant, and the latter on the reaction rate of the antioxidant with the oxidant relative to the reaction rate of the oxidant with the target lipid molecule.

The most comprehensive studies on the relative importance of small molecule antioxidants and protein thiols against lipid peroxidation in human plasma have used aqueous peroxyl radicals as oxidants. The group of Ingold and Burton measured the *total* radical-trapping *antioxidant parameter* (TRAP) of plasma by exposing diluted plasma in vitro to aqueous peroxyl radicals generated at a known and steady rate by thermal decomposition of the water-soluble azo-compound 2,2'-azobis(2-amidinopropane) hydrochloride (AAPH) (Wayner et al. 1985, 1987)

$$AAPH \xrightarrow{\Delta T} 2\ A\cdot \xrightarrow{O_2} 2\ AOO\cdot \qquad (1)$$

(A·, AAPH-derived carbon-centered radical; AOO·, AAPH-derived aqueous peroxyl radical). TRAP values were determined by measuring the so-called "induction period," i.e., the time period during which lipid peroxidation (measured as oxygen uptake) was suppressed. TRAP values in 45 plasma samples of healthy individuals were found to be 820 ± 148 μM (average value ± S.D.) (Wayner et al. 1987), i.e., each liter of plasma on an average had the capacity to scavenge 820 μmole of aqueous peroxyl radicals. TRAP correlated roughly with the sum of the individual contributing (radical-trapping) *antioxidant parameters* (CAP$_i$) of the four major plasma antioxidants: ascorbic acid, uric acid, protein thiols, and α-tocopherol. CAP$_i$ of each antioxidant was calculated by multiplying the concentration of the antioxidant in plasma (see Table 1) with the number of peroxyl radicals trapped by each molecule of the antioxidant (its so-called *n*-value). Thus: TRAP ≈ 1.7 [ascorbic acid] + 1.3 [uric acid] + 0.33 [protein-thiols] + 2.0 [α-tocopherol] (Wayner et al. 1987).

TABLE 2 INDIVIDUAL CONTRIBUTING (RADICAL-TRAPPING) ANTIOXIDANT PARAMETERS OF PLASMA ANTIOXIDANTS IN PERCENTAGE OF THE TOTAL RADICAL-TRAPPING ANTIOXIDANT PARAMETER OF HUMAN BLOOD PLASMA

Component	n-value	CAP_i [a] (%)	CAP_i [b] (%)
Uric acid	1.3	58 ±18	32.5
Protein thiols	0.33	21 ±10	15.6
Ascorbic acid	1.7/0.8[c]	14 ± 8	6.4
α-Tocopherol	2.0	7 ± 2	7.1
Bilirubin	3.67[d]	0	7.3
Lipids	1.0[e]	0	30.3

Reprinted, with permission, from Stocker and Frei (1991).
[a] As reported by Wayner et al. (1987).
[b] As determined from the results in Fig. 1 of Frei et al. (1988a) (Fig. 1 of this article).
[c] Although an n-value of 1.7 has been used by Wayner et al. (1987) for ascorbic acid in diluted plasma, we have used undiluted plasma samples where the corresponding n-value is ~0.8 (Wayner et al. 1986).
[d] The value of 3.67 takes into account that about 37% of albumin-bound bilirubin with an n-value of 1.9 is oxidized to biliverdin, which itself has an n-value of 4.7 (Stocker et al. 1987; Stocker and Ames 1987).
[e] We assume that during the phase of inhibited lipid peroxidation, each molecule of oxidized lipid has reacted with one peroxyl radical (no propagation of lipid peroxidation).

The quantitative contributions of the four major plasma antioxidants to TRAP in plasma from healthy individuals as determined by Wayner et al. (1987) and ourselves are shown in Table 2. Using the calculations suggested by the former group, we found considerable discrepancies between measured and calculated TRAP values when only including ascorbic acid, uric acid, protein thiols, and α-tocopherol in our calculations. These four antioxidants accounted for only about 63% of TRAP measured experimentally (Table 2). The remaining 37% could be attributed to three additional compounds being oxidized by reaction with aqueous peroxyl radicals: (1) Albumin-bound bilirubin also acts as an antioxidant with an n-value of about 1.9 (Stocker et al. 1987). (2) In addition, biliverdin, the oxidation product of bilirubin, can scavenge peroxyl radicals with an n-value of 4.7 (Stocker and Ames 1987). Together, bilirubin and biliverdin contribute about 7% to TRAP (Table 2). (3) Most importantly, we observed that lipids in plasma were oxidized to a significant extent during

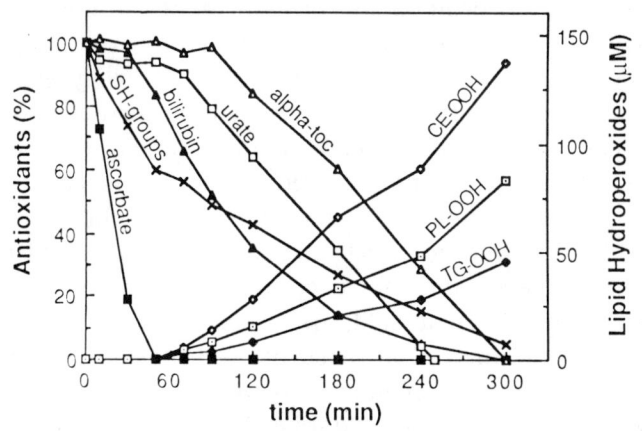

FIGURE 1 Antioxidant defenses and lipid peroxidation in human blood plasma exposed to the water-soluble radical initiator AAPH. Plasma was incubated at 37°C in the presence of 50 mM AAPH. The levels of the antioxidants ascorbate (initial concentration 72 μM), protein thiols (SH-groups, 425 μM), bilirubin (18 μM), urate (225 μM), and α-tocopherol (alpha-toc, 32 μM) are given as the percentage of the initial concentrations (left ordinate). The levels of the lipid hydroperoxides, triglyceride hydroperoxides (TG-OOH), cholesterol ester hydroperoxides (CE-OOH), and phospholipid hydroperoxides (PL-OOH) are given in μM concentrations (right ordinate). One experiment typical of four is shown. (Reprinted, with permission, from Frei et al. 1988a.)

the induction period (see Fig. 1) (Frei et al. 1988a), accounting for about 30% of TRAP (Table 2). Since this value represents substantial oxidative damage to lipids instead of antiperoxidative protection, the usefulness of the TRAP assay as an indicator of a person's antioxidant status is questionable. Thus, although the TRAP assay may measure the capacity of plasma to partially suppress lipid peroxidation, it does not provide information on the capacity of plasma antioxidants to prevent pathologically significant lipid peroxidative damage.

To investigate the *qualitative* importance of plasma antioxidants against lipid peroxidation, we have used a highly sensitive and selective assay to measure lipid hydroperoxides with a detection limit of about 10 nM neutral lipid hydroperoxides in plasma (Frei et al. 1988b; Yamamoto et al. 1990). We found that in plasma exposed to a constant flux of AAPH-derived peroxyl radicals, the sequence of antioxidant consumption was ascorbic acid, protein thiols > albumin-bound bilirubin > uric acid > α-tocopherol; peroxidation of plasma phospholipids, cholesterol esters, and triglycerides could only be detected after ascorbic acid had been depleted completely (Fig. 1) (Frei et al. 1988a). In a series of experiments, we subsequently demonstrated that the observed lag phase preceding detectable lipid peroxidation in plasma exposed to AAPH (Fig. 1) was indeed due to the antiperoxidative activity of ascorbic acid and that ascorbic acid was able to effectively spare bilirubin, uric acid, and α-tocopherol (Frei et al. 1989). It was concluded that ascorbic acid was the only endogenous antioxidant in plasma capable of fully protecting lipids against detectable peroxidative damage induced by aqueous peroxyl radicals and that all other major plasma antioxidants were less effective (Frei et al. 1990b). As indicated in Table 2 and illustrated in Figure 1, about one third of the aqueous peroxyl radicals generated from AAPH after complete consumption of ascorbic acid escaped the remaining plasma antioxidants, both in the aqueous phase and within the lipoproteins, and caused oxidative damage to lipids. It is not surprising that α-tocopherol and the other lipoprotein-associated antioxidants proved less effective than ascorbic acid against lipid peroxidation induced by aqueous peroxyl radicals: Lipid-soluble antioxidants scavenge chain-carrying lipid peroxyl radicals within lipoproteins, thereby preventing *propagation* of lipid peroxidation (reaction 3) after lipid peroxidation has been initiated (reaction 2), whereas ascorbic acid prevents *initiation* of lipid peroxidation by effectively scavenging AAPH-derived peroxyl radicals in the aqueous phase of plasma (reaction 4) before they can diffuse into and attack plasma lipids (reaction 2)

AOO• + LH + O_2 → AOOH + LOO•   (2)
LOO• + α-tocopherol → LOOH + α-tocopheroxyl radical   (3)
AOO• + ascorbic acid → AOOH + ascorbyl radical   (4)

(AOO•, AAPH-derived aqueous peroxyl radical; AOOH, corresponding hydroperoxide; LH, polyunsaturated fatty acyl side chain in lipoprotein lipid; LOO•, corresponding peroxyl radical; LOOH, corresponding hydroperoxide.)

In whole blood exposed to AAPH, the sequence of plasma antioxidant consumption was the same as in AAPH-exposed plasma (Niki et al. 1988). The period of effective antiperoxidative protection by ascorbic acid lasted about twice as long in whole blood as in plasma (B. Frei and B.N. Ames, unpubl.), possibly due to regeneration of plasma ascorbic acid from its oxidized form(s) by erythrocytes (Orringer and Roer 1979).

When a lipid-soluble (rather than a water-soluble) peroxyl radical initiator was used in plasma, lipid-soluble α-tocopherol belonged to the first line of antioxidant defense, together with ascorbic acid (Frei et al. 1990b). As soon as ascorbic acid had been depleted completely, bilirubin became oxidized, too. These findings indicate that ascorbic acid and bilirubin interact with α-tocopherol in the lipoproteins, when radicals are formed in the lipids. Indeed, there is in vitro evidence for synergistic antioxidant interaction between ascorbic acid and α-tocopherol in LDL (Sato et al. 1990) (see above), and between conjugated bilirubin and α-tocopherol in phospholipid liposomal membranes (Stocker and Peterhans 1989). Uric acid and protein thiols were not consumed in plasma exposed to lipid-soluble peroxyl radicals, confirming that radicals were produced solely in the plasma lipids. Lipid peroxidation occurred concomitantly with consumption of α-tocopherol (Frei et al. 1990b), and no lag phase preceding detectable lipid peroxidation was observed. Obviously, α-tocopherol and other lipoprotein-associated antioxidants are less effective in protecting lipids against lipid-soluble peroxyl radicals than is ascorbic acid against aqueous peroxyl radicals (see above, reactions 2-4).

RELATIVE EFFECTIVENESS OF SMALL MOLECULE ANTIOXIDANTS AND PROTEIN THIOLS IN PLASMA AGAINST LIPID PEROXIDATION INDUCED BY PATHOPHYSIOLOGICAL TYPES OF OXIDATIVE STRESS

AAPH-derived peroxyl radicals obviously are of no pathophysiological significance, and the same may be true for aqueous peroxyl radicals in general. Thus, the above results cannot be applied to human pathology without major reservations. Lipid-soluble peroxyl radicals are somewhat more relevant because they are formed in membranes and lipoproteins as intermediate products of lipid peroxidation, and by metal-ion-catalyzed decomposition of preformed lipid hydroperoxides (Halliwell and Gutteridge 1990). However, lipid-soluble peroxyl radicals appear not to be formed in vivo as primary oxidants that initiate lipid peroxidation, unlike the case in the in vitro experiments. Thus, studies exposing plasma to physiological or pathological types of oxidative stress, rather than aqueous or lipid-soluble peroxyl radicals, should provide more relevant data. Such in vitro studies indeed have been performed, using activated human neutrophils (Frei et al. 1988a; Cross et al. 1990), xanthine oxidase-derived oxidants (Radi et al. 1991; B. Frei and B.N. Ames, see below), cigarette smoke (Frei et al. 1991), and transition metal ions (Stocks et al. 1974; Chahboun et al. 1990). These types of oxidizing conditions probably are relevant to inflammatory-immune diseases (e.g., rheumatoid arthritis, autoimmune diseases), reoxygenation injury, respiratory distress syndrome, emphysema, lung cancer, idiopathic hemochromatosis, and a number of other human diseases (Hallliwell and Gutteridge 1989).

In their major findings, the in vitro studies using pathophysiological types of oxidative stress confirmed the results obtained with peroxyl radical-exposed plasma. Except for experiments using transition metal ions, where metal-binding proteins form the first line of antioxidant defense (see below), ascorbic acid always provided the most effective antiperoxidative protection. For example, in human plasma exposed to $\cdot O_2^-$ and H_2O_2 generated by the xanthine/xanthine oxidase system, first, ascorbic acid was consumed, and then formation of small, detectable amounts of cholesterol ester hydroperoxides and triglyceride hydroperoxides occurred (Fig. 2). Protein thiols were also oxidized rapidly to about 50%, whereas bilirubin and α-tocopherol were not consumed under these oxidative stress conditions. Uric acid concentrations increased, due to formation from xanthine upon reaction with xanthine oxidase. Rapid and partial oxidation of plasma thiols by hypoxanthine/xanthine oxidase was also reported by Radi et al. (1991). Since these authors used nonspecific and non-

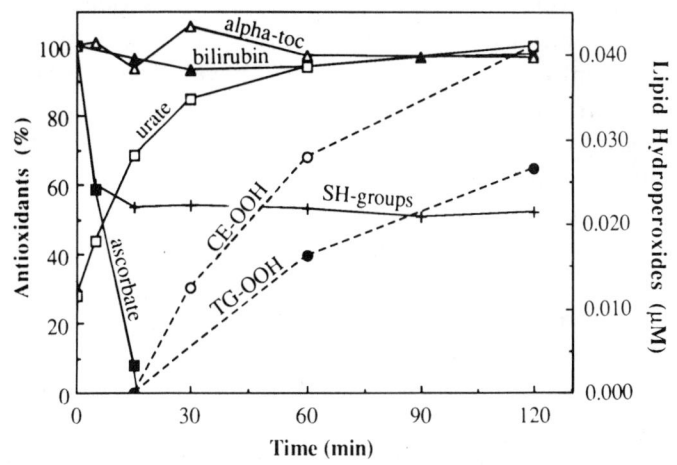

FIGURE 2 Antioxidant defenses and lipid peroxidation in human blood plasma exposed to the xanthine/xanthine oxidase system. Plasma was exposed at 37°C to 10 mU/ml of xanthine oxidase in the presence of 1.5 mM xanthine. The levels of the antioxidants ascorbate (initial concentration 93 µM), protein thiols (SH-groups, 498 µM), bilirubin (8.9 µM), and α-tocopherol (alpha-toc, 32 µM) are given as the percentage of the initial concentrations (left ordinate). The initial urate concentration was 252 µM and increased to 1.55 mM (100%) after 2 hr of incubation. The levels of triglyceride hydroperoxides (TG-OOH) and cholesterol ester hydroperoxides (CE-OOH) are given in µM concentrations (right ordinate). One experiment representative of three is shown.

sensitive indicators of lipid peroxidation (thiobarbituric acid reactive substances and conjugated dienes), they were unable to detect lipid peroxidation in xanthine-oxidase-exposed plasma, which we were able to detect (Fig. 2) with a more sensitive and specific lipid hydroperoxide assay (Frei et al. 1988b; Yamamoto et al. 1990).

On the basis of the measured rate of uric acid production (Fig. 2) and the known approximate ratios of uric acid to $\cdot O_2^-$ production (0.35) and uric acid to H_2O_2 production (1.0) by the xanthine/xanthine oxidase system (Link and Riley 1988), we estimated that in our experiments, only about 15% of the xanthine oxidase-derived oxidants were trapped by ascorbic acid and protein thiols or reacted with plasma lipids to form hydroperoxides. The other 85% of xanthine oxidase-derived

$\cdot O_2^-$ and H_2O_2 remained unaccounted for. Since plasma may contain small amounts of extracellular SOD and/or (contaminating) catalase activity (see above), we included diethyldithiocarbamate or sodium azide, strong inhibitors of SOD and catalase, in the incubation of plasma with xanthine and xanthine oxidase. We found that these inhibitors did not cause increased rates of antioxidant consumption or lipid peroxidation, supporting the view that removal of $\cdot O_2^-$ and H_2O_2 by SOD and catalase contributes little to antioxidant activities of extracellular fluids (Halliwell and Gutteridge 1986). Whether the bulk of xanthine oxidase-derived oxidants in plasma is scavenged by albumin, similar to scavenging of hypochlorite (Wasil et al. 1987), or whether other plasma proteins such as ceruloplasmin also contribute significantly (Goldstein et al. 1979), remains to be established. In whole blood, $\cdot O_2^-$ and H_2O_2 formed extracellularly might be taken up into erythrocytes, either by anion channels in the erythrocyte membrane (Lynch and Fridovich 1978) or by passive diffusion due to lipid-solubility (Halliwell and Gutteridge 1989), respectively, followed by reaction with intraerythrocytic antioxidant enzymes or small molecule antioxidants. Indeed, erythrocytes have been suggested to act as sinks for extraerythrocytic H_2O_2 (Agar et al. 1986). Similarly, copper-induced lipid peroxidation in plasma has been shown to be inhibited by hemolysate (Chahboun et al. 1990).

One important difference between experiments using AAPH and pathophysiological types of oxidative stress (xanthine oxidase, neutrophils, and cigarette smoke) was that the amounts of lipid hydroperoxides formed in plasma under the latter conditions were in the low micromolar range (see Fig. 2) (Frei et al. 1988a, 1991), i.e., much smaller than the amounts formed by incubation with AAPH (see Fig. 1). Most interestingly, moderate lipid peroxidation under pathophysiological oxidative stress caused pathophysiological changes in plasma LDL, whereas massive lipid peroxidation induced by AAPH did not cause such changes. For example, LDL of plasma containing about 0.7 μM lipid hydroperoxides following exposure to gas-phase cigarette smoke displayed increased electrophoretic mobility (Frei et al. 1991), whereas LDL isolated from AAPH-exposed plasma showed normal electrophoretic mobility (B.

Frei, unpubl.). It is known that lipid hydroperoxide-dependent increases in electrophoretic mobility of LDL, as well as other modifications of LDL that increase its atherogenicity, require breakdown of phospholipid hydroperoxides (Steinbrecher et al. 1984), probably by an intrinsic phospholipase A_2 activity of apoprotein B-100 (Parthasarathy and Barnett 1990). Since this phospholipase A_2 activity is itself inactivated by oxidative stress (Parthasarathy and Barnett 1990), it is possible that in AAPH-exposed plasma the phospholipid hydroperoxides formed in LDL are not hydrolyzed due to inactivation of phospholipase A_2, whereas under relatively mild pathophysiological oxidative stress conditions, enough phospholipase A_2 activity remains to stimulate phospholipid hydroperoxide-dependent atherogenic changes in LDL.

When the relative importance of antioxidants in plasma or serum is investigated in in vitro systems based on metal-catalyzed lipid peroxidation, the metal-binding proteins of plasma become of primary importance (Stocks et al. 1974). Except for uric acid, which can chelate iron and copper (Davies et al. 1986), the small molecule antioxidants appear to form a second line of defense under these conditions (Halliwell and Gutteridge 1990). Thus, transferrin, lactoferrin, ceruloplasmin, albumin, uric acid, hemopexin, or haptoglobin can be expected to form the first line of antioxidant defense in plasma when iron, copper, heme, or hemoglobin is released into it (see above). In this context it is interesting to note that ceruloplasmin and haptoglobin are acute-phase proteins, whose plasma concentrations are increased by up to about 50% and two- to threefold, respectively, in acute inflammatory plasma dysproteinemia (Kushner et al. 1981).

RELATIVE EFFECTIVENESS OF LIPID-SOLUBLE SMALL MOLECULE ANTIOXIDANTS AGAINST LIPID PEROXIDATION IN LDL

The antiperoxidative effectiveness of lipid-soluble antioxidants associated with LDL was first investigated by Esterbauer et al. (1989a,b). Using Cu^{++} to induce oxidative stress, it was shown that the antioxidants in isolated human LDL were progressively consumed in the sequence α,γ-tocopherol > lycopene > phytofluene > β-carotene. Lipid peroxidation, measured as diene conjugation, occurred with antioxidant consumption. This can be expected because scavenging of lipid peroxyl radicals by α-tocopherol, and other lipid-soluble, chain-breaking antioxidants in LDL, leads to lipid hydroperoxides according to reaction 3 (Esterbauer et al. 1989b). When the antioxidants in LDL had been depleted, lipid peroxidation entered its uninhibited propagation phase, leading to rapid accumulation of diene conjugates. Added ascorbic acid fully protected the endogenous antioxidants and lipids in LDL against Cu^{++}-induced oxidation (Esterbauer et al. 1989b), supporting the above data obtained with plasma. The strong antiperoxidative effects of ascorbic acid, and ascorbic acid's capability to spare LDL-associated antioxidants, have been confirmed recently by another group, using LDL exposed to Cu^{++} or human monocyte/macrophages (Jialal and Grundy 1991).

We have shown that in isolated LDL exposed to AAPH, ubiquinol-10 was consumed before α-tocopherol, β-carotene, and lycopene (Stocker et al. 1991). As measured by the lipid hydroperoxide assay mentioned above (Frei et al. 1988b; Yamamoto et al. 1990), the rate of radical-mediated formation of lipid hydroperoxides in LDL was low as long as ubiquinol-10 was present but increased rapidly after its consumption, even though more than 80% and 95% of endogenous carotenoids and α-tocopherol, respectively, were still present (Stocker et al. 1991). In the presence of ascorbic acid, there was an initial lag phase of detectable lipid peroxidation in LDL exposed to AAPH. Qualitatively similar results were obtained when peroxyl radicals were generated within LDL, or when LDL was exposed to oxidants produced by activated human neutrophils (Stocker et al. 1991). Thus, it appears that within LDL, ubiquinol-10 forms the first line of antiperoxidative defense, inhibiting LDL oxidation more efficiently than carotenoids and α-tocopherol. The notion that α-tocopherol, despite being by far the most concentrated lipid-soluble antioxidant in LDL, does not form the first line of antioxidant defense in this lipoprotein is further supported by recent evidence that the susceptibility of LDL to oxidation is not always related to its α-tocopherol content (Babiy et al. 1990; Knipping et al. 1990). Whether ubiquinol-10 levels better predict the ease with which different LDL preparations are oxidized remains to be investigated.

The reactivity of lipid-soluble antioxidants under another type of oxidative stress, viz. singlet oxygen, was shown to decrease in the order lycopene > α-carotene > β-carotene > lutein > bilirubin > α-tocopherol (Di Mascio et al. 1989). Taking the plasma concentrations of these antioxidants into consideration, the authors concluded that lycopene and bilirubin are qualitatively about equally important in quenching singlet oxygen in plasma, followed by β-carotene and α-tocopherol. However, another report, also based on plasma concentrations and quenching constants of the individual antioxidants, concluded that plasma proteins and uric acid quench most of the singlet oxygen, whereas carotenoids, bilirubin, tocopherols, and ascorbic acid make only a small contribution (<4%) (Kanofsky 1990).

CONCLUSIONS

The data reviewed in this chapter indicate that in extracellular fluids, where antioxidant enzymes are basically absent, ascorbic acid is of primary importance as a water-soluble antioxidant. Ascorbic acid protects against a plethora of pathophysiological types of oxidative stress. Thus, vitamin C should help prevent degenerative and other diseases in which oxidative stress plays a causative or exacerbating role. Clinical and epidemiological data indeed indicate that vitamin C may contribute to protection against atherosclerosis and certain types of cancer (Block et al. 1991; Riemersma et al. 1991). The in vivo relevance of observations made in vitro that ascorbic acid interacts with other antioxidants of extracellular fluids, such as α-tocopherol, uric acid, and protein thiols, remains to be established. Other open questions are concerned with scavenging of $\cdot O_2^-$, H_2O_2, and singlet oxygen in plasma and blood, and lack of oxidative (atherogenic) modification of plasma LDL by aqueous peroxyl radicals despite massive lipid peroxidation.

Within lipids, it now appears that α-tocopherol is not as effective as generally assumed, a finding that may be particularly important for the prevention of atherosclerosis. Although it is well documented that α-tocopherol can strongly suppress lipid peroxidation, the important question is whether this protection is also biologically or pathologically significant, i.e., sufficient to prevent disease. Recent in vitro data indicate that ubiquinol-10 is more effective than α-tocopherol and that there may well exist as yet unidentified lipid-soluble antioxidants in lipoproteins. Clinical and epidemiological studies on the antioxidant effectiveness of α-tocopherol in the diet may be complicated by the fact that the distribution of ubiquinone in foods is quite similar to that of α-tocopherol (Kamei et al. 1985). Thus, a reevaluation of many of these studies seems indicated.

REFERENCES

Agar, N.S., S.M.H. Sadrzadeh, P.E. Hallaway, and I.W. Eaton. 1986. Erythrocyte catalase. A somatic oxidant defense?. *Clin. Invest.* **77:** 319.

Ames, B.N. 1983. Dietary carcinogens and anticarcinogens. *Science* **221:** 1256.

Ames, B.N., R. Cathcart, E. Schwiers, and P. Hochstein. 1981. Uric acid provides an antioxidant defense in humans against oxidant- and radical-caused aging and cancer: A hypothesis. *Proc. Natl. Acad. Sci.* **78:** 6858.

Aruoma, O.I. and B. Halliwell. 1987. Superoxide-dependent and ascorbate-dependent formation of hydroxyl radicals from hydrogen peroxide in the presence of iron. Are lactoferrin and transferrin promoters of hydroxyl-radical generation? *Biochem. J.* **241:** 273.

Avissar, N., J.C. Whitin, P.Z. Allen, D.D. Wagner, P. Liegey, and H.J. Cohen. 1989. Plasma selenium-dependent glutathione peroxidase. Cell of origin and secretion. *J. Biol. Chem.* **264:** 15850.

Biedmond, P., H.G. van Eijk, A.J.G Swaak, and J.F. Koster. 1984. Iron mobilization from ferritin by superoxide derived from stimulated polymorphonuclear leukocytes: Possible mechanism in inflammation diseases. *J. Clin. Invest.* **73:** 1576.

Babiy, A.V., J.M. Gebicki, and D.R. Sullivan. 1990. Vitamin E content and low density lipoprotein oxidizability induced by free radicals. *Atherosclerosis* **81:** 175.

Block, G., D.E. Henson, and M. Levine. 1991. Vitamin C: Biologic functions and relation to cancer. *Nutr. Cancer* **15:** 249.

Cantin, A.M., S.L. North, R.C. Hubbard, and R.G. Crystal. 1987. Normal alveolar epithelial lining fluid contains high levels of glutathione. *J. Appl. Physiol.* **63:**152.

Cerutti, P.A. 1991. Oxidant stress and carcinogenesis. *Eur. J. Clin. Invest.* **21:** 1.
Chahboun, S., C. Tallineau, R. Pontcharraud, A. Guettier, and A. Piriou. 1990. Polyunsaturated fatty acid profiles and α-tocopherol levels in plasma and whole blood incubated with copper. Evidence of inhibition of lipoperoxidation in plasma by hemolysate. *Biochim. Biophys. Acta* **1042:** 324.
Chance, B., H. Sies, and A. Boveris. 1979. Hydroperoxide metabolism in mammalian organs. *Physiol. Rev.* **59:** 527.
Cross, C.E., T. Forte, R. Stocker, S. Louie, Y. Yamamoto, B.N. Ames, and B. Frei. 1990. Oxidative stress and abnormal cholesterol metabolism in patients with adult respiratory distress syndrome. *J. Lab. Clin. Med.* **115:** 396.
D'Aquino, M., C. Dunster, and R.L. Willson. 1989. Vitamin A and glutathione-mediated free radical damage: Competing reactions with polyunsaturated fatty acids and vitamin C. *Biochem. Biophys. Res. Commun.* **161:** 1199.
Davies, K.J.A., A. Sevanian, S.F. Muakkassah-Kelly, and P. Hochstein. 1986. Uric acid-iron ion complexes. A new aspect of the antioxidant functions of uric acid. *Biochem. J.* **235:** 747.
Di Mascio, P., S. Kaiser, and H. Sies. 1989. Lycopene as the most efficient biological carotenoid singlet oxygen quencher. *Arch. Biochem. Biophys.* **274:** 532.
Esterbauer, H., G. Jürgens, H. Puhl, and O. Quehenberger. 1989a. Role of oxidatively modified LDL in atherogenesis. In *Medical biochemical and chemical aspects of free radicals* (ed. O. Hayaishi et al.), p. 1203. Elsevier, Amsterdam.
Esterbauer, H., G. Striegl, H. Puhl, S. Oberreither, M. Rotheneder, M. El-Saadani, and G. Jürgens. 1989b. The role of vitamin E and carotenoids in preventing oxidation of low density lipoprotein. *Ann. N.Y. Acad. Sci.* **570:** 254.
Frei, B., L. England, and B.N. Ames. 1989. Ascorbate is an outstanding antioxidant in human blood plasma. *Proc. Natl. Acad. Sci.* **86:** 6377.
Frei, B., M.C. Kim, and B.N. Ames. 1990a. Ubiquinol-10 is an effective lipid-soluble antioxidant at physiological concentrations. *Proc. Natl. Acad. Sci.* **87:** 4879.
Frei, B., R. Stocker, and B.N. Ames. 1988a. Antioxidant defenses and lipid peroxidation in human blood plasma. *Proc. Natl. Acad. Sci.* **85:** 9748.
Frei, B., T. Forte, B.N. Ames, and C.E. Cross. 1991. Gas-phase oxidants of cigarette smoke induce lipid peroxidation and changes in lipoprotein properties in human blood plasma: Protective effects of ascorbic acid. *Biochem. J.* **277:** 133.
Frei, B., Y. Yamamoto, D. Niclas, and B.N. Ames. 1988b. Evaluation of an isoluminol chemiluminescence assay for the detection of hydroperoxides in human blood plasma. *Anal. Biochem.* **175:** 120.

Frei, B., R. Stocker, L. England, and B.N. Ames. 1990b. Ascorbate: The most effective antioxidant in human blood plasma. In *Antioxidants in therapy and preventive medicine* (ed. I. Emerit and B. Halliwell), p. 155. Plenum Press, New York.

Goldstein, I.M., H.B. Kaplan, H.S. Edelson, and G. Weissman. 1979. Ceruloplasmin. A scavenger of superoxide anion radicals. *J. Biol. Chem.* **254:** 4040.

Goth, L. 1991. A simple method for determination of serum catalase activity and revision of reference range. *Clin. Chim. Acta* **196:** 143.

Grootveld, M., J.D. Bell, B. Halliwell, O.I. Aruoma, A. Bomford, and P.J. Sadler. 1989. Non-transferrin-bound iron in plasma or serum from patients with idiopathic hemochromatosis. Characterization by high performance liquid chromatography and nuclear magnetic resonance spectroscopy. *J. Biol. Chem.* **264:** 4417.

Gutteridge, J.M.C. 1983. Antioxidant properties of caeruloplasmin towards iron- and copper-dependent oxygen radical formation. *FEBS Lett.* **157:** 37.

———. 1984. Copper-phenanthroline-induced site-specific oxygen-radical damage to DNA. Detection of loosely bound trace copper in biological fluids. *Biochem. J.* **218:** 983.

———. 1986. Antioxidant properties of the proteins caeruloplasmin, albumin and transferrin. A study of their activity in serum and synovial fluid from patients with rheumatoid arthritis. *Biochim. Biophys. Acta* **869:** 119.

Gutteridge, J.M.C., D.A. Rowley, and B. Halliwell. 1981. Superoxide-dependent formation of hydroxyl radicals in the presence of iron salts. Detection of "free" iron in biological systems by using bleomycin-dependent degradation of DNA. *Biochem. J.* **199:** 263.

Halliwell, B. 1988. Albumin—An important extracellular antioxidant? *Biochem. Pharmacol.* **37:** 569.

Halliwell, B. and J.M.C. Gutteridge. 1985. Oxygen radicals and the nervous system. *Trends Neurosci.* **8:** 22.

———. 1986. Oxygen free radicals and iron in relation to biology and medicine: Some problems and concepts. *Arch. Biochem. Biophys.* **246:** 501.

———. 1989. *Free radicals in biology and medicine*, 2nd edition. Clarendon Press, Oxford.

———. 1990. The antioxidants of human extracellular fluids. *Arch. Biochem. Biophys.* **280:** 1.

Jacobs, A., F.R.C. Path, and M. Worwood. 1975. Ferritin in serum. Clinical and biochemical implications. *N. Engl. J. Med.* **292:** 951.

Jialal, I. and S.M. Grundy. 1991. Preservation of the endogenous antioxidants in low density lipoprotein by ascorbate but not probucol during oxidative modification. *J. Clin. Invest.* **87:** 597.

Kamei, M., T. Fujita, T. Kanbe, K. Sasaki, K. Oshiba, S. Otani, I. Matsui-Yuasa, and S. Morisawa. 1985. The distribution and content of ubiquinone in foods. *Int. J. Vitam. Nutr. Res.* **56:** 57.

Kanofsky, J.R. 1990. Quenching of singlet oxygen by human plasma. *Photochem. Photobiol.* **51:** 299.

Kappus, H. 1985. Lipid peroxidation: Mechanisms, analysis, enzymology and biological relevance. In *Oxidative stress: Oxidants and antioxidants* (ed. H. Sies), p. 273. Academic Press, London.

Karlsson, K. and S.L. Marklund. 1987. Heparin-induced release of extracellular superoxide dismutase to human blood plasma. *Biochem. J.* **242:** 55.

Knipping, G., M. Rotheneder, G. Striegl, and H. Esterbauer. 1990. Antioxidants and resistance against oxidation of porcine LDL subfractions. *J. Lipid Res.* **31:** 1965.

Kushner, I., H. Gewurz, and M.D. Benson. 1981. C-reactive protein and the acute phase response. *J. Lab. Clin. Med.* **97:** 739.

Lentner, C., ed. 1981. *Geigy scientific tables,* vol. 1. Ciba-Geigy, Basel.

———. 1984. *Geigy scientific tables,* vol. 3. Ciba-Geigy, Basel.

Link, E.M. and P.A Riley. 1988. Role of hydrogen peroxide in the cytotoxicity of the xanthine/xanthine oxidase system. *Biochem. J.* **249:** 391.

Lynch, R.E. and I. Fridovich. 1978. Permeation of the erythrocyte stroma by superoxide radical. *J. Biol. Chem.* **253:** 4697.

Maddipati, K.R. and L.J. Marnett. 1987. Characterization of the major hydroperoxide-reducing activity of human plasma. *J. Biol. Chem.* **262:** 17398.

Maguire, J.J., D.S. Wilson, and L. Packer. 1989. Mitochondrial electron transport-linked tocopheroxyl radical reduction. *J. Biol. Chem.* **264:** 21462.

Maples, K.R. and R.P. Mason. 1988. Free radical metabolite of uric acid. *J. Biol. Chem.* **263:** 1709.

Niki, E., Y. Yamamoto, M. Takahashi, K. Yamamoto, Y. Yamamoto, E. Komuro, M. Miki, H. Yasuda, and M. Mino. 1988. Free radical-mediated damage of blood and its inhibition by antioxidants. *J. Nutr. Sci. Vitaminol.* **34:** 507.

Orringer, E.P. and M.E.S. Roer. 1979. An ascorbate-mediated transmembrane-reducing system of the human erythrocyte. *J. Clin. Invest.* **63:** 53.

Parthasarathy, S. and J. Barnett. 1990. Phospholipase A_2 activity of low density lipoprotein: Evidence for an intrinsic phospholipase A_2 activity of apoprotein B-100. *Proc. Natl. Acad. Sci.* **87:** 9741.

Radi, R., K.M. Bush, T.P. Cosgrove, and B.A. Freeman. 1991. Reaction of xanthine oxidase-derived oxidants with lipid and protein of human plasma. *Arch. Biochem. Biophys.* **286:** 117.

Riemersma, R.A., D.A. Wood, C.C.A. MacIntyre, R.A. Elton, K.F. Gey, and M.F. Oliver. 1991. Risk of angina pectoris and plasma concentrations of vitamins A, C, and E and carotene. *Lancet* **337:** 1.

Sato, K., E. Niki, and H. Shimasaki. 1990. Free radical-mediated chain oxidation of low density lipoprotein and its synergistic inhibition by vitamin E and vitamin C. *Arch. Biochem. Biophys.* **279:** 402.

Schöneich, C., K.-D. Asmus, U. Dillinger, and F. von Bruchhausen. 1989. Thiyl radical attack on polyunsaturated fatty acids: A possible route to lipid peroxidation. *Biochem. Biophys. Res. Commun.* **161:** 113.

Schwartz, C.J., A.J. Valente, E.A. Sprague, J.L. Kelley, and R.M. Nerem. 1991. The pathogenesis of atherosclerosis: An overview. *Clin. Cardiol.* **14:** I-1.

Specter, R. 1977. Vitamin homeostasis in the central nervous system. *N. Engl. J. Med.* **296:** 1393.

Steinberg, D., S. Parthasarathy, T.E. Carew, J.C. Khoo, and J.L. Witztum. 1989. Beyond cholesterol. Modifications of low-density lipoprotein that increase its atherogenicity. *N. Engl. J. Med.* **320:** 915.

Steinbrecher, U.P., S. Parthasarathy, D.S. Leake, J.L. Witztum, and D. Steinberg. 1984. Modification of low density lipoprotein by endothelial cells involves lipid peroxidation and degradation of low density lipoprotein phospholipids. *Proc. Natl. Acad. Sci.* **81:** 3883.

Stocker, R. and B.N. Ames. 1987. Potential role of conjugated bilirubin and copper in the metabolism of lipid peroxides in bile. *Proc. Natl. Acad. Sci.* **84:** 8130.

Stocker, R. and B. Frei. 1991. Endogenous antioxidant defences in human blood plasma. In *Oxidative stress: Oxidants and antioxidants* (ed. H. Sies) p. 213. Academic Press, London.

Stocker, R. and E. Peterhans. 1989. Synergistic interaction between vitamin E and the bile pigments bilirubin and biliverdin. *Biochim. Biophys. Acta* **1002:** 238.

Stocker, R., V.W. Bowry, and B. Frei. 1991. Ubiquinol-10 protects human low density lipoprotein more efficiently against lipid peroxidation than does α-tocopherol. *Proc. Natl. Acad. Sci.* **88:** 1646.

Stocker, R., A.N. Glazer, and B.N. Ames. 1987. Antioxidant activity of albumin-bound bilirubin. *Proc. Natl. Acad. Sci.* **84:** 5918.

Stocks, J., J.M.C. Gutteridge, R.J. Sharp, and T.L. Dormandy. 1974. The inhibition of lipid autoxidation by human serum and its relation to serum proteins and α-tocopherol. *Clin. Sci. Mol. Med.* **47:** 223.

Takahashi, K., N. Avissar, J. Whitin, and H. Cohen. 1987. Purification and characterization of human plasma glutathione peroxidase: A selenoglycoprotein distinct from the known cellular enzyme. *Arch. Biochem. Biophys.* **256:** 677.

Wasil, M., B. Halliwell, D.C.S. Hutchison, and H. Baum. 1987. The antioxidant action of human extracellular fluids. Effect of human serum and its protein components on the inactivation of α_1-antiproteinase by hypochlorous acid and by hydrogen peroxide. *Biochem. J.* **243:** 219.

Wayner, D.D.M., G.W. Burton, and K.U. Ingold. 1986. The

antioxidant efficiency of vitamin C is concentration-dependent. *Biochim. Biophys. Acta* **884:** 119.

Wayner, D.D.M., G.W. Burton, K.U. Ingold, and S.J. Locke. 1985. Quantitative measurement of the total, peroxyl radical-trapping antioxidant capability of human blood plasma by controlled peroxidation. The important contribution made by plasma proteins. *FEBS Lett.* **187:** 33.

Wayner, D.D.M., G.W. Burton, K.U. Ingold, L.R.C. Barclay, and S.J. Locke. 1987. The relative contributions of vitamin E, urate, ascorbate and proteins to the total peroxyl radical-trapping antioxidant activity of human blood plasma. *Biochim. Biophys. Acta* **924:** 408.

Yamamoto, Y., B. Frei, and B.N. Ames. 1990. Assay of lipid hydroperoxides using high-performance liquid chromatography with isoluminol chemiluminescence detection. *Methods Enzymol.* **186:** 371.

DNA Damage by Oxygen-derived Species: Its Mechanism, and Measurement Using Chromatographic Methods

B. Halliwell[1] and O.I. Aruoma[2]
[1]Division of Pulmonary Medicine, Davis Medical Center
University of California, Sacramento, California 95817
[2]Department of Biochemistry, King's College, University of London
London WC2R 2LS, United Kingdom

It is well established that aerobes constantly produce small amounts of oxygen-derived species, such as superoxide radical ($\cdot O_2^-$), hydrogen peroxide (H_2O_2), and hypochlorous acid (HOCl), the latter being generated by the enzyme myeloperoxidase in neutrophils (for reviews, see di Guiseppi and Fridovich 1984; Halliwell and Gutteridge 1989; Weiss 1989). Exposure of living organisms to background levels of ionizing radiation leads to homolytic fission of oxygen-hydrogen bonds in water to produce highly reactive hydroxyl radicals, $\cdot OH$ (for review, see Von Sonntag 1987). Hydroxyl radicals can also be generated when H_2O_2 comes into contact with certain transition metal ion chelates, especially those of iron and copper. In general, the reduced forms of these metal ions (Fe^{++}, Cu^+) produce $\cdot OH$ at a faster rate upon reaction with H_2O_2 than the oxidized forms (Fe^{+++}, Cu^{++}), and so reducing agents such as $\cdot O_2^-$ and ascorbic acid can often accelerate $\cdot OH$ generation by metal ion/H_2O_2 mixtures (Halliwell and Gutteridge 1990a).

Aerobes have evolved antioxidant defenses to protect themselves against the oxygen-derived species generated in vivo. These defenses include enzymes (such as superoxide dismutases, catalase, and glutathione peroxidases), low-molecular-mass agents (such as α-tocopherol and ascorbic acid), and proteins that bind metal ions in forms unable to accelerate free radical reactions (Sies 1991; di Guiseppi and Fridovich 1984; Halliwell and Gutteridge 1990a,b). *Oxidative stress* results when oxygen-derived species are not adequately removed. This can happen if antioxidants are depleted and/or if the formation of oxygen-derived species is increased beyond the ability of the defenses to cope with them (Sies 1991).

Subjecting cells to oxidative stress can result in severe metabolic dysfunctions, including peroxidation of membrane lipids, depletion of nicotinamide nucleotides, rises in intracellular free Ca^{++} ions, cytoskeletal disruption, and DNA damage. The latter is often measured as formation of single-strand breaks, double-strand breaks, or chromosomal aberrations. Indeed, DNA damage has been almost invariably observed in a wide range of mammalian cell types exposed to oxidative stress in a number of different ways (Table 1). Oxidative stress (Larrick and Wright 1990) and DNA damage (Zimmerman et al. 1989) also occur when some mammalian cells are exposed to tumor necrosis factor. Oxidative stress may additionally play some role in the carcinogenicity of asbestos, cigarette smoke (see, e.g., Kiyosawa et al. 1990), and certain metals, such as nickel.

TABLE 1 METHODS USED TO SUBJECT CELLS TO OXIDATIVE STRESS THAT HAS PRODUCED INCREASED INTRACELLULAR DNA DAMAGE

Elevated O_2 concentrations
Exposure to activated phagocytic cells
Exposure to "redox cycling" drugs (e.g., alloxan, paraquat, menadione)
Exposure to cigarette smoke
Exposure to ozone
Exposure to ionizing radiation
Direct addition of hydrogen peroxide
Exposure to "autoxidizing" agents (e.g., dihydroxyfumarate)
Exposure to xanthine oxidase[a] plus its substrates (xanthine, hypoxanthine)

See Halliwell and Aruoma (1991).
[a]Care must be taken in the use of commercial xanthine oxidase, which is often heavily contaminated with proteases and other material directly injurious to cells.

POSSIBLE MECHANISMS OF DNA DAMAGE INDUCED BY OXIDATIVE STRESS

Why does oxidative stress cause DNA damage? In the case of externally generated oxygen-derived species (e.g., when cells are incubated with H_2O_2, activated phagocytes, or xanthine oxidase plus its substrates), damage is usually inhibited by adding catalase, showing that H_2O_2 is needed. Superoxide dismutase (SOD) does not usually inhibit much, which could mean either that $\cdot O_2^-$ is not involved in the DNA damage, or that SOD does not enter cells easily. That the latter interpretation is correct in at least one cell system is shown by the observations that SOD can protect hepatocytes from the toxicity of H_2O_2 or t-butyl hydroperoxide under conditions where SOD can enter the cells (Kyle et al. 1988; Nakae et al. 1990).

It has been known for many years that neither $\cdot O_2^-$ nor H_2O_2 causes any strand breakage in DNA, if the reaction mixture is carefully freed of transition metal ions (see, e.g., Rowley and Halliwell 1983). Our more recent work (Aruoma et al. 1989a,b, 1991) has confirmed this inability of $\cdot O_2^-$ or H_2O_2 at physiologically relevant concentrations to damage DNA, by looking for chemical changes in the purine and pyrimidine bases (see below). Hence, DNA damage by oxidative stress in vivo is unlikely to involve direct attack of $\cdot O_2^-$ or H_2O_2 upon the DNA.

Two explanations of the DNA damage have been advanced (Fig. 1). First, it is possible that the damage is due to $\cdot OH$ radical formation, a proposal first clearly stated by Mello-Filho et al. (1984). Thus, it is envisaged that H_2O_2, which crosses biological membranes easily, can penetrate to the nucleus and react with iron and/or copper ions to form $\cdot OH$. Because of the high reactivity of $\cdot OH$ and its resultant inability to diffuse significant distances within the cell (Halliwell and Gutteridge 1990a), this mechanism is only feasible if the $\cdot OH$ is generated from H_2O_2 by reaction with metal ions bound *upon or very close to* the DNA. One possibility is that these metal ions are always present bound to the DNA in vivo. For example, copper ions are thought to be present in chromosomes (Lewis and Laemmli 1982; Dijkwel and Wenink 1986; Prutz et al. 1990), and copper ions are very effective in promoting H_2O_2-depen-

A. Fenton Chemistry

Oxidative stress → •OH generation upon DNA by reaction of H_2O_2 with transition metal ions already bound to DNA → Strand Breakage / DNA Base Modification / Deoxyribose Fragmentation

Oxidative stress → release of catalytic copper or iron ions within the cell → binding of ions to DNA → (•OH generation box above)

B. Nuclease Activation

Oxidative stress → Inactivation of Ca^{2+}-binding by endoplasmic reticulum. Inhibition of plasma membrane Ca^{2+}-extrusion systems. Release of Ca^{2+} from mitochondria. → DNA Fragmentation (no base modification)

↓ rise in intracellular free Ca^{2+} → endonuclease activation → (DNA Fragmentation box above)

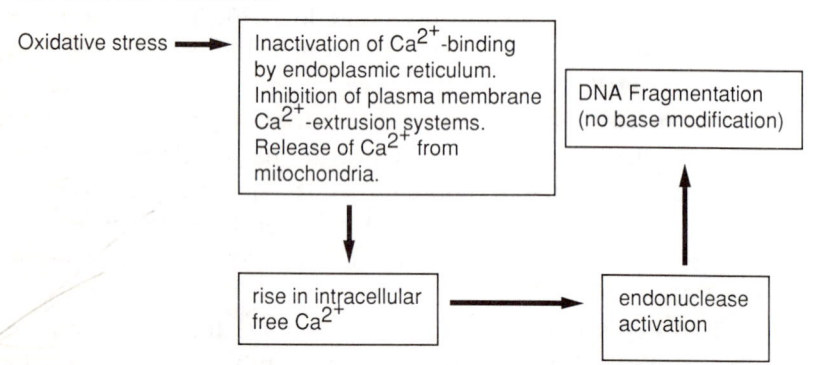

FIGURE 1 Hypotheses to explain DNA damage resulting from exposing cells to oxidative stress.

dent damage to isolated DNA and to DNA within chromatin in vitro (see below). A second possibility (suggested by Halliwell 1987) is that the metal ions are released within the cell as a result of oxidative stress, and then bind to the DNA. Thus, just as oxidative stress causes rises in intracellular free Ca^{++} (Orrenius et al. 1989), it may cause rises in intracellular free iron and/or copper ions by interfering with normal intracellular sequestration mechanisms. Some of these released ions may then bind to DNA and make it a target for oxidative damage.

A second explanation of the ability of oxidative stress to

cause DNA damage is that the stress triggers a series of metabolic events within the cell that lead to activation of nuclease enzymes, which cleave the DNA backbone. Much has been written recently (for discussions, see Orrenius et al. 1989; Farber 1990) about the suggestion that oxidative stress causes rises in intracellular free Ca^{++}, which fragment DNA by activating Ca^{++}-dependent endonucleases in a mechanism resembling that of apoptosis (programmed cell death) (see Wyllie 1980). An example of apoptosis is the killing of immature thymocytes by glucocorticoid hormones, which activate a cell-destructive process that apparently involves DNA fragmentation by a Ca^{++}-dependent nuclease.

These two mechanisms (DNA damage by •OH or by activation of nucleases) are not mutually exclusive, i.e., they could both take place (Fig. 1). Indeed, there is evidence consistent with the existence of both mechanisms. Their relative importance may depend on the cell type used and on how the oxidative stress is imposed (Halliwell and Aruoma 1991). For example, chelating agents that bind iron ions into chelates unable to generate •OH (such as desferrioxamine, desferrithiocin, and phenanthroline) can often protect cells against DNA damage and other toxic effects of oxidative stress (Mello-Filho et al. 1984; Imlay and Linn 1988; Halliwell and Aruoma 1991). The effects of desferrioxamine are variable, since in general it does not cross cell membranes readily, although it appears to enter some cell types (such as hepatocytes) more readily than it enters others. The evidence for a role played by metabolic changes in the DNA damage produced in cells as a result of oxidative stress is also strong (Birnboim 1988; Larsson and Cerutti 1989; Orrenius et al. 1989). Menadione and other quinones (which "redox cycle" within cells to give •O_2^- and H_2O_2) appear to produce DNA strand breaks in hepatocytes by Ca^{++}-dependent activation of an endonuclease. DNA damage could be inhibited by preventing the rise in Ca^{++} using Ca^{++} chelators. Oxidative stress can also sometimes activate and/or cause changes in the subcellular location of protein kinase C. Cantoni et al. (1989) found that the Ca^{++}-chelator quin 2 inhibited H_2O_2-induced DNA strand breakage in CHO cells, although it did not inhibit iron ion-dependent •OH generation from H_2O_2 in vitro under their reaction conditions (its effect on copper ion-dependent •OH formation was not examined). Of course, even if transition metal ion-quin 2 complexes are capable of catalyzing •OH formation, the chelator could still protect by removing metal ions from the vicinity of the DNA, so that any •OH generated no longer attacks this molecule (Halliwell and Gutteridge 1990a). It is also possible that chelators such as desferrioxamine and phenanthroline interfere with changes in cell Ca^{++} metabolism in response to oxidative stress. It is clear that attempting to elucidate the mechanism of DNA damage in the nucleus of cells subjected to oxidative stress by adding free radical scavengers or metal ion chelators to the surrounding media is unlikely to give unambiguous answers.

THE PHYSIOLOGICAL IMPORTANCE OF DNA DAMAGE INDUCED BY OXIDATIVE STRESS

Why is it important to understand the mechanism of DNA damage by oxidative stress? Oxidative stress, imposed by a variety of mechanisms (including increased O_2 concentrations) has been convincingly shown to be mutagenic to bacterial and mammalian cells (for review, see Halliwell and Aruoma 1991). For example, *Escherichia coli* mutants lacking SOD activity show greatly enhanced rates of spontaneous mutation (Touati 1989). Moraes et al. (1990) studied the pattern of mutations obtained in a gene of a shuttle plasmid when simian cells transfected with this plasmid were exposed to H_2O_2. Both single base changes and deletions were observed. The majority of base changes were at GC base pairs, the GC → AT base transition being predominant. Treatment of the plasmid with H_2O_2 in vitro before transfection did not produce an increased number of mutations (unless iron ions were added), consistent with the inability of H_2O_2 to react directly with DNA. McBride et al. (1991) found similar results when single-stranded M13mp2 DNA was incubated with Fe^{++} ions under aerobic conditions and then transfected into *E. coli*.

Mutations induced by oxidative stress may lead to cancer. Ionizing radiation is well known to be both mutagenic and carcinogenic. Since much of the cell damage caused by such radiation involves •OH production by homolytic fission of the oxygen-hydrogen bonds in water, •OH can probably be classified as a complete carcinogen. Base-pair changes and some frameshifts are the commonest mutations observed in cells exposed to ionizing radiation (for review, see Breimer 1988, 1990). Chemical changes in the DNA bases, single- and double-strand breaks, and enhanced expression of certain proto-oncogenes have also been described (Von Sonntag 1987). However, the precise relationship between these different events and the development of cancer is uncertain. Thus, the chemical changes in DNA may themselves somehow lead to cancer (for discussion, see Floyd 1990). An unrepaired lesion in DNA may be bypassed in an error-prone fashion. Resynthesis of DNA after excision repair may conceivably introduce errors.

There are many steps between a healthy cell and a malignant tumor. Cancer biologists have often referred to at least three stages: initiation (an irreversible change in DNA), promotion (probably involving changes in gene expression), and progression (further changes in DNA leading to the eventual production of a malignant tumor). Both Zimmerman and Cerutti (1984) and Weitzman et al. (1985) showed that a clone of C3H mouse fibroblasts exposed to activated human neutrophils or to hypoxanthine plus xanthine oxidase underwent malignant transformation. Nassi-Calo et al. (1989) showed that H_2O_2 also transformed these cells, an action prevented by the chelating agent o-phenanthroline. The ability of oxidative stress to induce transformation has also been shown in human lung fibroblasts (Weitberg and Corvese 1990). Weitzman et al. (1988) reported that DNA isolated from C3H mouse fibroblasts that had been transformed by exposure to activated neutrophils could sometimes transform another cell line, NIH-3T3, when the DNA was transfected into the latter cells.

Although most attention has been paid in the literature to the action of oxygen-derived species as promoters of carcinogenesis, their ability to damage DNA and produce alterations in gene expression implies that they could be involved in all stages of carcinogenesis. Indeed, it has been argued (see, e.g., Totter 1980; Ames 1989) that continuous oxidative damage to DNA by free radical mechanisms is a significant cause of cancer in humans. Of course, DNA damage resulting from oxidative stress (or from any other mechanism) need not necessarily lead to cancer. Low levels of damage may be efficiently repaired with a minimal risk of error (Breimer 1991). High levels of oxidative stress may lead to cell death, so that initiated cells do not remain in the organism. Thus, an intermediate level of oxidative stress is most likely to predispose to malignancy. It is interesting to note the association of chronic inflammation (involving phagocytic production of $\cdot O_2^-$ and H_2O_2) with malignancy in such human diseases as ulcerative colitis, Crohn's disease, and reflux esophagitis. Cerutti et al. (1989) showed that one difference between a clone of mouse epidermal cells that was promotable by xanthine/xanthine oxidase and a nonpromotable clone was that the latter had lower levels of SOD and catalase and was more sensitive to killing by oxygen-derived species. Thus, increased antioxidant defenses, by protecting against cell death resulting from oxidative stress, may sometimes conceivably, and ironically, lead to increased cancer.

PROBING THE MECHANISM OF DNA DAMAGE IN CELLS EXPOSED TO OXIDATIVE STRESS: THE PRINCIPLES

We have already commented that it is difficult to gain information about the mechanism of oxidative stress-induced DNA damage by using antioxidants and scavengers. Another means of implicating free radicals as damaging agents is to use what we have called a "fingerprint" approach: If a free radical produces a unique pattern of chemical change in a biological molecule, then observation of the same pattern in vivo is evidence that the radical attacked that molecule. The damage pattern *must* be unique to that radical. Therefore, we set out to characterize the chemical changes produced in DNA by different oxygen-derived species, using the technique of gas chromatography/mass spectrometry with selected ion monitoring, largely developed for work with DNA by Dr. Miral Dizdaroglu (1991). We concentrated on damage to the purine (adenine, guanine) and pyrimidine (cytosine, thymine) bases, because sugar-damage products are much less chemically distinctive (Von Sonntag 1987). We found, as expected, that •O$_2^-$ and H$_2$O$_2$ at physiologically relevant concentrations do not themselves cause any base damage in DNA (Aruoma et al. 1989a,b, 1991).

Studies by Dizdaroglu and others (for review, see Dizdaroglu 1991) had already shown that •OH reacts in a multiplicity of ways with all four DNA bases. Thus, •OH can add on to guanine residues at C4, C5, and C8 positions to give hydroxyguanine radicals that can do various things. For example, addition of •OH to C8 of guanine produces a radical that can be reduced to 8-hydroxy-7,8-dihydroguanine, oxidized to 8-hydroxyguanine (8-OH-Gua), or can undergo ring opening followed by one electron reduction and protonation to give 2,6-diamino-4-hydroxy-5-formamidopyrimidine, usually abbreviated as FapyGua. Figure 2 shows the structures of some of these products. Similarly, •OH can add on to C4, C5, or C8 of adenine. Among other fates, the C8 •OH adenine radical can be converted into 8-hydroxyadenine (8-OH-Ade) by oxidation

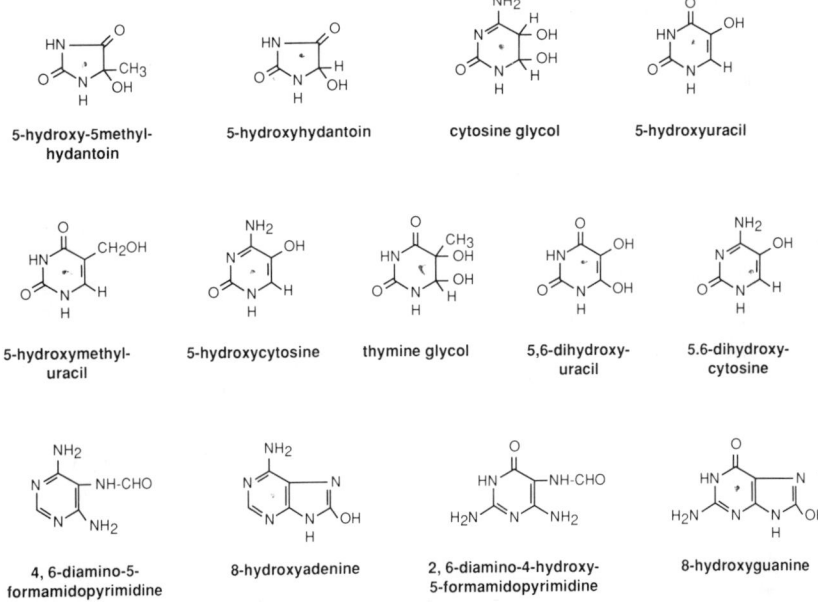

FIGURE 2 Some of the end products that result from attack of hydroxyl radicals upon the bases of DNA.

or can undergo ring opening followed by one-electron reduction to give 5-formamido-4,6-diaminopyrimidine (FapyAde). Pyrimidines are also attacked by •OH to give multiple products. Thus, thymine can form *cis* and *trans* thymine glycols (5,6-dihydroxy-6-hydrothymines), 5-hydroxy-5-methylhydantoin, 5,6-dihydrothymine, and 5-hydroxymethyluracil. Cytosine can form several products, including cytosine glycol and 5,6-dihydroxycytosine (Fig. 2). In addition, •OH generation within whole cells or isolated chromatin can result in formation of cross-links between DNA bases and amino acid residues in nuclear proteins. Thus, thymine-tyrosine, thymine-aliphatic amino acid, and cytosine-tyrosine links have been identified in isolated calf thymus chromatin subjected to gamma-irradiation (for review, see Dizdaroglu 1991).

Molecular biologists have examined the likely physiological effects of these various lesions in DNA. 8-OH-Gua (and, by inference, 8-OH-Ade) may lead to mutations by inducing misreading of the base itself and possibly of the adjacent bases (Breimer 1991; Shibutani et al. 1991). Thymine glycol may have some mutagenic action, and it can be lethal if not removed from the DNA before replication. Ring-fragmented bases are thought to block DNA replication. Abasic sites, which can also result from attack of •OH, can be mutagenic as well (Breimer 1991).

It is clear that •OH produces multiple changes in DNA, whereas •O_2^- and H_2O_2 have no effect, but the situation with other oxygen-derived species is less clear-cut at present. Singlet oxygen is able to produce limited strand breakage in isolated DNA, and its ability to modify the DNA bases is also limited. Thus, M. Dizdaroglu and H. Sies (pers. comm.) found small amounts of 8-OH-Gua and FapyGua but no other significant base changes in DNA exposed to singlet O_2 generated by the thermal decomposition of an endoperoxide. Thus, singlet O_2 certainly does not induce the extensive pattern of DNA base modification produced by •OH. Peroxyl and alkoxyl radicals might preferentially react with guanine, whereas HOCl would be expected to chlorinate –NH_2 groups. More work is required to characterize these reactions in detail, but we doubt that any species other than •OH produces the extensive pattern of base modification shown in Figure 2. Several authors have reported that peroxidizing lipids damage DNA but, in interpreting the data, one must bear in mind that peroxidizing lipids produce a range of reactive oxygen species including •OH, H_2O_2, singlet oxygen, peroxyl radicals, and alkoxyl radicals (Gutteridge and Halliwell 1990), and the exact contributions of these species to the DNA damage observed need to be determined. Lipid peroxides also decompose at body temperature in the presence of transition metal ions to give a huge range of products, including carbonyl compounds, such as malondialdehyde (MDA) and the unsaturated aldehyde 4-hydroxy-2-*trans*-nonenal, which has been shown to be mutagenic to mammalian cells (Canonero et al. 1990). If these aldehydes are generated in the vicinity of DNA, they may be able to combine with it to form distinctive products. Thus, MDA reacts with adenine, cytosine, and guanine (Stone et al. 1990a,b), and a guanine-MDA adduct has been identified in human urine (Hadley and Draper 1990). The product of reaction of hydroxynonenal with deoxyguanosine has also been characterized (Sodum and Chung 1988).

Humans are constantly exposed to background levels of ionizing radiation, which will generate some •OH in vivo. This radical may also arise by reaction of metal ions with H_2O_2 in vivo. Thus, it is not surprising to find that repair systems have evolved to remove at least some of the lesions in DNA that can result from attack of •OH (for review, see Breimer 1991).

Some of the modified DNA bases shown in Figure 2, and their nucleosides (base-deoxyribose), have been detected in the urine of humans and other mammals. Thus, 8-OH-Ade, 7-methyl-8-hydroxyguanine, thymine glycol, thymidine glycol, hydroxymethyluracil, 8-OH-Gua, and 8-hydroxydeoxyguanosine have been detected in mammalian urine (Ames 1989; Stilwell et al. 1989; Fraga et al. 1990). The presence of these products in urine suggests that oxidative damage to the DNA bases does occur in vivo and that repair systems are active to cleave modified bases from DNA. However, it is possible that some of the modified bases excreted originate from the diet or from the metabolism of gut flora. In addition, it is possible that DNA released from dead and dying cells within an organism undergoes rapid oxidative damage (since cell disruption can increase free radical reactions; see Halliwell and Gutteridge

1984). Hence, one must be cautious in using the amounts of modified DNA bases excreted from the body as an index of the extent of repair of oxidative DNA damage in *healthy* cells.

METHODS OF MEASURING PRODUCTS OF FREE RADICAL ATTACK ON THE DNA BASES

High Performance Liquid Chromatography

The development of an HPLC technique (Kasai et al. 1986), coupled with highly sensitive electrochemical detection (Floyd et al. 1986), for the measurement of 8-hydroxydeoxyguanosine has led to a series of pioneering studies in which measurement of this product has been used to gain information about free radical damage to DNA in intact cells and whole organisms (Ames 1989; Floyd 1990; Fraga et al. 1990). For example, the amount of 8-OH-Gua in the DNA from certain subpopulations of rat liver mitochondria was found to be considerably higher than in rat liver nuclear DNA, leading to proposals about the role of mitochondria in aging and carcinogenesis (Richter 1988). Exposure of numerous cell types to oxidative stress has been reported to increase the 8-OH-Gua content of their DNA (Kasai et al. 1986; Floyd 1990).

These studies have certainly produced qualitative evidence for oxidative damage to DNA in vivo, although care must be used in interpreting the data (Halliwell and Aruoma 1991). For example, there are mechanisms of forming 8-OH-Gua in DNA that may not involve oxygen radicals (Kohda et al. 1987). One must also be extremely cautious in attempting to use measurement of 8-OH-Gua (or any other single product) as a *quantitative* measure of DNA base damage by oxygen-derived species. We have already seen that when •OH attacks DNA bases, radicals are formed that can react in various ways, depending on the experimental conditions used. Thus, attack of •OH on guanine can lead to formation of 8-OH-Gua by oxidation of the C8 •OH adduct radical, but this radical can lead to other products as well (such as FapyGua), depending on the reaction conditions. Hence, different amounts of 8-OH-Gua can result from attack of the same amount of •OH on guanine in DNA. It follows that changes in 8-OH-Gua levels do not necessarily reflect changes in the amount of free radical attack upon DNA. To take some examples, iron ion-dependent systems generating •OH led to substantial formation of FapyGua as well as 8-OH-Gua in DNA (Aruoma et al. 1989a,b), whereas systems containing copper ions and H_2O_2 greatly favored 8-OH-Gua over FapyGua production (Aruoma et al. 1991). When isolated, mammalian chromatin was irradiated in aqueous suspension, the relative amounts of 8-hydroxypurines and formamidopyrimidines generated depended on the environment provided by the gases used to saturate the aqueous solution. For example, the presence of oxygen favored formation of 8-hydroxypurines (Gajewski et al. 1990). In Table 2, we summarize some of the results obtained. Products derived from pyrimidines can similarly be affected by changes in reaction conditions. If base-damage products are continuously formed in DNA in vivo, then changes in the ef-

TABLE 2 HOW REACTION CONDITIONS CAN ALTER THE END PRODUCTS DERIVED FROM ATTACK OF HYDROXYL RADICALS GENERATED BY DIFFERENT SYSTEMS ON PURINE BASES IN THE DNA OF ISOLATED CHROMATIN

Systems used to generate •OH	Ratios of 8-OH-Ade to FapyAde	8-OH-Gua to FapyGua
H_2O_2 metal ions: Air saturated solutions		
H_2O_2/Fe^{+++}/ascorbate	1.1	5.9
H_2O_2/Fe^{+++}-EDTA	4.2	8.6
H_2O_2/Fe^{+++}-EDTA/ascorbate	0.5	2.2
H_2O_2/Fe^{+++}-NTA	1.5	8.3
H_2O_2/Fe^{+++}-NTA/ascorbate	1.2	5.3
H_2O_2/Cu^{++}	18.6	48.4
H_2O_2/Cu^{++}/ascorbate	11.1	31.5
Ionizing radiation: Solutions saturated with		
argon	0.55	0.57
air	1.8	3.5
nitrous oxide	0.8	0.75
nitrous oxide/oxygen	3.4	4.5

(NTA) Nitrilotriacetic acid. Results were obtained with mammalian chromatin in aqueous suspension. Calculations courtesy of Dr. M. Dizdaroglu. Data abstracted from Gajewski et al. (1990) and Aruoma et al. (1989a,b, 1991).

ficiency of repair systems (e.g., free radical damage to the repair enzymes) could also cause modified bases to accumulate in DNA in vivo. Enzymes and other proteins are an important cellular target of damage by oxidative stress (Von Sonntag 1987; Orrenius et al. 1989).

Gas Chromatography/Mass Spectrometry

A complete characterization of damage to DNA by oxygen-derived species can be achieved by the technique of gas chromatography/mass spectrometry (GC/MS) (Dizdaroglu 1991), which may be applied to DNA itself or to DNA-protein complexes such as chromatin. The DNA or chromatin is hydrolyzed (usually by acid) and the products are converted to volatile derivatives, which are separated by gas chromatography and identified by mass spectrometry. High sensitivity of detection can be achieved by operating the mass spectrometer in the selected ion monitoring (SIM) mode. In this mode, the mass spectrometer is set to monitor several ions derived by fragmentation of a particular product during the time at which this product is expected to emerge from the GC column. The GC/MS-SIM technique is being used in the authors' laboratory to examine the mechanism by which DNA is damaged in cells subjected to oxidative stress. Thus, if damage is due to •OH generation, then products characteristic of •OH attack should be detected (Fig. 2), as has been observed in the DNA from murine hybridoma cells treated with H_2O_2 (Dizdaroglu et al. 1991b) and in DNA from primate tracheal epithelial cells exposed to ozone or to cigarette smoke (O.I. Aruoma et al., unpubl.). In contrast, cleavage of the DNA backbone by the action of nucleases should leave the purines and pyrimidines unaltered (Fig. 1). For studies on DNA modification, extraction of chromatin from cells for analysis is preferable to extraction of DNA, since it minimizes the loss of extensively fragmented DNA and of DNA that has become covalently cross-linked to protein. The results of GC/MS analysis of modified DNA bases are usually expressed as nanomoles of modified bases per milligram of DNA. However, it is easy to convert these data into the actual number of bases modified. Thus, dividing the figure of nmole bases per milligrams of DNA by 3.14 (or multiplying by 0.32) gives the number of modified bases per thousand DNA base pairs.

The apparently unique pattern of DNA damage produced by attack of •OH has been used not only to study the role of •OH in causing DNA damage in cells exposed to oxidative stress, but also to solve several other biochemical problems.

1. Mixtures of copper ions with H_2O_2 and •O_2^- or ascorbate do considerable damage to DNA and proteins. Although this was attributed to •OH (for review, see Halliwell and Gutteridge 1990a), several authors persistently disputed the formation of •OH. Examination of the pattern of DNA base damage clearly showed that •OH was involved (Aruoma et al. 1991). An interesting consequence of these experiments was the discovery that copper ions mediate much more damage to DNA bases by H_2O_2 (± ascorbate) than do iron ions in comparable reaction mixtures (Aruoma et al. 1991; Dizdaroglu et al. 1991a). In Table 3, we present some illustrative data. Copper ions also favor the formation of 8-OH-Gua in DNA (Table 3). This could be because copper ions bind to DNA in the vicinity of guanine and "site-direct" •OH formation to that base, because copper ions react faster than iron ions with H_2O_2 to form •OH (Halliwell and Gutteridge 1989), and because the copper ions may facilitate

TABLE 3 OXIDATIVE DAMAGE TO ISOLATED DNA: A COMPARISON OF IRON AND COPPER IONS

System used	nmole modified bases/mg DNA produced	
	total bases	8-OH-Gua
DNA alone	2.32 ± 0.19	1.02 ± 0.09
Plus Cu^{++}/H_2O_2	14.30 ± 0.39	9.02 ± 0.12
Plus Fe^{+++}/H_2O_2	3.63 ± 0.24	1.29 ± 0.09
Plus Cu^{++}/H_2O_2/ascorbate	83.2 ± 8.84	48.20 ± 6.80
Plus Fe^{+++}/H_2O_2/ascorbate	8.15 ± 0.84	2.14 ± 0.24

Calf thymus DNA was exposed to H_2O_2 alone, or H_2O_2 plus ascorbate, in the presence of added Cu^{++} or Fe^{+++} ions at pH 7.4. No extra DNA damage over the control (DNA alone) was seen if metal ions were not added.

Results are mean ± S.D. from three experiments each performed in triplicate. For further details, see Aruoma et al. (1991) and Dizdaroglu et al. (1991a).

the conversion of the hydroxylated guanine radical into 8-OH-Gua rather than other products such as FapyGua.
2. A copper ion-phenanthroline complex (Cu [phen]$_2$) cleaves DNA in the presence of reducing agents. The proposal of Gutteridge and Halliwell (1982) that this cleavage is due to "site-specific" formation of •OH radicals upon DNA was confirmed by examining the pattern of base modification and showing it to be characteristic of attack by •OH (Dizdaroglu et al. 1990). Formation of H_2O_2 in copper ion/phenanthroline/ascorbic acid (or mercaptoethanol) reaction mixtures has been directly demonstrated by O.I. Aruoma (unpubl.) using a peroxidase-based system (Aruoma et al. 1989c). Addition of extra H_2O_2 increases Cu[phen]$_2$-dependent base modification and deoxyribose breakdown in DNA.
3. Gutteridge (1985) was the first to demonstrate experimentally, using the deoxyribose assay, that certain *ferric* chelates react with H_2O_2 to form •OH, in a process inhibitable by SOD. However, some scientists have been skeptical. Therefore, we studied the mechanism of damage to DNA by various ferric chelates in the presence of H_2O_2. The pattern of base damage observed confirmed the involvement of •OH (Aruoma et al. 1989b). Ferric nitrilotriacetic acid (NTA) produced the most DNA base damage in the presence of H_2O_2. It is disturbing to note that NTA, previously a common constituent of detergents, is now widely distributed in the environment (for discussion, see Aruoma and Halliwell 1991).
4. The antitumor antibiotic bleomycin cleaves DNA in the presence of O_2, ferric ions and ascorbate, or ferric ions and H_2O_2. It is thought that the DNA-cleaving species is a bleomycin-oxo-iron complex rather than •OH. In support of this, GC/MS-SIM studies showed that very little base modification occurs to DNA in either of these reaction systems (Gajewski et al. 1991).

Other recent uses of GC/MS-SIM have included identification of several DNA base damage products in neoplastic liver tissue from fish (Malins and Haimanot 1990) and the demonstration that the damage done to pure DNA by exposure to activated human neutrophils is probably due to •OH generated from H_2O_2 and $\cdot O_2^-$ in the presence of iron ions contaminating the reaction mixture (Jackson et al. 1989).

We believe that the use of GC/MS-SIM to "fingerprint" DNA damage is one way forward in investigating the role of reactive oxygen species in damaging DNA in vivo and assessing the contribution that such species make to the mechanism of action of carcinogens and to the increased cumulative risk of cancer with age (Ames 1989).

ACKNOWLEDGMENTS

We are grateful to the Medical Research Council, Arthritis and Rheumatism Council, Cancer Research Campaign, and Association for International Cancer Research for research support. Professor John Gutteridge and Drs. Lars Breimer and Miral Dizdaroglu are particularly thanked for many helpful discussions. We thank Dr. Miral Dizdaroglu and Professor Helmut Sies for providing unpublished information. Dr. Aruoma was a visiting scientist (1988–89) at the National Institute of Standards and Technology, Gaithersburg, Maryland.

REFERENCES

Ames, B.N. 1989. Endogenous oxidative DNA damage, aging and cancer. *Free Radical Res. Commun.* **7:** 121.

Aruoma, O.I. and B. Halliwell. 1991. DNA damage and free radicals. *Chem. Br.* (Feb), p.149.

Aruoma, O.I., B. Halliwell, and M. Dizdaroglu. 1989a. Iron ion-dependent modification of bases in DNA by the superoxide radical-generating system hypoxanthine/xanthine oxidase. *J. Biol. Chem.* **264:** 13024.

Aruoma, O.I., B. Halliwell, E. Gajewski, and M. Dizdaroglu. 1989b. Damage to the bases in DNA induced by hydrogen peroxide and ferric ion chelates. *J. Biol. Chem.* **264:** 20509.

―――. 1991. Copper-ion-dependent damage to the bases in DNA in the presence of hydrogen peroxide. *Biochem. J.* **273:** 601.

Aruoma, O.I., B. Halliwell, B.M. Hoey, and J. Butler. 1989c. The antioxidant action of N-acetylcysteine. *Free Radicals Biol. Med.* **6:** 593.

Birnboim, H.C. 1988. A superoxide anion induced DNA strand-break metabolic pathway in human leukocytes: Effect of vanadium. *Biochem. Cell. Biol.* **66:** 374.

Breimer, L.H. 1988. Ionizing radiation-induced mutation. *Br. J Cancer* **57:** 6.

———. 1990. Molecular mechanisms of oxygen radical carcinogenesis and mutagenesis. The role of DNA base damage. *Mol. Carcinog.* **3:** 188.

———. 1991. Repair of DNA damage induced by reactive oxygen species. *Free Radical Res. Commun.* **14:** 159.

Canonero, R., A. Martelli, U.R. Marinari, and G. Brambilla. 1990. Mutation induction in Chinese hamster lung V79 cells by five alk-2-enals produced by lipid peroxidation. *Mutat. Res.* **244:** 153.

Cantoni, O., P. Sestili, F. Cattabeni, G. Bellomo, S. Pou, M. Cohen, and P. Cerutti. 1989. Calcium chelator quin 2 prevents hydrogen-peroxide-induced DNA breakage and cytotoxicity. *Eur. J. Biochem.* **181:** 209.

Cerutti, P., R. Larsson, G. Krupitza, D. Muehlmatter, D. Crawford, and P. Amstad. 1989. Pathophysiological mechanisms of active oxygen. *Mutat. Res.* **214:** 81.

di Guiseppi, G. and I. Fridovich. 1984. The toxicology of molecular oxygen. *CRC Crit. Rev. Toxicol.* **12:** 315.

Dijkwel, P.A. and P.W. Wenink. 1986. Structural integrity of the nuclear matrix; differential effects of thiol agents and metal chelators. *J. Cell Sci.* **84:** 53.

Dizdaroglu, M. 1991. Chemical determination of free radical-induced damage to DNA. *Free Radicals Biol. Med.* **10:** 225.

Dizdaroglu, M., O.I. Aruoma, and B. Halliwell. 1990. Modification of bases in DNA by copper ion-1,10-phenanthroline complexes. *Biochemistry* **29:** 8447.

Dizdaroglu, M., G. Rao, B. Halliwell, and E. Gajewski. 1991a. Damage to the DNA bases in mammalian chromatin by hydrogen peroxide in the presence of ferric and cupric ions. *Arch. Biochem. Biophys.* **285:** 317.

Dizdaroglu, M., Z. Nackerdien, B.C. Chao, E. Gajewski, and G. Rao. 1991b. Chemical nature of *in vivo* DNA base damage in hydrogen peroxide-treated mammalian cells. *Arch. Biochem. Biophys.* **285:** 388.

Farber, J.L. 1990. The role of calcium in lethal cell injury. *Chem. Res. Toxicol.* **3:** 503.

Floyd, R.A. 1990. The role of 8-hydroxyguanine in carcinogenesis. *Carcinogenesis* **11:** 1447.

Floyd, R.A., J.J. Watson, P.K. Wong, D.H. Altmiller, and R.C. Rickard. 1986. Hydroxyl-free radical adduct of deoxyguanosine: Sensitive detection and mechanisms of formation. *Free Radical Res. Commun.* **1:** 163.

Fraga, C.C., M.K. Shigenaga, J.W. Park, P. Degan, and B.N. Ames. 1990. Oxidative damage to DNA during aging: 8-Hydroxy-2'-deoxyguanosine in rat organ DNA and urine. *Proc. Natl. Acad. Sci.* **87:** 4533.

Gajewski, E., O.I. Aruoma, M. Dizdaroglu, and B. Halliwell. 1991. Bleomycin-dependent damage to the bases in DNA is a minor side-reaction. *Biochemistry* **30:** 2444.

Gajewski, E., G. Rao, Z. Nackerdien, and M. Dizdaroglu. 1990. Modification of DNA bases in mammalian chromatin by radiation-generated free radicals. *Biochemistry* **29:** 7876.

Gutteridge, J.M.C. 1985. Superoxide dismutase inhibits the superoxide-driven Fenton reaction at two different levels: Implications for a wider protective role. *FEBS Lett.* **185:** 19.

Gutteridge, J.M.C. and B. Halliwell. 1982. The role of superoxide and hydroxyl radicals in the degradation of DNA and deoxyribose induced by a copper-phenanthroline complex. *Biochem. Pharmacol.* **31:** 2801.

———. 1990. The measurement and mechanism of lipid peroxidation in biological systems. *Trends Biochem. Sci.* **15:** 129.

Hadley, M. and H.H. Draper. 1990. Isolation of a guanine-malondialdehyde adduct from rat and human urine. *Lipids* **25:** 81.

Halliwell, B. 1987. Oxidants and human disease; some new concepts. *FASEB J.* **1:** 358.

Halliwell, B. and O.I. Aruoma. 1991. DNA damage by oxygen-derived species. Its mechanism and measurement in mammalian systems. *FEBS Lett.* **281:** 9.

Halliwell, B. and J.M.C. Gutteridge. 1984. Lipid peroxidation, oxygen radicals, cell damage and antioxidant therapy. *Lancet* **1:** 1396.

———. 1989. *Free radicals in biology and medicine*, 2nd edition. Clarendon Press, Oxford, England.

———. 1990a. Role of free radicals and catalytic metal ions in human disease. *Methods Enzymol.* **186:** 1.

———. 1990b. The antioxidants of human extracellular fluids. *Arch. Biochem. Biophys.* **280:** 1.

Imlay, J.A. and S. Linn. 1988. DNA damage and oxygen radical toxicity. *Science* **240:** 1302.

Jackson, J.H., E. Gajewski, I.U. Schraufstatter, P.A. Hyslop, A.F. Fuciarelli, C.G. Cochrane, and M. Dizdaroglu. 1989. Damage to the bases in DNA induced by stimulated human neutrophils. *J. Clin. Invest.* **84:** 1644.

Kasai, H., P.F. Crain, Y. Kuchino, S. Nishimura, A. Ootsuyama, and H. Tanooka. 1986. Formation of 8-hydroxyguanine moiety in cellular DNA by agents producing oxygen radicals and evidence for its repair. *Carcinogenesis* **7:** 1849.

Kiyosawa, T., M. Suko, H. Okudaira, K. Murata, T. Miyamoto, M.H. Chung, H. Kasai, and S. Nishimura. 1990. Cigarette smoke induces formation of 8-hydroxydeoxyguanosine, one of the oxidative DNA damages in human peripheral leukocytes. *Free Radical Res. Commun.* **11:** 23.

Kohda, T., M. Tada, A. Hakura, H. Kasai, and Y. Kawazoe. 1987.

Formation of 8-hydroxyguanine residues in DNA treated with 4-hydroxyamino quinoline 1-oxide and its related compounds in the presence of seryl-AMP. *Biochem. Biophys. Res. Commun.* **149:** 1141.

Kyle, M.E., D. Nakae, I. Sakaida, S. Miccadei, and J.L. Farber. 1988. Endocytosis of superoxide dismutase is required in order for the enzyme to protect hepatocytes from the cytotoxicity of hydrogen peroxide. *J. Biol. Chem.* **263:** 3784.

Larrick, J.W. and S.C. Wright. 1990. Cytotoxic mechanism of tumor necrosis factor alpha. *FASEB J.* **4:** 3215.

Larsson, R. and P. Cerutti. 1989. Translocation and enhancement of phosphotransferase activity of protein kinase C following exposure of mouse epidermal cells to oxidants. *Cancer Res.* **49:** 5627.

Lewis, C.D. and U.K. Laemmli. 1982. Higher order metaphase chromosome structure: Evidence for metalloprotein interactions. *Cell* **29:** 171.

Malins, D.C. and R. Haimanot. 1990. 4,6-Diamino-5-formamidopyrimidine, 8-hydroxyguanine and 8-hydroxyadenine in DNA from neoplastic liver of English sole exposed to carcinogens. *Biochem. Biophys. Res. Commun.* **173:** 614.

McBride, T.J., B.D. Preston, and L.A. Loeb. 1991. Mutagenic spectrum resulting from DNA damage by oxygen radicals. *Biochemistry* **30:** 207.

Mello-Filho, A.C., R.E. Hoffman, and R. Meneghini. 1984. Cell killing and DNA damage by hydrogen peroxide are mediated by intracellular iron. *Biochem. J.* **218:** 273.

Moraes, E.C., S.M. Keyse, and R.M. Tyrrell. 1990. Mutagenesis by hydrogen peroxide treatment of mammalian cells: A molecular analysis. *Carcinogenesis* **11:** 283.

Nakae, D., H. Yoshiji, T. Amanuma, T. Kinugasa, J.L. Farber, and Y. Konishi. 1990. Endocytosis-independent uptake of liposome-encapsulated superoxide dismutase prevents the killing of cultured hepatocytes by tert-butyl hydroperoxide. *Arch. Biochem. Biophys.* **279:** 315.

Nassi-Calo, L., A.C. Mello-Filho, and R. Meneghini. 1989. O-phenanthroline protects mammalian cells from hydrogen peroxide-induced gene mutation and morphological transformation. *Carcinogenesis* **10:** 1055.

Orrenius, S., D.J. McConkey, G. Bellomo, and P. Nicotera. 1989. Role of Ca^{++} in toxic cell killing. *Trends Pharmacol. Sci.* **10:** 281.

Prutz, W.A., J. Butler, and E.J. Land. 1990. Interaction of copper (I) with nucleic acids. *Int. J. Radiat. Biol.* **58:** 215.

Richter, C. 1988. Do mitochondrial DNA fragments promote cancer and aging? *FEBS Lett.* **241:** 1.

Rowley, D.A. and B. Halliwell. 1983. DNA damage by superoxide-generating systems in relation to the mechanism of action of the anti-tumor antibiotic adriamycin. *Biochim. Biophys. Acta* **761:** 86.

Shibutani, S., M. Takeshita, and A.P. Grollman. 1991. Insertion of specific bases during DNA synthesis past the oxidation-damaged base 8-oxodG. *Nature* **349**: 431.

Sies, H, ed. 1991. *Oxidative stress, oxidants and antioxidants.* Academic Press, New York.

Sodum, R.S. and F.L. Chung. 1988. Stereoselective formation of *in vitro* nucleic acid adducts by 2,3 epoxy-4-hydroxynonanal. *Cancer Res.* **48**: 320.

Stilwell, W.G., H.X. Xu, J.A. Adkins, J.S. Wishnok, and S.R. Tannenbaum. 1989. Analysis of methylated and oxidized purines in urine by capillary gas chromatography-mass spectrometry. *Chem. Res. Toxicol.* **2**: 94.

Stone, K., M. Ksebati, and L.J. Marnett. 1990a. Identification of adducts formed by reaction of malondialdehyde with adenosine. *Chem. Res. Toxicol.* **3**: 33.

Stone, K., A. Uzieblo, and L.J. Marnett. 1990b. Studies of the reaction of malonaldehyde with cytosine nucleosides. *Chem. Res. Toxicol.* **3**: 467.

Totter, J.R. 1980. Spontaneous cancer and its possible relation to oxygen metabolism. *Proc. Natl. Acad. Sci.* **77**: 1763.

Touati, D. 1989. The molecular genetics of superoxide dismutase in *E. coli. Free Radical Res. Commun.* **8**: 1.

Von Sonntag, C. 1987. *The chemical basis of radiation biology.* Taylor and Francis, London.

Weiss, S.J. 1989. Tissue destruction by neutrophils. *N. Engl. J. Med.* **320**: 365.

Weitberg, A.B. and D. Corvese. 1990. Translocation of chromosomes 16 and 18 in oxygen radical-transformed human lung fibroblasts. *Biochem. Biophys. Res. Commun.* **169**: 70.

Weitzman, S.A., A.B. Weitberg, E.P. Clark, and T.P. Stossel. 1985. Phagocytes as carcinogens: Malignant transformation produced by human neutrophils. *Science* **227**: 1231.

Weitzman, S., C. Schmeichel, P. Turk, C. Stevens, S. Toloma, and N. Bouck. 1988. Phagocyte-mediated carcinogenesis. DNA from phagocyte-transformed C3H 10T1/2 cells can transform NIH/3T3 cells. *Ann. N.Y. Acad. Sci.* **551**: 103.

Wyllie, A.H. 1980. Glucocorticoid-induced thymocyte apoptosis is associated with endogenous endonuclease activation. *Nature* **284**: 555.

Zimmerman, R. and P. Cerutti. 1984. Active oxygen acts as a promoter of carcinogenesis in C3H/10T1/2/C18 fibroblasts. *Proc. Natl. Acad. Sci.* **81**: 2085.

Zimmerman, R.J., A. Chan, and S.A. Leadon. 1989. Oxidative damage in murine tumor cells treated in vitro by recombinant human tumor necrosis factor. *Cancer Res.* **49**: 1644.

Possible Protective Mechanisms of Tumor Necrosis Factors against Oxidative Stress

G.H.W. Wong,[1] A. Kamb,[2] L.A. Tartaglia,[1] and D.V. Goeddel[1]
[1]Department of Molecular Biology, Genentech, Inc.
South San Francisco, California 94080
[2]Department of Biochemistry, University of California
San Francisco, California 94143

Tumor necrosis factor-α (TNF-α) was originally identified by its ability to cause necrosis of tumors in animals (Carswell et al. 1975), but it has proved subsequently to possess a variety of other properties. It plays a role in immunoregulation and in antimicrobial and antiviral defense (Goeddel et al. 1986; Old 1990; Fiers 1991; Neta et al. 1992; Wong et al. 1992a,c). TNF-α is thought to be associated with cachexia (Beutler and Cerami 1986), although recent data have demonstrated that TNF-α does not, by itself, result in a sustained cachetic effect (Mullen et al. 1990; Fiers 1991). TNF-α is produced by a number of cell types, including activated macrophages, lymphocytes, mast cells, and glial cells (Wong and Goeddel 1989; Neta et al. 1992). TNF-β (also called lymphotoxin) binds the same receptors as does TNF-α and mediates a similar spectrum of physiological responses (Goeddel et al. 1986; Fiers 1991). TNF-β is only produced by activated T or B lymphocytes but not by activated macrophages or mast cells. Human TNF-α and TNF-β, encoded by linked genes located on the short arm of chromosome 6 near the major histocompatibility complex (MHC) region (Spies et al. 1986), are identical at 28% of their residues (Goeddel et al. 1986).

Cellular responses to TNF are diverse. Some tumor cells are killed by TNF alone, but most cells are not (Sugarman et al. 1985; Goeddel et al. 1986). Insensitivity is not generally due to

Molecular Biology of Free Radical Scavenging Systems.
Copyright 1992 Cold Spring Harbor Laboratory Press 0-87969-000-0/92 $3.00 + 00

a lack of receptors, because essentially all cells express both types of receptors: a 55-kD protein (TNF-R1) and a 75-kD protein (TNF-R2) (Fiers 1991; Loetscher et al. 1991). The two receptors share about 30% sequence identity in their extracellular, ligand-binding domains and show a similar level of homology with the extracellular domain of several other receptors, including the nerve growth factor receptor (Fiers 1991; Loetscher et al. 1991). However, the intracellular domains of TNF-R1 and TNF-R2 are not similar to each other, nor are they similar to any other protein so far reported (Lewis et al. 1991; Loetscher et al. 1991). The lack of homology between TNF-R1 and TNF-R2 suggests that the intracellular signals transduced by these two receptors differ from each other. This model is supported by the observation that the two receptors signal distinct cellular responses (Tartaglia et al. 1991).

TNF triggers a series of events inside cells, including the induction of certain genes and the formation of highly reactive molecules such as superoxide anions ($\cdot O_2^-$) (Matsuyama and Ziff 1986; Meier et al. 1989; Radeke et al. 1990), hydroxyl radicals ($\cdot OH$) (Yamauchi et al. 1989), hydrogen peroxide (H_2O_2) (Hoffman and Weinberg 1987; Radeke et al. 1990; Tiku et al. 1990), and nitric oxide (NO) (Drapier et al. 1988; Kilbourn et al. 1990). These molecules can damage proteins (Prinsze et al. 1990), DNA, and lipids (Boveris 1977; Cross et al. 1987; Comporti 1989). In this review, we discuss the somewhat paradoxical relationship between TNF and oxidative damage. We propose several mechanisms by which TNF may protect against oxidative stress and suggest an explanation for its opposite effects on normal cells and abnormal cells.

PROTECTION CONFERRED BY TNF

TNF is a component of the body's system of defense against infection by pathogens, such as fungi, bacteria, protozoa, and viruses (Old 1990; Fiers 1991; Neta et al. 1992; Wong et al. 1992a). It appears to act partly through generating toxic free radicals. For example, TNF can stimulate neutrophils and macrophages, which in turn release $\cdot O_2^-$ and NO into their surroundings, bombarding any neighboring microbes with a stream of highly reactive and damaging molecules (Shalaby et al. 1985; Berkow et al. 1987; Drapier et al. 1988). Other cells, such as fibroblasts (Meier et al. 1989), mesangial cells (Radeke et al. 1990), chondrocytes (Tiku et al. 1990), and the human T cell line HuT-78 (Fig. 1), may also produce oxygen free radicals in response to TNF. It will be of interest to examine whether these reactive radicals induced by TNF can trigger cellular gene expression and other biological activities mediated by TNF.

TNF protects against viral disease in several ways. First, it can induce in cells a resistance to infection by a range of viruses (Wong and Goeddel 1986). Second, TNF selectively kills

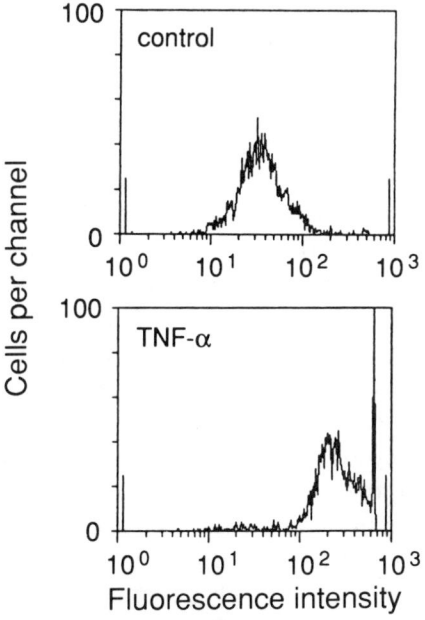

FIGURE 1 TNF-α induces the production of free radicals in HuT-78 cells. 2'7'-dichlorofluorescin diacetate (DCF-DA) was used to detect the intracellular production of free radicals as described previously (Himmelfarb et al. 1991). DCF-DA can cross the cellular membranes and is trapped in cells after deacetylation and becomes fluorescent after oxidation by free radicals. HuT-78 cells (1 × 10^6/ml) were incubated with 50 μM of DCF-DA for 15 min at 37°C before addition of TNF-α (0.1 μg/ml) for 5 min. The fluorescent levels were quantitated by fluorescence-activated cell sorter.

cells that are already infected with viruses (Aderka et al. 1985; Koff and Fann 1986; Wong and Goeddel 1986; Wong et al. 1988). Third, it enhances the expression of MHC antigens (Collins et al. 1986) together with viral antigens (Folks et al. 1987), thus increasing the recognition and destruction of virus-infected cells by cytotoxic T cells. Fourth, TNF dramatically enhances the efficacy of the known antiviral proteins, interferon-α, -β (Wong and Goeddel 1987, 1988a), and -γ (Wong and Goeddel 1986, 1988a). Finally, it sensitizes virus-infected cells to heat and other oxidative stresses (Wong et al. 1991). The antiviral activity of TNF has also been demonstrated in vivo (Sambhi et al. 1991). It is possible that the antiviral properties of TNF also depend partly on intracellular production of free radicals.

Despite the action of TNF in stimulating free radical production, TNF pretreatment can actually protect animals from some insults that appear to be associated with the generation of free radicals (Fig. 2). These include radiation (Neta et al. 1988; Slordal et al. 1989), cytotoxic drugs (Slordal et al. 1990), ischemia reperfusion (L. Eddy and G.H.W. Wong, in prep.), endotoxin shock (Sheppard et al. 1989), hyperoxia (Tsan et al. 1990a), and the alopecia that results from administration of the anticancer drug cytosine arabinoside (Jimenez et al. 1991). TNF has other protective effects: It protects mice from contracting certain autoimmune diseases such as systemic lupus erythematosus (Jacob et al. 1990b) and type I diabetes (Jacob et al. 1990a), and it protects against subsequent large doses of TNF that would otherwise be lethal (Patton et al. 1987). Thus, TNF may induce a state of resistance in the body's tissues against a wide variety of insults. Indeed, antibodies that deplete endogenous TNF from the circulation render animals more sensitive to radiation (Neta et al. 1991), cytotoxic drugs (Moreb et al. 1990), and experimental peritonitis (Alexander et al. 1991) and increase stress hyperthermia (Long et al. 1990b). Antibodies against TNF also block the well-known radioprotective (Neta et al. 1991) and chemoprotective (Futami et al. 1990) actions, as well as the fever-inducing activity (Long et al. 1990a) of lipopolysaccharide (LPS). This suggests that a major portion of LPS-induced protection in vivo is partly mediated by TNF.

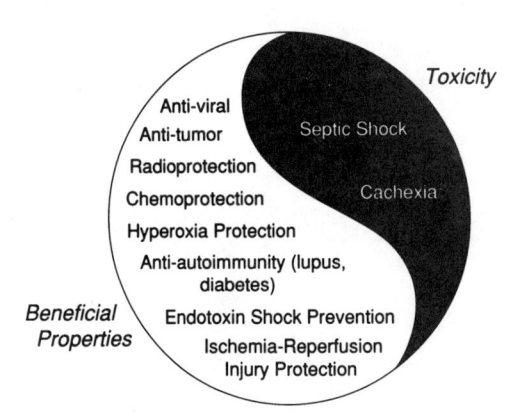

FIGURE 2 Biological activities of TNF-α in vivo.

In addition to the protective benefits observed in vivo, TNF pretreatment has protective effects in vitro. Pretreatment of cells with TNF protects against damage caused by radiation and heat (Wong et al. 1991, 1992b), H_2O_2 (G.H.W. Wong, unpubl.), and agents that generate free radicals, such as paraquat (Warner et al. 1991), adriamycin, and menadione (G.H.W. Wong, unpubl.). These results suggest that TNF may act directly on target cells to protect them from oxidative damage.

INDUCTION OF PROTECTIVE PROTEINS BY TNF

Experiments involving the translation inhibitor cycloheximide (CHI) imply that the protective effects of TNF may be mediated in part by proteins that protect cells against insults, including TNF itself (Wallach 1984). Most cells are not damaged by TNF alone; they become sensitive to TNF killing when treated with TNF plus CHI, presumably because the constitutive synthesis or induction of protective proteins is inhibited. For example, cells of the human lung carcinoma cell line A549 are normally resistant to TNF killing but can be rendered sensitive by addition of CHI (Wong et al. 1992c). If A549 cells are treated with TNF before exposure to CHI plus TNF, they are no longer susceptible to killing (Fig. 3). This suggests that TNF induces proteins that protect against the potential cytotoxic actions of TNF. Interestingly, A549 cells that are treated with CHI alone become partially resistant to subsequent challenge with TNF

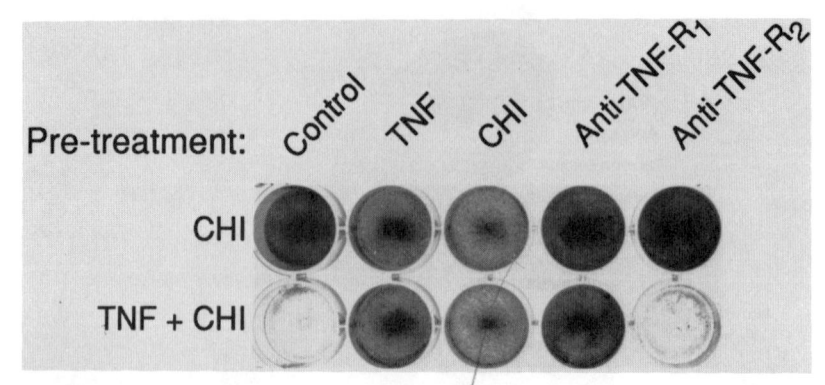

FIGURE 3 Effect of pretreatment of A549 cells with various agents on subsequent killing with TNF-α + CHI. A549 cells (3 x 10^5/ml) were incubated with either TNF-α (0.1 µg/ml), CHI (50 µg/ml), anti-TNF-R1 (1:100), or anti-TNF-R2 (1:100) for 1/2 hr before exposure to serum-free RPMI-1640 medium containing CHI (100 µg/ml) in the presence or absence of TNF-α (1 µg/ml) for 24 hr. Cells were stained with crystal violet as described previously (Kramer and Carver 1986).

plus CHI (Fig. 3). One explanation is that TNF may act through labile protein mediators to kill cells. Possible candidates for such mediators are proteases (Ruggiero et al. 1987), lysosomal enzymes (Watanabe et al. 1988), phospholipase A2 (Neale et al. 1988), and ADP-ribose polymerase (Lichtenstein et al. 1991), because inhibitors of these enzymes block TNF killing.

INDUCTION OF MANGANOUS SUPEROXIDE DISMUTASE

Because TNF induces expression of a large number of proteins, it is not easy to determine which of these proteins contribute to TNF's protective properties. However, the association between TNF activity and oxygen free radicals provides one clue. Of the enzymes known to be directly involved in oxygen free radical metabolism, the only one whose synthesis is induced by TNF is manganous superoxide dismutase (MnSOD) (Wong and Goeddel 1988c; Asoh et al. 1989; Kawaguchi et al. 1990; Shaffer et al. 1990; Valentine and Nick 1990; Visner et al. 1990). MnSOD, localized mainly to mitochondria, detoxifies superoxide radicals (Weisiger and Fridovich 1973). Other antioxidant enzymes, such as catalase, glutathione reductase, glutathione peroxidase, and CuZnSOD, are not induced by

TNF (Wong et al. 1992c). Induction of MnSOD mRNA is rapid (detectable within 1 hr and requiring a very low level of TNF [0.1 ng/ml]), is general (occurs in most cells), is direct (does not require new protein synthesis), and is physiologically relevant (occurs also in vivo [Wong and Goeddel 1988c; Tsan et al. 1990b]).

In addition to TNF-α, TNF-β and interleukin-1 (IL-1) also induce MnSOD (Wong and Goeddel 1988c; Masuda et al. 1988). Like TNF-α, pretreatment of mice with TNF-β also results in a significant increase in survival following irradiation (Wong et al. 1992b). Pretreatment of A549 cells with TNF-β or IL-1 protects against killing by CHI plus TNF (Wong et al. 1992c). Furthermore, IL-1 also protects animals from cytotoxic drug-induced alopecia (Jimenez et al. 1991), radiation (Neta et al. 1986), hyperoxia (Tsan et al. 1990a), endotoxin shock (Alexander et al. 1991), and ischemic reperfusion (Brown et al. 1990). Interferon-γ (IFN-γ) induces MnSOD in some cells (Morikawa et al. 1990; Harris et al. 1991) and displays a synergistic interaction with low doses of either TNF-α or IL-1 in MnSOD induction (Harris et al. 1991).

IDENTIFICATION OF MNSOD AS A PROTECTIVE PROTEIN

The induction of MnSOD appears to represent a significant contribution to the protective activity of TNF. TNF-sensitive cells such as the ME-180 cell line express much lower levels of MnSOD than TNF-resistant T24 cells (Wong and Goeddel 1988c). However, the levels of other antioxidant enzymes such as catalase, CuZnSOD, and glutathione peroxidase are similar in these two cell types. ME-180 and 293 cell lines can be converted to partial resistance to TNF killing by artificial overexpression of MnSOD (Wong et al. 1989). Conversely, cells that are resistant to TNF can be rendered more sensitive by overexpression of MnSOD antisense RNA, which greatly diminishes the cellular content of MnSOD protein (Wong et al. 1989). In addition, 293 cell lines that overexpress MnSOD are more resistant to radiation damage than parental cells, whereas the 293 cell lines that overexpress antisense MnSOD RNA are more sensitive (Wong et al. 1992b). These cell lines that have been engineered to express either higher or lower levels of MnSOD than the parental line display no significant differences in expression of other antioxidant enzymes. Thus, the variation in sensitivity to TNF appears to depend solely on alterations in MnSOD expression. Overexpression of MnSOD in other cell types also increases resistance to TNF, anticancer drugs, and ionizing radiation (Hirose et al. 1992) or to paraquat-mediated toxicity (St. Clair et al. 1991).

Why is MnSOD expression regulated by TNF? Accelerated mitochondrial oxidative metabolism not accompanied by induction of MnSOD has been suggested to cause oxidative injury in kidney at an early stage of diabetes (Asayama et al. 1989). Recent results suggest that MnSOD levels must lie within a certain range in order to provide protection from oxidative stress such as reperfusion injury (Omar and McCord 1990). Levels of MnSOD that are higher than optimal not only fail to protect cells from damage, but actually worsen the damage (Omar and McCord 1990). The basis for this effect is not known. However, one possibility is that low levels of $\cdot O_2^-$ in mitochondria may benefit cells; for example, by destroying other reactive free radicals (Omar and McCord 1990). Thus, cells may require a basal level of $\cdot O_2^-$ production to remain healthy. This hypothesis may explain why MnSOD levels are normally low in the body but are induced rapidly by TNF, which is produced in response to episodes of trauma such as inflammation, infection, hyperoxia, and exposure to radiation. Furthermore, the hypothesis predicts that MnSOD is labile so that its level can quickly return to normal values. Indeed, the half-life of MnSOD appears to be much shorter (50 min) than that of CuZnSOD (8 hr) (L.W. Oberley, pers. comm.). Another possible explanation for regulation of MnSOD is that oxygen free radicals influence cell growth. It has been shown that fibroblast proliferation is stimulated by low levels of $\cdot O_2^-$ but is inhibited at higher levels (Murrell et al. 1990). Thus, MnSOD expression may indirectly control cell growth through its involvement in $\cdot O_2^-$ dismutation. This may explain why some rapidly growing tumor cells have lower levels of MnSOD than do normal cells (Oberley et al. 1989). It also raises the possibility that MnSOD is a member of the growing family of tumor suppressor genes.

In certain cells, the induction of protective proteins such as MnSOD by TNF may be defective. In specific tumor cell lines (G.H.W. Wong, unpubl.) and in virus-infected cells (Wong et al. 1991), TNF treatment does not lead to heightened MnSOD expression. Moreover, TNF does not protect these cells from oxidative insults (Wong et al. 1991). On the contrary, these cells are actually sensitized (Wong et al. 1991). For example, HIV-infected HuT-78 cells have lower levels of MnSOD than uninfected cells and are more sensitive to heat, radiation (Wong et al. 1991), and adriamycin (Fig. 4) than uninfected cells. TNF also does not induce MnSOD in HIV-infected HuT-78 cells, and it sensitizes them to heat, radiation (Wong et al. 1991), and adriamycin (Fig. 4). TNF-β but not IL-1β acts similarly (G.H.W. Wong, unpubl.). HuT-78 cells engineered to overexpress MnSOD are still susceptible to HIV infection; however, these HIV-infected cells display increased resistance to heat and radiation compared to the infected parental line that

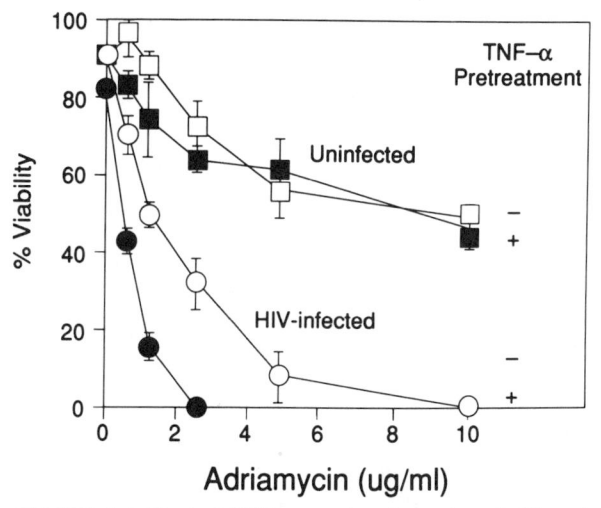

FIGURE 4 Effect of TNF-α pretreatment on killing by adriamycin. HuT-78 cells with or without HIV infection were pretreated with TNF-α (0.1 μg/ml) for 24 hr before challenge with adriamycin (5 μg/ml) for 48 hr. The viability of cells was examined by staining with trypan blue. (Open boxes) uninfected HuT-78 cells; (filled boxes) TNF-pretreated HuT-78 cells; (open circles) HIV-infected HuT-78 cells; (filled circles) TNF-treated HIV-infected HuT-78 cells. Each point represents the mean of triplicate samples ± S.D.

does not express high levels of MnSOD (Wong et al. 1991). Overexpression of CuZnSOD has no effect (Wong et al. 1991). TNF still sensitizes HIV-infected cells that overexpress high levels of MnSOD or CuZnSOD to heat and radiation (Wong et al. 1991).

INDUCTION OF MNSOD IS MEDIATED BY TNF-R1

The mechanism of induction of gene expression by TNF is still obscure. The amino acid sequences of the intracellular domains of both TNF-R1 and TNF-R2 provide no hint as to the signal transduced by TNF binding. Clearly, signal transduction does not require internalization of TNF (Englemann et al. 1990; Tartaglia et al. 1991). Instead, induction of at least several TNF activities appears to depend on direct activation of TNF receptors because antibodies that bind specifically to TNF-R1 mimic TNF (Englemann et al. 1990; Tartaglia et al. 1991). Recently, induction of MnSOD in mouse 3T3 cells (Tartaglia et al. 1991) and human WI-38 fibroblasts (Fig. 5) was shown to be mediated through TNF-R1. Both polyclonal and mixed monoclonal antibodies against TNF-R1 but not TNF-R2 induce high levels of MnSOD mRNA (Fig. 5). However, all 12 monoclonal antibodies to TNF-R1 had no effect when tested individually (G.H.W. Wong, unpubl.). Paradoxically, TNF-R1 is also the receptor that mediates cytotoxicity. Antihuman TNF-R1 antibodies kill A549 cells in the presence of CHI, and they also protect A549 from subsequent killing by TNF plus CHI or anti-TNF-R1 antibodies plus CHI (Fig. 3). Interestingly, agonistic effects of antimouse TNF-R1 could also be observed on rat fibroblast (Rat-1) cells (G.H.W. Wong, unpubl.). Antibodies directed against TNF-R2 do not display any of these effects. Instead, polyclonal antibodies directed against the murine TNF-R2 stimulate proliferation of mouse thymocytes and the mouse T cell line CT-6 (Tartaglia et al. 1991). However, in human foreskin (Englemann et al. 1990) and WI-38 lung fibroblast (G.H.W. Wong, unpubl.) cells, anti-TNF-R1 but not anti-TNF-R2 antibodies induce cell proliferation. Additional biological activities mediated by TNF-R1 but not TNF-R2 include the production of oxygen free radicals, lac-

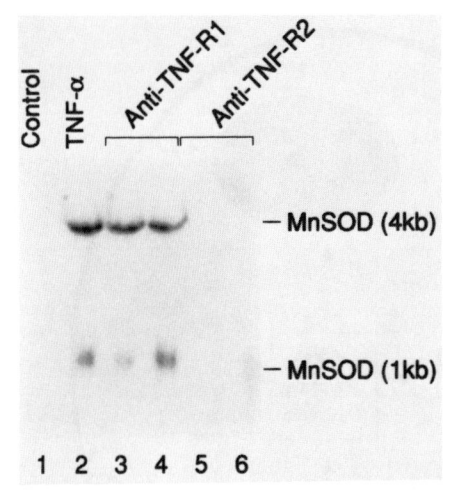

FIGURE 5 Regulation of MnSOD mRNA in human WI-38 fibroblasts by anti-TNF-R1 but not by anti-TNF-R2. Total RNA was extracted from cells pretreated with either human TNF-α (0.1 µg/ml) or indicated antibodies for 15 hr. The antibodies used were rabbit polyclonal antibodies against human TNF-R1 (1:100, lane 3); mouse monoclonal antibodies (a mixture of 12 different types, 0.1 µg/ml of each, lane 4) against human TNF-R1; rabbit polyclonal antibodies against human TNF-R2 (1:100, lane 5), and mouse monoclonal antibodies (a mixture of 10 different types, 0.1 µg/ml of each, lane 6) against human TNF-R2.

tate production, enhanced expression of MHC class I and class II, ICAM expression, and antiviral activities (G.H.W. Wong et al., in prep.).

There are three possible pathways through which TNF-R1 may signal induction of MnSOD (Fig. 6). First, TNF or anti-TNF-R1 antibody binds to TNF-R1 to cause receptor aggregation and thereby stimulates production of classic intracellular messengers that act directly on nuclear gene expression. Of particular interest, the induction of MnSOD by TNF or by anti-TNF-R1 agonistic antibodies is blocked by lipoxygenase but not cyclo-oxygenase inhibitors (G.H.W. Wong et al., in prep.). Second, intracellular free radicals may mediate at least a portion of the signaling process. However, agents that generate free radicals in cells, such as radiation and hyperoxia, induce only very low levels of MnSOD mRNA in the absence of TNF production (Fig. 7). In vivo expression of MnSOD can be in-

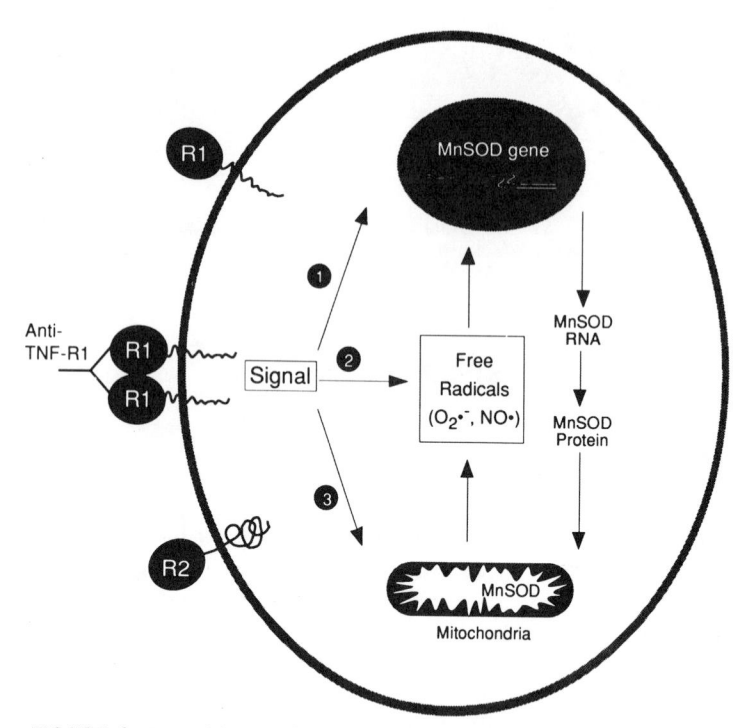

FIGURE 6 Possible mechanisms that underlie induction of MnSOD by TNF. Antibody directed against TNF-R1 (or TNF) binds to TNF-R1, causing the receptors to aggregate and activate the transduction process. The intracellular signal may be of three types: (1) an effector that directly activates MnSOD gene transcription in the nucleus; (2) an effector that triggers the production of free radicals within the cytoplasm that in turn activate MnSOD gene transcription; or (3) an effector that acts on mitochondria which in turn signals transcription of the MnSOD gene, possibly through increased free radical generation.

duced by radiation (Oberley et al. 1987), by hyperoxia (Fleming and Gitlin 1989), or by vincristine (Johnke et al. 1991), but these treatments may induce TNF and IL-1 (Oxholm et al. 1988; Wong et al. 1992b). Thus, it is possible that MnSOD induction in vivo caused by oxidative stress is mediated by TNF and/or IL-1. Furthermore, treatment of A549 cells with H_2O_2 (Wong and Goeddel 1988c), paraquat, or menadione (G.H.W. Wong and L.A. Tartaglia, unpubl.) results in very low levels of MnSOD induction. It remains to be examined whether free

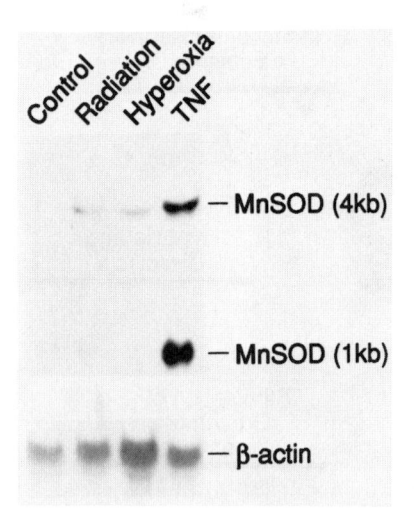

FIGURE 7 Regulation of MnSOD by radiation and hyperoxia in A549 cells. RNA was extracted from A549 cells that had been treated with radiation (40 Gy), 95% O_2 for 3 hr (hyperoxia), or TNF-α (0.1 μg/ml) for 3 hr.

radicals induced by TNF contribute in initiating changes in gene expression. Third, TNF or anti-TNF-R1 antibody may transmit a signal to mitochondria which in turn effects expression of the nuclear MnSOD gene, possibly through mitochondrial production of free radicals (Fig. 6).

POSSIBLE MECHANISMS OF PROTECTION AGAINST OXIDATIVE STRESS INDUCED BY TNF

Several possible mechanisms may explain TNF's protective actions, particularly those directed against damage caused by oxygen free radicals. Below we present several plausible candidates (Fig. 8) that may underlie cellular protection conferred not only by TNF but also by IL-1. It should be noted that the protective effects of TNF and IL-1 in vivo are not necessarily direct. Other cytokines may also be involved in radioprotection such as D-factor (also called leukemic inhibitory factor) (Wong et al. 1992b), whose synthesis can be induced by TNF and IL-1 (Wetzler et al. 1991). Interestingly, TNF and D-factor act syner-

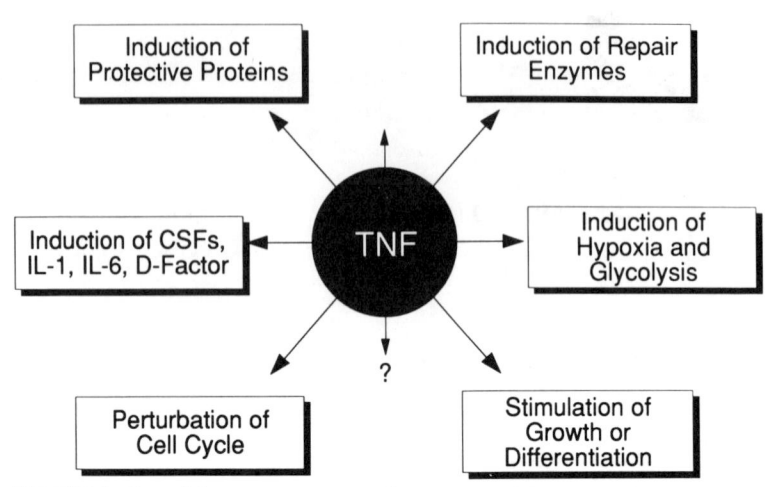

FIGURE 8 Possible protective mechanisms of TNF-α in vivo.

gistically to protect mice from lethal doses of radiation (Wong et al. 1992b). Other growth factors, such as fibroblast growth factor (Haimovitz-Friedman et al. 1991), epidermal growth factor (Kwok and Sutherland 1991), or hematopoietic growth factor (Zucali et al. 1988), may also enhance the protective effects of TNF observed in vivo.

Induction of Protective Proteins

Because TNF-treated cells take some time to become resistant to oxidative insults, it is reasonable to hypothesize that TNF induces protective proteins. MnSOD is one such protein. Although this enzyme is known to detoxify superoxide anions in mitochondria, it may also have other functions. For instance, it may scavenge free radicals other than $\cdot O_2^-$, such as NO. In addition, it may function in organelles other than mitochondria. Overexpression of MnSOD in transgenic tobacco plants (Bowler et al. 1991) and in *Escherichia coli* (Gruber et al. 1990) reduces cellular damage that results from oxygen free radicals. Furthermore, expression of MnSOD in cell lines appears to counteract toxicity caused by TNF (Wong et al. 1989) and other agents such as heat and radiation that generate free radicals (Wong et al. 1991, 1992b). MnSOD expression is essential but not sufficient to confer full protection from TNF and other oxidative insults. Additional protective proteins induced by TNF (e.g., acute phase proteins, metallothioneins [Wong and Goeddel 1988b], ferritin, and possibly heat shock proteins) may also contribute to TNF-mediated resistance. Recent studies have shown that cell lines expressing high levels of HER-2 (also known as c-erbB-2) oncogene and ADP-ribose polymerase (Lichtenstein et al. 1991) or plasminogen activator inhibitor type-2 (Kumar and Baglioni 1991) appear to be more resistant to TNF killing. It will be of interest to compare the types of genes induced by TNF with those induced by ionizing radiation (Woloschak et al. 1990; Weichselbaum et al. 1991) or other oxidative insults.

Induction of Repair

A second mechanism of protection against oxidative insults could involve repair rather than prevention of damage. TNF may activate processes in the cell that remedy damage caused by free radicals (Boveris 1977; Freeman and Crapo 1982; Cross et al. 1987). Because agents that inhibit DNA repair sensitize cells to TNF killing (Coffman et al. 1989), TNF may protect cells by induction of enzymes involved in DNA repair. TNF has been shown to increase ADP ribosylation in both TNF-sensitive and TNF-resistant cells (Lichtenstein et al. 1991). Because inhibition of ADP ribosylation inhibits DNA repair (Berger et al. 1979; Ben-Hur 1984), it is possible that TNF stimulates DNA repair partly by induction of ADP ribosylation. TNF may also stimulate general tissue repair by triggering production of growth factors. Indeed, TNF appears to induce the production of colony stimulating factors that promote stem cell division (Broudy et al. 1986). This process may facilitate the repopulation of lymphoid and myeloid compartments from bone marrow progenitors.

Effects on the Cell Cycle

A third possibility is that TNF perturbs the cell cycle in a way that permits cells to withstand stress more effectively (Warren et al. 1990). It is known that cell cycle stage is an important determinant of cellular resistance to oxidative stress and can influence the response of cells to radiation (Denekamp 1986). Thus, agents that alter the distribution of cells among the stages of the cell cycle may influence the sensitivity of cells to radiation and possibly other stresses. TNF affects proliferation in some cell types. It causes growth arrest in certain cells, an effect that appears to be necessary for protection against TNF killing (Belizario and Dinarello 1991) and against radiation damage in vivo (Warren et al. 1990). TNF causes differentiation of certain cells, leading to the suggestion that the radioprotective effect of TNF is partly due to this perturbation of the cell cycle (Broxmeyer et al. 1986; Takeda et al. 1986).

Effects on Glycolysis

Another possibility is that TNF influences cellular energy metabolism to enable cells to cope with oxidative challenges. It has been shown that tumor cell killing by TNF is inhibited by anaerobic conditions (Matthews et al. 1987) and that mitochondria are damaged by TNF treatment (Matthews 1983; Lancaster et al. 1989). Following exposure to TNF, cells greatly increase glucose utilization and lactate production (Lee et al. 1987; Taylor et al. 1988; Tredget et al. 1988). These are signatures of a switch from aerobic to glycolytic metabolism and are accompanied by other changes both in vivo and in vitro. The Krebs cycle is reduced to a fraction of its previous output, but at least in some cells, the synthesis of ATP actually increases. Thus, it appears that cells respond to TNF by switching their energy metabolism away from efficient oxidative metabolism and toward relatively inefficient anaerobic metabolism, while maintaining or increasing energy output. Consistent with this trend toward decreased oxygen utilization, TNF also causes constriction of capillaries, thereby reducing oxygen availability to tissues (Kallinowski et al. 1989; Edwards et al. 1991) and hence increasing the radioresistance of tissue. These changes may have the effect of relieving mitochondria from their normal burden of oxygen free radicals (Boveris 1977; Freeman and Crapo 1982; Comporti 1989), thus enabling them to deal better with exogenously induced free radicals. There is evidence that hypoxic cells, for example those in the core region of solid tumors, are more resistant to radiation (Suit and Shalek 1963). By analogy, TNF protection may be due partially to its induction of a hypoxic state (see Fig. 8).

TNF: A PARADOX

An attempt to find a unifying explanation for protection against oxidative stress mediated by TNF is frustrated by the apparent paradoxes that TNF phenomena present. These paradoxes include: (1) TNF protects cells against oxidative damage while inducing formation of free radicals; (2) TNF stimulates division in some cells, possibly through oxygen free radicals (e.g., fibroblasts and thymocytes), while it protects generally against oxidative damage; (3) TNF shuts down respiratory metabolism while producing oxygen free radicals; and (4) TNF inhibits respiratory metabolism in mitochondria, presumably removing the source of damaging $\cdot O_2^-$, but it also induces the expression of mitochondrial MnSOD.

Some of these apparent contradictions may have simple resolutions. For example, the timing of events in cells may be a critical factor. TNF may induce a rapid burst of respiratory activity in cells, followed by a slower transition to glycolysis. In addition, some of the free radicals generated by TNF may not originate in mitochondria, but rather, in the cytoplasm as products of oxidation-reduction events, or in other organelles (Boveris 1977). Other paradoxes may disappear when mechanisms are examined in detail in specific cell types. In view of the pleiotropy of TNF-induced responses, it is possible, even likely, that protection mediated by TNF does not have a single basis in all cells. Different cell types may respond to TNF in different ways, perhaps using one of the possible mechanisms or a combination of mechanisms to achieve similar results.

Regardless of how peculiar some of the protective properties of TNF appear, one aspect is now emerging with clarity: A primary target of TNF-induced protection is the mitochondrion (see Fig. 6). Mitochondria, it seems, are especially sensitive to oxidative damage (Mehrotra et al. 1991). If the metabolic formation of free radicals is reduced, the mitochondria are perhaps the primary beneficiary. Their exposure to free radicals is vastly reduced. Furthermore, MnSOD is a mitochondrial enzyme. Thus, cells treated with TNF not only inhibit production of oxygen free radicals in mitochondria, but also increase expression of an enzyme that detoxifies $\cdot O_2^-$. Added to these factors is a diminished supply of oxygen brought about by capillary constriction caused by TNF (Kallinowski et al. 1989; Redl et al. 1990; Edwards et al. 1991). The net result in cells is drastic reduction in the concentration of oxygen free radicals in mitochondria. The reasons that such extreme protective measures are applied to mitochondria remain to be worked

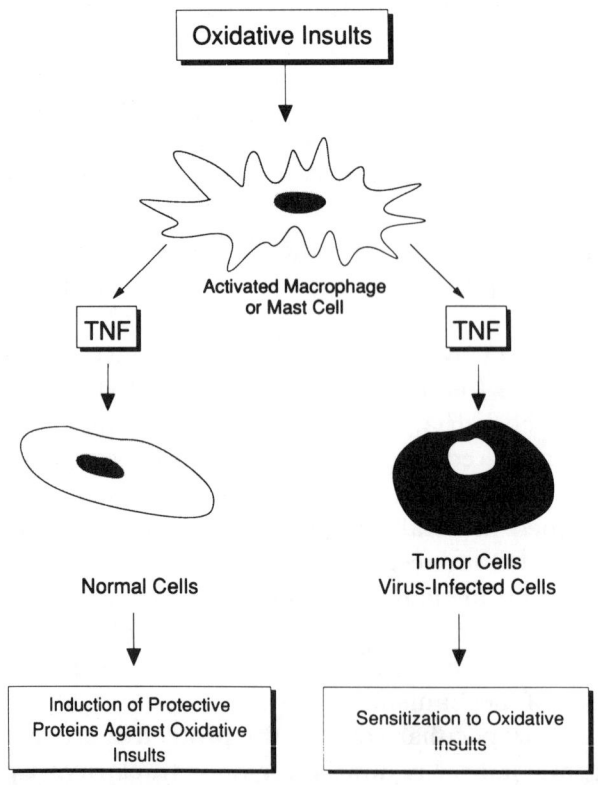

FIGURE 9 Differential effects of TNF-α on the response of normal or abnormal cells to oxidative insults. Oxidative insults trigger the release of TNF-α from macrophages or mast cells (lymphocytes or other cell types) or TNF-β from activated lymphocytes. TNF-α (or TNF-β) induces a state of resistance in normal cells to oxidative stress, but it sensitizes tumor cells and virus-infected cells.

out. One explanation may be the inability of mitochondria to repair oxidative damaged DNA or other cellular components.

SUMMARY

TNF presents several puzzles. It appears to have substantial protective powers, and yet it is not tolerated well at high levels by the host (Beutler and Cerami 1986). Thus, its expression increases transiently during episodes of infection or oxidative trauma such as fever and radiation exposure (Hallahan et al. 1989). We suggest that TNF induces in normal cells protective mechanisms that enable them to avoid (and/or repair) damage from oxidative stress (Fig. 9). However, certain abnormal cells, including specific tumor types and virus-infected cells, are deficient in these protective measures and are more sensitive to oxidative insults. In fact, in many cases, TNF further sensitizes these abnormal cell types. Thus, TNF may place a deliberate stress on cells that normal cells can overcome. In the process, they gain increased resistance to subsequent oxidative insults (Fig. 9). In the words of F. Nietzsche, "That which does not kill me, makes me stronger."

ACKNOWLEDGMENTS

We thank the manufacturing group at Genentech for providing pure recombinant human cytokines; Louis Tamayo for art work; and Elizabeth van Genderen for help in preparing the manuscript.

REFERENCES

Aderka, D., D. Novick, T. Hahn, D.G. Fischer, and D. Wallach. 1985. Increase of vulnerability to lymphotoxin in cells infected by vesicular stomatitis virus and its further augmentation by interferon. *Cell. Immunol.* **92:** 218.

Alexander, H.R., G.M. Doherty, D.L. Fraker, M.I. Block, J.A. Swendenborg, and J.A. Norton. 1991. Human recombinant interleukin-1α; Protection against the lethality of endotoxin and experimental sepsis in mice. *J. Surg. Res.* **50:** 421.

Asayama, K., H. Hayashibe, K. Dobashi, T. Niitsu, A. Miyao, and K.

Kato. 1989. Antioxidant enzyme status and lipid peroxidation in various tissues of diabetic and starved rats. *Diabetes Res.* **12:** 85.

Asoh, K.-I., Y. Watanabe, H. Mizoguchi, M. Mawatari, M. Ono, K. Kohno, and M. Kuwano. 1989. Induction of manganese superoxide dismutase by tumor necrosis factor in human breast cancer MCF-7 cell line and its TNF-resistant variant. *Biochem. Biophys. Res. Commun.* **162:** 794.

Belizario, J.E. and C.A. Dinarello. 1991. Interleukin 1, interleukin 6, tumor necrosis factor, and transforming growth factor β increase cell resistance to tumor necrosis factor cytotoxicity by growth arrest in the G_1 phase of the cell cycle. *Cancer Res.* **51:** 2379.

Ben-Hur, E. 1984. Involvement of poly(ADP-ribose) in the radiation response of mammalian cells. *Int. J. Radiat. Biol. Relat. Stud. Phys. Chem. Med.* **46:** 659.

Berger, N.A., G.W. Sikorski, S.J. Petzoid, and K.K. Kurohara. 1979. Association of poly(ADP-ribose) synthesis with DNA damage and repair in normal human lymphocytes. *J. Clin. Invest.* **63:** 1164.

Berkow, R.L., D. Wang, J.W. Dodson, and T.H. Howard. 1987. Enhancement of neutrophil superoxide production by preincubation with recombinant human tumor necrosis factor. *J. Immunol.* **139:** 3783.

Beutler, B. and A. Cerami. 1986. Cachectin and tumor necrosis factor as two sides of the same biological coin. *Nature* **320:** 584.

Boveris, A. 1977. Mitochondrial production of superoxide radical and hydrogen peroxide. *Adv. Exp. Med. Biol.* **78:** 67.

Bowler, C., L. Slooten, S. Vandenbranden, R. De Rycke, J. Botterman, C. Sybesma, M. Van Montagu, and D. Inze. 1991. Manganese superoxide dismutase can reduce cellular damage mediated by oxygen radicals in transgenic plants. *EMBO J.* **10:** 1723.

Broudy, V.C., K. Kaushansky, G.M. Segal, J.M. Harlan, and J.W. Adamson. 1986. Tumor necrosis factor type α stimulates human endothelial cells to produce granulocyte macrophage colony-stimulating factor. *Proc. Natl. Acad. Sci.* **83:** 7467.

Brown, J.M., C.W. White, L.S. Terada, M.A. Grosso, P.F. Shanley, D.W. Mullvin, A. Banerjee, G.J.R. Whitman, A.H. Harken, and J.E. Repine. 1990. Interleukin 1 pretreatment decreases ischemia/reperfusion injury. *Proc. Natl. Acad. Sci.* **87:** 5026.

Broxmeyer, H.E., D.E. Williams, L. Lu, S. Cooper, S.L. Anderson, G.S. Beyer, R. Hoffman, and B.Y. Rubin. 1986. The suppressive influences of human tumor necrosis factor on bone marrow hematopoietic progenitor cells from normal donors and patients with leukemia: Synergism of tumor necrosis factor and interferon-γ. *J. Immunol.* **136:** 4487.

Carswell, E.A., L.J. Old, R.L. Kassel, S. Green, N. Fiore, and B. Williamson. 1975. An endotoxin-induced serum factor that causes necrosis of tumors. *Proc. Natl. Acad. Sci.* **72:** 3666.

Coffman, F.D., L.M. Green, A. Godwin, and C.F. Ware. 1989.

Cytotoxicity mediated by tumor necrosis factor in variant subclones of the ME-180 cervical carcinoma line: Modulation by specific inhibitors of DNA topoisomerase II. *J. Cell Biochem.* **39:** 95.

Collins, T., L.A. Lapierre, W. Fiers, J.L. Strominger, and J.S. Pober. 1986. Recombinant human tumor necrosis factor increases mRNA levels and surface expression of HLA-A, B antigens in vascular endothelial cells and dermal fibroblasts *in vitro. Proc. Natl. Acad. Sci.* **83:** 446.

Comporti, M. 1989. Three models of free radical-induced cell injury. *Chem. Biol. Interact.* **72:** 1.

Cross, C.E., B. Halliwell, E.T. Borish, W.A. Pryor, B.N. Ames, R.L. Saul, J.M. McCord, and D. Harman. 1987. Oxygen radicals and human disease. *Ann. Intern. Med.* **107:** 526.

Denekamp, J. 1986. Cell kinetics and radiation biology. *Int. J. Radiat. Biol. Relat. Stud. Phys. Chem. Med.* **49:** 357.

Drapier, J-C., J. Wietzerbin, and J.B. Hibbs, Jr. 1988. Interferon-γ and tumor necrosis factor induce the L-arginine-dependent cytotoxic effector mechanism in murine macrophages. *Eur. J. Immunol.* **18:** 1587.

Edwards, H.S., J.C.M. Bremner, and I.J. Stratford. 1991. Induction of hypoxia in the KHT sarcoma by tumour necrosis factor and flavone acetic acid. *Int. J. Radiat. Biol. Relat. Stud. Phys. Chem. Med.* **59:** 419.

Englemann, H., H. Holtmann, C. Brakebusch, Y.S. Avni, I. Sarov, Y. Nophar, E. Hadas, O. Leitner, and D. Wallach. 1990. Antibodies to a soluble form of a tumor necrosis factor (TNF) receptor have TNF-like activity. *J. Biol. Chem.* **265:** 14497.

Fiers, W. 1991. Tumor necrosis factor characterization at the molecular, cellular and *in vivo* level. *FEBS Lett.* **285:** 199.

Fleming, R.E. and J.D. Gitlin. 1989. Regulation of manganese superoxide dismutase gene expression during hyperoxia. *Pediatr. Res.* **25:** (Abstr.).

Folks, T.M., J. Justement, A. Kinter, C.A. Dinarello, and A.S. Fauci. 1987. Cytokine-induced expression of HIV-1 in a chronically infected promonocyte cell line. *Science* **238:** 800.

Freeman, B.A. and J.D. Crapo. 1982. Biology of disease. Free radicals and tissue injury. *Lab. Invest.* **47:** 412.

Futami, H., R. Jansen, M.J. MacPhee, J. Keller, K. McCormick. D.L. Longo, J.J. Oppenheim, F.W. Ruscetti, and R.H. Wiltrout. 1990. Chemoprotective effects of recombinant human IL-1α in cyclophosphamide-treated normal and tumor-bearing mice. *J. Immunol.* **145:** 4121.

Goeddel, D.V., B.B. Aggarwal, P.W. Gray, D.W. Leung, G.E. Nedwin, M.A. Palladino, J.S. Patton, D. Pennica, H.M. Shepard, B.J. Sugarman, and G.H.W. Wong. 1986. Tumor necrosis factors: Gene structure and biological activities. *Cold Spring Harbor Symp. Quant. Biol.* **51:** 597.

Gruber, M.Y., B.R. Glick, and J.E. Thompson. 1990. Cloned manganese superoxide dismutase reduces oxidative stress in *Escherichia coli* and *Anacystis nidulans*. *Proc. Natl. Acad. Sci.* **87:** 2608.

Haimovitz-Friedman, A., I. Vlodavsky, A. Chaudhuri, L. Witte, and Z. Fuks. 1991. Autocrine effects of fibroblast growth factor in repair of radiation damage in endothelial cells. *Cancer Res.* **51:** 2552.

Hallahan, D.E., D.R. Spriggs, M.A. Beckett, D.W. Kufe, and R.R. Weichselbaum. 1989. Increased tumor necrosis factor α mRNA after cellular exposure to ionizing radiation. *Proc. Natl. Acad. Sci.* **86:** 10104.

Harris, C.A., K.S. Derbin, B. Hunte-McDonough, M.R. Krauss, K.T. Chen, D.M. Smith, and L.B. Epstein. 1991. Manganese superoxide dismutase is induced by INF-γ in multiple cell types. Synergistic induction by INF-γ and tumor necrosis factor or IL-1. *J. Immunol.* **147:** 149.

Himmelfarb, J., J.M. Lazarus, and R. Hakim. 1991. Reactive oxygen species production by monocytes and polymorphonuclear leukocytes during dialysis. *Am. J. Kidney Dis.* **17:** 271.

Hirose, K., D.L. Longo, J.J. Oppenheim, and K. Matsushima. 1992. Overexpression of mitochondrial manganese superoxide dismutase confers resistance on tumor cells to interleukin 1, tumor necrosis factor, selected anti-cancer drugs and ionizing radiation. *FASEB J.* (in press).

Hoffman, M. and J.B. Weinberg. 1987. Tumor necrosis factor-α induces increased hydrogen peroxide production and Fc receptor expression, but not increased Ia antigen expression by peritoneal macrophages. *J. Leukocyte Biol.* **42:** 704.

Jacob, C.O., S. Aiso, S.A. Michie, H.O. McDevitt, and H. Acha-Orbea. 1990a. Prevention of diabetes in nonobese diabetic mice by tumor necrosis factor (TNF): Similarities between TNF-α and interleukin 1. *Proc. Natl. Acad. Sci.* **87:** 968.

Jacob, C.O., Z. Fronek, G.D. Lewis, M. Koo, J.A. Hansen, and H.O. McDevitt. 1990b. Heritable major histocompatibility complex class II-associated differences in production of tumor necrosis factor α: Relevance to genetic predisposition to systemic lupus erythematosus. *Proc. Natl. Acad. Sci.* **87:** 1233.

Jimenez, J.J., G.H.W. Wong, and A.A. Yunis. 1991. Interleukin 1 protects from cytosine arabinoside-induced alopecia in the rat model. *FASEB J.* **5:** 2456.

Johnke, R.M., D.P. Loven, R.S. Abernathy, M.J. Bennett, and S.A. Murphy. 1991. Marrow antioxidant enzyme activity in tumor-bearing and non-tumor-bearing mice following vincristine treatment. *Int. J. Radiat. Oncol. Biol. Phys.* **20:** 369.

Kallinowski, F., C. Schaefer, G. Tyler, and P. Vaupel. 1989. *In vivo* targets of human tumour necrosis factor-α: Blood flow, oxygen consumption and growth of isotransplanted rat tumours. *Br. J. Cancer* **60:** 555.

Kawaguchi, T., A. Takeyasu, K. Matsunobu T. Uta, M. Ishizawa, K. Suzuki, T. Nishiura, M. Ishikawa, and N. Taniguchi. 1990. Stimulation of manganese-superoxide dismutase expression by tumor necrosis factor-alpha: Quantitative determination of manganese-SOD protein levels in TNF-resistant and sensitive cells by ELISA. *Biochem. Biophys. Res. Commun.* **171:** 1378.

Kilbourn, R.G., S.S. Gross, A. Jubran, J. Adams, O.W. Griffith, R. Levi, and R.F. Lodato. 1990. N^G-methyl-L-arginine inhibits tumor necrosis factor-induced hypotension: Implications for the involvement of nitric oxide. *Proc. Natl. Acad. Sci.* **87:** 3629.

Koff, W.C. and A.V. Fann. 1986. Human tumor necrosis factor-α kills herpesvirus-infected but not normal cells. *Lymphokine Res.* **5:** 215.

Kramer, S.M. and M.E. Carver. 1986. Serum-free *in vitro* bioassay for the detection of tumor necrosis factor. *J. Immunol. Methods* **93:** 201.

Kumar, S. and C. Baglioni. 1991. Protection from tumor necrosis factor-mediated cytolysis by overexpression of plasminogen activator inhibitor type-2*. *J. Biol. Chem.* **266:** 20960.

Kwok, T.T. and R.M. Sutherland. 1991. Epidermal growth factor modification of radioresistance related to cell-cell interactions. *Int. J. Radiat. Oncol. Biol. Phys.* **20:** 315.

Lancaster, J.R., Jr., S.M. Laster, and L.R. Gooding. 1989. Inhibition of target cell mitochondrial electron transfer by tumor necrosis factor. *FEBS Lett.* **248:** 169.

Lee, M.D., A. Zentella, P.H. Pekala, and A. Cerami. 1987. Effect of endotoxin-induced monokines on glucose metabolism in the muscle cell line L6. *Proc. Natl. Acad. Sci.* **84:** 2590.

Lewis, M., L.A. Tartaglia, A. Lee, G.L. Bennett, G.C. Rice, G.H.W. Wong, E.Y. Chen, and D.V. Goeddel. 1991. Cloning and expression of cDNA for two distinct murine tumor necrosis factor receptors demonstrate one receptor is species specific. *Proc. Natl. Acad. Sci.* **88:** 2830.

Lichtenstein, A., J.F. Gera, J. Andrews, J. Berenson, and C.F. Ware. 1991. Inhibitors of ADP-ribose polymerase decrease the resistance of HER2/neu-expressing cancer cells to the cytotoxic effects of tumor necrosis factor. *J. Immunol.* **146:** 2052.

Loetscher, H., M. Steinmetz, and W. Lesslauer. 1991. Tumor necrosis factor: Receptors and inhibitors. *Cancer Cells* **3:** 221.

Long, N.C., S.L. Kunkel, A.J. Vander, and M.J. Kluger. 1990a. Antiserum against tumor necrosis factor enhances lipopolysaccharide fever in rats. *Am. J. Physiol.* **258:** R332.

Long, N.C., A.J. Vander, S.L. Kunkel, and M.J. Kluger. 1990b. Antiserum against tumor necrosis factor increases stress hyper-

thermia in rats. *Am. J. Physiol.* **258:** R591.
Masuda, A., D.L. Longo, Y. Kobayashi, E. Appella, J.J. Oppenheim, and K. Matsushima. 1988. Induction of mitochondrial manganese superoxide dismutase by interleukin-1. *FASEB J.* **2:** 3087.
Matsuyama, T. and M. Ziff. 1986. Increased superoxide anion release from human endothelial cells in response to cytokines. *J. Immunol.* **137:** 3295.
Matthews, N. 1983. Anti-tumour cytotoxin produced by human monocytes: Studies on its mode of action. *Br. J. Cancer* **48:** 405.
Matthews, N., M.L. Neale, S.K. Jackson, and J.M. Stark. 1987. Tumour cell killing by tumour necrosis factor: Inhibition by anaerobic conditions, free-radical scavengers and inhibitors of arachidonate metabolism. *Immunology* **62:** 153.
Mehrotra, S., P. Kakkar, and P.N. Viswanathan. 1991. Mitochondrial damage by active oxygen species *in vitro*. *J. Free Radicals Biol. Med.* **10:** 277.
Meier, B., H.H. Radeke, S. Selle, M. Younes, H. Sies, K. Resch, and G.G. Habermehl. 1989. Human fibroblasts release reactive oxygen species in response to interleukin-1 or tumour necrosis factor-α. *Biochem. J.* **263:** 539.
Moreb, J., J.R. Zucali, and S. Rueth. 1990. The effects of tumor necrosis factor-α on early human hematopoietic progenitor cells treated with 4-hydroperoxycyclophosphamide. *Blood* **76:** 681.
Morikawa, K., R. Morikawa, J.J. Killion, D. Fan, and I.J. Fidler. 1990. Isolation of human colon carcinoma cells for resistance to a single interferon associated with cross-resistance to multiple recombinant interferons: Alpha, beta, and gamma. *J. Natl. Cancer Inst.* **82:** 517.
Mullen, B.J., R.B.S. Harris, J.S. Patton, and R.J. Martin. 1990. Recombinant tumor necrosis factor-α chronically administered in rats: Lack of cachectic effect. *Proc. Soc. Exp. Biol. Med.* **193:** 318.
Murrell, G.A.C., M.J.O. Francis, and L. Bromley. 1990. Modulation of fibroblast proliferation by oxygen free radicals. *Biochem. J.* **265:** 659.
Neale, M.L., R.A. Fiera, and N. Matthews. 1988. Involvement of phospholipase A2 activation in tumour cell killing by tumour necrosis factor. *Immunology* **64:** 81.
Neta, R., S.D. Douches, and J.J. Oppenheim. 1986. Interleukin-1 is a radioprotector. *J. Immunol.* **136:** 2483.
Neta, R., J.J. Oppenheim, and S.D. Douches. 1988. Interdependence of the radioprotective effects of human recombinant IL-1, TNF, G-CSF, and murine recombinant G-CSF. *J. Immunol.* **140:** 108.
Neta, R., T. Sayers, and J.J. Oppenheim. 1992. Relationship of tumor necrosis factor to interleukins. In *Tumor necrosis factor: Structure, function, and mechanism of actions* (ed. J. Vilcek et al.), p. 499. Marcel Dekker, New York.
Neta, R., J.J. Oppenheim, R.D. Schreiber, R. Chizzonite, G.D. Ledney,

and T.J. MacVittie. 1991. Role of cytokines (interleukin 1, tumor necrosis factor and transforming growth factor β) in natural and lipopolysaccharide-enhanced radioresistance. *J. Exp. Med.* **173:** 1177.

Oberley, L.W., D. Kasemset-St. Clair, A.P. Autor, and T.D. Oberley. 1987. Increase in manganese superoxide dismutase activity in the mouse heart after X-irradiation. *Arch. Biochem. Biophys.* **254:** 69.

Oberley, L.W., M.L. McCormick, E. Sierra-Rivera, and D. Kasemset-St. Clair. 1989. Manganese superoxide dismutase in normal and transformed human embryonic lung fibroblasts. *J. Free Radicals Biol. Med.* **6:** 379.

Old, L.J. 1990. Tumor necrosis factor. In *TNF: Structure, mechanism action, role in disease and therapy* (ed. B. Bonavida and G. Granger), p. 1. Karger, New York.

Omar, B.A. and J.M. McCord. 1990. The cardioprotective effect of Mn-superoxide dismutase is lost at high doses in the postischemic isolated rabbit heart. *J. Free Radicals Biol. Med.* **9:** 473.

Oxholm, A., P. Oxholm, B. Straberg, and K. Bendtzen. 1988. Immuno-histological detection of interleukin 1-like molecules and tumour necrosis factor in human epidermis before and after UVB-irradiation *in vivo. Br. J. Dermatol.* **118:** 369.

Patton, J.S., P.M. Peters, J. McCabe, D. Crase, S. Hansen, A.B. Chen, and D. Liggitt. 1987. Development of partial tolerance to the gastrointestinal effects of high doses of recombinant tumor necrosis factor-α in rodents. *J. Clin. Invest.* **80:** 1587.

Prinsze, C., T.M.A.R. Dubbelman, and J. Van Steveninck. 1990. Protein damage, induced by small amounts of photodynamically generated singlet oxygen or hydroxyl radicals. *Biochim. Biophys. Acta* **1038:** 152.

Radeke, H.H., B. Meier, N. Topley, J. Floge, G.G. Habermehl, and K. Resch. 1990. Interleukin 1-α and tumor necrosis factor-α induce oxygen radical production in mesangial cells. *Kidney Int.* **37:** 767.

Redl, H., G. Schlag, and H. Lamche. 1990. TNF- and LPS-induced changes of lung vascular permeability: Studies in un-anesthetised sheep. *Circ. Shock* **31:** 183.

Ruggiero, V., S.E. Johnson, and C. Baglioni. 1987. Protection from tumor necrosis factor cytotoxicity by protease inhibitors. *Cell. Immunol.* **107:** 317.

Sambhi, S.K., M.R.J. Kohonen-Corish, and I.A. Ramshaw. 1991. Local production of tumor necrosis factor encoded by recombinant vaccinia virus is effective in controlling viral replication *in vivo. Proc. Natl. Acad. Sci.* **88:** 4025.

Shaffer, J.B., C.P. Treanor, and P.J. Del Vecchio. 1990. Expression of bovine and mouse endothelial cell antioxidant enzymes following TNF-α exposure. *J. Free Radicals Biol. Med.* **8:** 497.

Shalaby, M.R., B.B. Aggarwal, E. Rinderknecht, L.P. Svedersky, B.S. Finkle, and M.A. Palladino, Jr. 1985. Activation of human polymorphonuclear neutrophil functions by interferon-γ and tumor necrosis factors. *J. Immunol.* **135:** 2069.

Sheppard, B.C., D.L. Fraker, and J. A. Norton. 1989. Prevention and treatment of endotoxin and sepsis lethality with recombinant human tumor necrosis factor. *Surgery* **106:** 156.

Slordal, L., D.J. Warren, and M.A.S. Moore. 1990. Protective effects of tumor necrosis factor on murine hematopoiesis during cycle-specific cytotoxic chemotherapy. *Cancer Res.* **50:** 4216.

Slordal, L., M.O. Muench, D.J. Warren, and M.A.S. Moore. 1989. Radioprotection by murine and human tumor-necrosis factor: Dose-dependent effects on hematopoiesis in the mouse. *Eur. J. Haematol.* **43:** 428.

Spies, T., C.C. Morton, S.A. Nedospasov, W. Fiers, D. Pious, and J.L. Strominger. 1986. Genes for the tumor necrosis factors α and β are linked to the human major histocompatibility complex. *Proc. Natl. Acad. Sci.* **83:** 8699.

St. Clair, D.K., T.D. Oberley, and Y.-S. Ho. 1991. Overproduction of human Mn-superoxide dismutase modulates paraquat-mediated toxicity in mammalian cells. *FEBS Lett.* **293:** 199.

Sugarman, B.J., B.B. Aggarwal, P.E. Hass, I.S. Figari, M.A. Palladino, Jr., and H.M. Shepard. 1985. Recombinant human tumor necrosis factor-α: Effects on proliferation of normal and transformed cells in vitro. *Science* **230:** 943.

Suit, H.D. and R.J. Shalek. 1963. Response of spontaneous mammary carcinoma of the C3H mouse to X-irradiation given under conditions of local tissue anoxia. *J. Natl. Cancer Inst.* **31:** 497.

Takeda, K., S. Iwamoto, H. Sugimoto, T. Takuma, N. Kawatani, M. Noda, A. Masaki, H. Morise, H. Arimura, and K. Konno. 1986. Identity of differentiation inducing factor and tumour necrosis factor. *Nature* **323:** 338.

Tartaglia, L.A., R.F. Weber, I.S. Figari, C. Reynolds, M.A. Palladino, Jr., and D.V. Goeddel. 1991. The two different receptors for tumor necrosis factor mediate distinct cellular responses. *Proc. Natl. Acad. Sci.* **88:** 9292.

Taylor, D.J., R.J. Whitehead, J.M. Evanson, D. Westmacott, M. Feldmann, H. Bertfield, M.A. Morris, and D.E. Woolley. 1988. Effect of recombinant cytokines on glycolysis and fructose 2, 6-bisphosphate in rheumatoid synovial cells in vitro. *Biochem. J.* **250:** 111.

Tiku, M.L., J.B. Liesch, and F.M. Robertson. 1990. Production of hydrogen peroxide by rabbit articular chondrocytes. Enhancement by cytokines. *J. Immunol.* **145:** 690.

Tredget, E.E., Y.M. Yu, S. Zhong, R. Burini, S. Okusawa, J.A. Gelfand, C.A. Dinarello, V.R. Young, and J.F. Burke. 1988. Role of interleukin 1 and tumor necrosis factor on energy metabolism in rabbits. *Am. J. Physiol.* **18:** E760.

Tsan, M.-F., J.E. White, T.A. Santana, and C.Y. Lee. 1990a. Tracheal

insufflation of tumor necrosis factor protects rats against oxygen toxicity. *J. Appl. Physiol.* **68:** 1211.
Tsan, M.-F., J.E. White, C. Treanor, and J.B. Shaffer. 1990b. Molecular basis for tumor necrosis factor-induced increase in pulmonary superoxide dismutase activities. *Am. J. Physiol.* **259:** L506.
Valentine, J., and H. Nick. 1990. Regulation of manganous and copper-zinc superoxide dismutase messenger RNA in cultured rat intestinal epithelial cells. *FASEB J.* **J4:** 7.
Visner, G.A., W.C. Dougall, J.M. Wilson, I.A. Burr, and H.S. Nick. 1990. Regulation of manganese superoxide dismutase by lipopolysaccharide, interleukin-1 and tumor necrosis factor: Role in the acute inflammatory response. *J. Biol. Chem.* **265:** 2856.
Wallach, D. 1984. Preparations of lymphotoxin induce resistance to their own cytotoxic effect. *J. Immunol.* **132:** 2464.
Warner, B.B., M.S. Burhans, J.C. Clark, and J.R. Wispe. 1991. Tumor necrosis factor-alpha increases Mn-SOD expression: Protection against oxidant injury. *Am. J. Physiol.* **260:** L296.
Warren, D.J., L. Slordal, and M.A.S. Moore. 1990. Tumor-necrosis factor induces cell cycle arrest in multipotential hematopoietic stem cells: A possible radioprotective mechanism. *Eur. J. Hematol.* **45:** 158.
Watanabe, N., Y. Niitsu, H. Neda, H. Sone, N. Yamauchi, M. Maeda, and I. Urushizaki. 1988. Cytocidal mechanism of TNF: Effects of lysosomal enzyme and hydroxyl radical inhibition on cytotoxicity. *Immunopharmacol. Immunotoxicol.* **10:** 109.
Weichselbaum, R.R., D.E. Hallahan, V. Sukhatme, A. Dritschilo, M.L. Sherman, and D.W. Kufe. 1991. Biological consequences of gene regulation after ionizing radiation exposure. *J. Natl. Cancer. Inst.* **83:** 480.
Weisiger, R.A. and I. Fridovich. 1973. Superoxide dimutase. Organelle specificity. *J. Biol. Chem.* **248:** 3582.
Wetzler, M., M. Talpaz, D.G. Lowe, G. Baiocchi, J.U. Gutterman, and R. Kurzrock. 1991. Constitutive expression of leukemia inhibitory factor RNA by human bone marrow stromal cells and modulation by IL-1, TNF-α, and TNF-β. *Exp. Hematol.* **19:** 347.
Woloschak, G.E., C.-M. Chang-Liu, P.S. Jones, and C.A. Jones. 1990. Modulation of gene expression in Syrian hamster embryo cells following ionizing radiation. *Cancer Res.* **50:** 339.
Wong, G.H.W. and D.V. Goeddel. 1986. Tumour necrosis factors α and β inhibit virus replication and synergize with interferons. *Nature* **323:** 819.
———. 1987. Tumour necrosis factors have antiviral activity. In *The biology of the interferon system* (ed. K. Cantell and H. Schellekens), p. 273. Martinus-Nijhoff, Boston.
———. 1988a. Biological activities and production of TNF-α. In *Monokines and other non-lymphocytic cytokines* (ed. M.C.

Powanda), p. 251. A.R. Liss, New York.

———. 1988b. Tumor necrosis factors: Modulation of synthesis and biological activities. In *Lymphocyte activation and differentiation* (ed. J.C. Maini and J. Dornand), p. 217. de Gruyter, New York.

———. 1988c. Induction of manganous superoxide dismutase by tumor necrosis factor: Possible protective mechanism. *Science* **242:** 941.

———. 1989. Tumor necrosis factor. In *The human monocyte review* (ed. G.L. Asherson and M. Zembala.), p. 195. Academic Press, London.

Wong, G.H.W., A. Kamb, and D.V. Goeddel. 1992a. Antiviral properties of TNF. In *Tumor necrosis factors: The molecules and their emerging role in medicine* (ed. B. Beutler), p. 371. Raven Press, New York.

Wong, G.H.W., R. Neta, and D.V. Goeddel. 1992b. Protective roles of MnSOD, TNF-α, TNF-β and D-factor (LIF) in radiation injury. In *Eicosanoids and other bioactive lipids in cancer, inflammation and radiation injury* (ed. S. Nigam et al.). Kluwer Academic, Boston. (In press.)

Wong, G.H.W., J.H. Elwell, L.W. Oberley, and D.V. Goeddel. 1989. Manganous superoxide dismutase is essential for cellular resistance to cytotoxicity of tumor necrosis factor. *Cell* **58:** 923.

Wong, G.H.W., J.F. Krowka, D.P. Stites, and D.V. Goeddel. 1988. In vitro anti-human immunodeficiency virus activities of tumor necrosis factor α and interferon-γ. *J. Immunol.* **140:** 120.

Wong, G.H.W., T. McHugh, R. Weber, and D.V. Goeddel. 1991. Tumor necrosis factor α selectively sensitizes human immunodeficiency virus-infected cells to heat and radiation. *Proc. Natl. Acad. Sci.* **88:** 4372.

Wong, G.H.W., A. Kamb, J.H. Elwell, L.W. Oberley, and D.V. Goeddel. 1992c. MnSOD induction by TNF and its protective role. In *Tumor necrosis factors: The molecules and their emerging role in medicine* (ed. B. Beutler), p. 473. Raven Press, New York.

Yamauchi, N., H. Kuriyama, N. Watanabe, H. Neda, M. Maeda, and Y. Niitsu. 1989. Intracellular hydroxyl radical production induced by recombinant human tumor necrosis factor and its implication in the killing of tumor cells in vitro. *Cancer Res.* **49:** 1671.

Zucali, J.R., H.E. Broxmeyer, M.A. Gross, and C. A. Dinarello. 1988. Recombinant human tumor necrosis factor α and β stimulate fibroblasts to produce hematopoietic growth factors in vivo. *J. Immunol.* **140:** 840.

Regulation of Bacterial Catalase Synthesis

P.C. Loewen
Department of Microbiology, University of Manitoba
Winnipeg, Manitoba, Canada R3T 2N2

All catalases have the common role of dismutating H_2O_2 to O_2 and H_2O, but the activity resides in a structurally diverse group of proteins. *Micrococcus lysodeikticus* (Herbert and Pinsent 1948) and *Proteus mirabilis* (Jouve et al. 1984) produce catalases that closely resemble the classic eukaryotic enzyme with four protoheme IX groups associated with a tetramer of approximately 60-kD subunits. *Escherichia coli* (Loewen and Switala 1986) and *Bacillus subtilis* (Loewen and Switala 1988) produce catalases containing six heme d-isomer prosthetic groups in a hexameric structure of larger subunits. *E. coli* (Claiborne and Fridovich 1979) and *Rhodopseudomonas capsulata* (Hochman and Shemesh 1987) produce a bifunctional catalase-peroxidase enzyme with larger than normal subunits. *Lactobacillus plantarum* (Kono and Fridovich 1983) and *Thermoleophilum album* (Allgood and Perry 1986) produce catalases that contain one manganese per subunit rather than a heme prosthetic group associated in hexameric or tetrameric structures, respectively. Finally, one of the three catalases of *Klebsiella pneumoniae* (Goldberg and Hochman 1989a) exists in dimer form. In addition to this diversity in structures, most bacteria produce two catalases, with *K. pneumoniae* (Goldberg and Hochman 1989b) being an exception with three. The two catalases in *E. coli*, labeled hydroperoxidase I (HPI) (Claiborne and Fridovich 1979) and hydroperoxidase II (HPII) (Claiborne et al. 1979), have been most extensively studied and serve as the focus of this review.

HPI is a bifunctional catalase-peroxidase containing two protoheme IX groups associated with a tetramer of identical 80-kD subunits (Claiborne and Fridovich 1979). The deduced amino acid sequence of the HPI subunit (Triggs-Raine et al. 1988) has revealed significant similarity with the sequence of a peroxidase from *Bacillus stearothermophilus* (Loprasert et al. 1989) but not to other catalases, suggesting that it may have evolved as a peroxidase rather than a catalase. HPII is a monofunctional catalase with six heme d isomers (Chiu et al. 1989) associated in a hexameric structure of 84.2-kD subunits (Loewen and Switala 1986). Despite this dissimilarity in structure from classic catalases, the deduced amino acid sequence of the HPII subunit has revealed significant similarity with the sequences of known eukaryotic catalases (von Ossowski et al. 1991), suggesting that it has evolved as a true catalase.

The presence of two catalases in *E. coli* complicated early studies on the regulation of catalase synthesis because of the inability to ascribe changes in the overall catalase levels to one or the other of the two enzymes. For example, glucose was reported to lower catalase levels (Hassan and Fridovich 1978; Yoshpe-Purer et al. 1977), hydrogen peroxide was reported to induce catalase synthesis (Finn and Codon 1975; Yoshpe-Purer et al. 1977; Richter and Loewen 1981), and catalase levels were reported to increase as cells grew into stationary phase (Finn and Condon 1975), but it was not clear whether one or both of the catalases were changing in response to the stimuli. The isolation of mutants lacking one or the other of the two catalases made it possible to show that it was HPI that responded to H_2O_2 and that HPII was synthesized as the cells entered stationary phase (Fig. 1) (Loewen et al. 1985a). Because the primary role of catalase is to rid the cell of the strong oxidant H_2O_2 before it causes unwanted, possibly destructive, reactions or gives rise to the even more reactive hydroxyl radical •OH, the increase in HPI in response to H_2O_2 in the medium is an easily understood protective response to an oxidative stress. On the other hand, the rationale for why catalase HPII synthesis increases as cells enter stationary phase and not in response to H_2O_2 is not as clear but is probably a reflection of the need for protection against H_2O_2 that may arise under nongrowth conditions. Thus, not only are the two enzymes quite different structurally and genetically, but also the systems controlling their synthesis respond to different stimuli and involve different mechanisms.

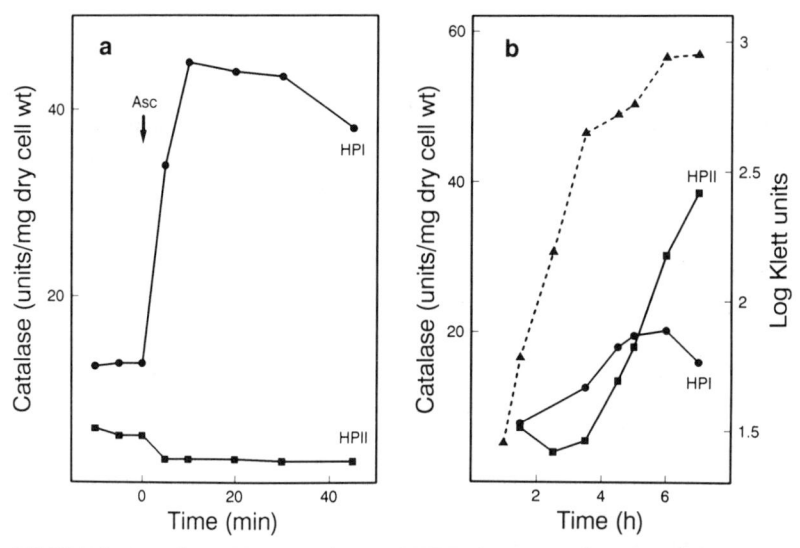

FIGURE 1 Catalase HPI (circle) and HPII (box) synthesis in response to H_2O_2 (from ascorbate added at 0 time) in *a* and to growth through exponential phase into stationary phase in *b*. Strain UM120 (*katE*::Tn*10*) was used for the determination of HPI, and strain UM202 (*katG*::Tn*10*) was used for the determination of HPII. The growth curve in *b* is shown by the dashed line. (Data modified from Loewen et al. 1985a.)

REGULATION OF CATALASE HPI SYNTHESIS

The subunit of HPI is encoded by *katG*, which has been mapped at 89.2 minutes on the *E. coli* chromosome (Loewen et al. 1985b). The gene has been cloned, physically characterized, and sequenced (Loewen et al. 1983; Triggs-Raine and Loewen 1987; Triggs-Raine et al. 1988), revealing an open reading frame of 2181 bp encoding an 80-kD subunit. Putative –10 (TATCGT) and –35 (TTATAA) sequences upstream (Triggs-Raine et al. 1988) of the transcription start site (Tartaglia et al. 1989) differed from the σ^{70} consensus promoter sequence, raising the possibility that an activating protein might be required for transcription initiation.

Cultures of both *E. coli* and *Salmonella typhimurium* treated with low doses of H_2O_2 respond with the synthesis of HPI, as already noted, but also with the synthesis of 33 other proteins that make the cell more resistant to killing by higher, normally lethal, levels of the same oxidant. Of the 34 proteins induced by H_2O_2, 9, including HPI and an alkyl hydroperoxidase (encoded by *ahpC* and *ahpF*), have been shown to be directly under the control of the *oxyR* gene product (Christman et al. 1985; Morgan et al. 1986). The fact that the levels of *katG* mRNA were increased some 50-fold in a *S. typhimurium oxyR1*-containing mutant that allows constitutive overexpression of the 9 proteins (Morgan et al. 1986) confirmed that the effect of OxyR protein on *katG* was at the level of transcription.

Sequence analysis of *oxyR* revealed an open reading frame encoding a 34-kD protein having significant sequence similarity with a family of bacterial regulator proteins including the *E. coli* LysR, IlvY, CysB proteins; the *S. typhimurium* MetR and CysB proteins; the *Rhizobium* NodD protein; the *Enterobacter cloacae* AmpR protein; and the *Pseudomonas aeruginosa* TrpI protein (Christman et al. 1989). Like OxyR, which activates transcription from *katG* and *ahpC* promoters, members of this family of regulatory proteins act principally as transcriptional activators, although some also act as repressors. Indeed, OxyR was also found to negatively autoregulate its synthesis, with a fivefold higher level of β-galactosidase expression from an *oxyR::lacZ* fusion (Christman et al. 1989) being found in a Δ*oxyR*-containing mutant as compared to a wild-type strain. Identification of the OxyR-binding site in the *oxyR* promoter region showed that it overlapped the +1 to −10 region, thereby preventing RNA polymerase initiation, while at the same time activating a putative promoter oriented in the opposite direction. The sequences of the OxyR-binding sites in the *oxyR*, *katG*, and *ahpCF* promoters have been compared (Tartaglia et al. 1989), revealing too little sequence similarity to allow identification of the OxyR-binding requirements but indicating that OxyR just overlaps part of the −35 regions (ATTA in all three cases), possibly facilitating RNA polymerase binding through a direct protein-protein interaction (Fig. 2). Mutations linked to either *katG* or *ahpC*, which suppress H_2O_2 sensitivity in Δ*oxyR*-containing mutants, have been isolated and characterized (Greenberg and Demple 1988). Those linked to *katG* overproduced catalase HPI to impart peroxide resistance, implicating a change in the *katG* promoter that enhanced RNA polymerase binding in the absence of OxyR. Sur-

```
                                                  -35
ahpC    aatcGGGTTGTTAGTTAACGCTTATTGATTTGATAATGGAAACGCATTAGccgaatcagcaa
                            T
                            ↑                   -35
katG    caacaatATGTAAGATCTCAACTATCGCATCCGTGGATTAATTCAATTATAActtctctctaa

                                              -35
oxyR    catcGCCACGATAGTTCATGGCGATAGGTAGAATAGCAATGAACGATTAtccctatcaagcat
                                    -10                          -35
```

FIGURE 2 Alignment of the OxyR-binding sites in the *ahpC*, *katG*, and *oxyR* promoters. The −35 regions for transcription to the right (away from the *oxyR* gene) are indicated above the sequence, and the −10 for transcription into the *oxyR* gene is shown below. The single C to T change in the *katG* promoter that enhances *oxyR*-independent transcription of *katG* is shown above the *katG* sequence. (Data modified from Tartaglia et al. [1989] and Storz et al. [1990].)

prisingly, all the *katG*-linked mutations were the result of a single C to T change at −58, approximately in the middle of the OxyR-binding site and well removed from the RNA polymerase-binding site (J.T. Greenberg et al., unpubl.). The mutation does not generate an obvious new promoter nor does it change the known RNA polymerase-binding site, suggesting that there may be another activator protein which can activate *katG* expression in the absence of OxyR. One possible source of a candidate for this role might be a putative activator for another, as yet uncharacterized, regulon that enhances the expression of some of the 26 other proteins in response to peroxide (Storz et al. 1990). A more detailed definition of the OxyR-binding sites to be found by studying the interaction in various mutants will help to define the important sites for OxyR-DNA interaction.

The fact that *oxyR* transcription did not increase as a result of H_2O_2 treatment indicated that the inducing effect of OxyR came about as a result of changes in the existing OxyR protein and not as a result of increases in OxyR levels. The oxidation state of OxyR was found to be the determining factor in transcription enhancement, with oxidized OxyR being an effective activator of transcription and the reduced form being ineffective (Storz et al. 1990). Furthermore, the protein could be taken through several reduction-oxidation cycles and still retain the ability to activate transcription. There was little difference in the affinities of reduced and oxidized OxyR for the promoters, but the binding footprints on the DNA were somewhat different, suggesting that it was a change in conformation brought on by the change in oxidation state that was responsible for enhancing transcription.

OxyR protein is therefore directly activated by the metabolic stimulus, an oxidant, to become the transcriptional activator, making it both the sensor of oxidative stress and the mediator of enhanced transcription of genes whose products are part of the protective response. A reactive species, which is generated from dissolved oxygen in the medium or from the addition of H_2O_2 to the medium, interacts with OxyR to change its oxidation state, causing a conformational change that affects the way in which the protein interacts with its target promoters. Consequently, the *oxyR* regulon response to oxidative stress is one of the few mechanisms for which the translation of an environmental stress into transcriptional control has been defined, and *E. coli* uses this response, in part, to survive the harmful metabolic effects of peroxide and possibly other oxidants in the medium. The importance of catalase HPI in this response is indicated by the fact that overproduction of HPI alone (Greenberg and Demple 1988) was sufficient to impart peroxide resistance. The other 33 proteins, 8 controlled by *oxyR* and 25 by some other mechanism(s), must also have protective roles, but, except for the alkylhydroperoxidase from the *ahp* genes, none has been characterized. The picture is further complicated by the overlap of control mechanisms, with common proteins being among those induced by redox-cycling agents (such as paraquat and menadione), by hydrogen peroxide (Greenberg and Demple 1989), and by heat shock (Morgan et al. 1986). *E. coli* has evolved a very intricate and complex system of overlapping responses to oxidative stresses, and catalase HPI is an important product of this protective response.

REGULATION OF CATALASE HPII SYNTHESIS

The subunit of catalase HPII is encoded by *katE*, which has been mapped at 37.8 minutes on the *E. coli* chromosome (Loewen 1984). The gene has been cloned, physically characterized, and sequenced (Mulvey et al. 1988; von Ossowski et al. 1991), revealing an open reading frame of 2259 bp encoding an 84.2-kD subunit. The pattern of HPII synthesis, including elevated levels in stationary phase and during growth on TCA cycle intermediates, was independent of *oxyR* (Christman et al. 1985) but required an active *katF* gene mapping at 59 minutes on the chromosome (Loewen and Triggs 1984). Cloning and sequence analysis (Mulvey et al. 1988; Mulvey and Loewen 1989) of the *katF* gene revealed a 1086-bp open reading frame encoding a 41.5-kD protein with striking sequence similarity to known sigma transcription factors, suggesting that its role in the activation of expression from *katE* involved facilitating RNA polymerase binding to the *katE* promoter.

It seemed unlikely that a sigma factor would have evolved for the transcription of just one gene, and subsequent work has confirmed that there are a number of other loci controlled by *katF*. The gene, *nur*, was originally identified by its phenotype of conferring resistance to near-ultraviolet (NUV) radiation, and mutants deficient in *nur* were found to be more sensitive to H_2O_2 and to NUV radiation (Sammartano et al. 1986). Genetic analysis revealed that *nur* was an allele of *katF*, and it was confirmed that *katF* mutations were sensitive to NUV (Sammartano et al. 1986) whereas *katE*-containing mutants were not, indicating that *katF* must be influencing the expression of other genes involved in NUV resistance. A number of loci, including several involved in DNA repair, *recA* (Carlsson and Carpenter 1980), *uvrA* (Sammartano et al. 1986), *polA* (Sammartano et al. 1986), and *xthA* (Demple et al. 1983), have been implicated in the repair of NUV damage, and of these, *xthA* has been shown to be under the control of *katF* (Sak et al. 1989) to the extent that mutations in *katF* caused a reduction in *xthA* expression from both chromosomal and plasmid encoded genes. Unexpectedly, the levels of exonuclease III (encoded by *xthA*) increased throughout exponential phase and dropped as the cells entered stationary phase. This is quite unlike the pattern of expression of HPII (Fig. 1), which remains low throughout exponential phase and increases substantially as the cells enter stationary phase. If this pattern of *xthA* expression is confirmed, and *katF* is indeed regulating both *xthA* and *katE* despite the different patterns of in vivo expression, it implies the involvement of additional factors to mediate the effect of the KatF protein on the two promoters. More recently, evidence has been presented that *appR* is very likely an allele of *katF* (Touati et al. 1991), which implies that *appA*, encoding an acid phosphatase, is another gene controlled by *katF*.

The correlation between HPII synthesis and entry into stationary phase has suggested a possible link between *katE-katF* expression and the expression of genes responding to starvation stress. Up to 55 proteins are synthesized as part of the starvation response, and they result in the cell's becoming thermotolerant, resistant to hydrogen peroxide, resistant to osmotic stress, and resistant to acid stress (for a review of the starvation stress response, see Matin 1990). A search for mutations affecting the transition of bacteria to starvation conditions resulted in the identification of a locus named *csi-2* (for *c*arbon-*s*tarvation-*i*nduced) that controlled the synthesis of at least 16 proteins (Lange and Hengge-Aronis 1991a). Inactivation of the locus by Tn*10* or *lacZ* insertion mutagenesis resulted in reduced glycogen synthesis, an inability to develop thermotolerance or H_2O_2 resistance, and an inability to induce *appA* expression in stationary phase. Like *appR*, *csi-2* was shown to be an allele of *katF*, and growth of a *csi2::lacZ*-containing strain into stationary phase resulted in the induction of β-galactosidase synthesis, confirming that its expression phenotype was very similar to the phenotype of HPII synthesis in wild-type cells (Lange and Hengge-Aronis 1991a). The importance of *katF* in the starvation response has been reinforced with the observation that expression of 32 carbon starvation proteins, including some encoded by *pex* genes, is controlled by *katF* and that starvation-induced cross-protection against heat shock, osmotic shock, oxidative stress, and acid stress did not occur in a *katF* mutant (McCann et al. 1991).

A significant morphological change in cells occurs as they grow into stationary phase. Rod-shaped, exponential phase cells become spherical-shaped as a result of starvation or growth into stationary phase, but the change does not occur in *katF*-containing mutants (Lange and Hengge-Aronis 1991b), thereby implicating *katF* in the regulation of morphological changes. One gene involved in mediating morphological change is *bolA* (Aldea et al. 1989), which encodes a small regulatory protein required for the synthesis of the penicillin-binding protein PBP6, a carboxy peptidase involved in peptidoglycan synthesis for the cell wall. Like HPII, its synthesis is turned on in wild-type cells growing into stationary phase, and expression from the *bolA* promoter was prevented in a *katF::Tn10*-containing mutant (Lange and Hengge-Aronis 1991b). Because only 32 of a possible 50–80 starvation proteins respond to *katF*, not all proteins turned on in stationary phase are controlled by KatF, and this probably includes the *mcb* operon (Genilloud et al. 1989), which, despite its expression in stationary phase (Connell et al. 1987), is only weakly affected by *katF* (Lange and Hengge-Aronis 1991b).

The sequences from a number of genes expressed in stationary phase and apparently controlled by *katF* are now available, making it possible to search for consensus sequences favored by KatF protein acting as a possible sigma factor (von Ossowski et al. 1991; Lange and Hengge-Aronis 1991b). Included in the comparison shown in Figure 3 are the promoter regions from *katE* (von Ossowski et al. 1991), *xthA* (Saporito et al. 1988), *bolA* (Aldea et al. 1989), *ftsQ* (Aldea et al. 1990), and *mcbA* (Genilloud et al. 1989). The first four have putative –35

Consensus	GTTAAGC	———	ACGTCC	———	A
katE	GTTtAGC	—15 bp—	ACGTCC	—6 bp—	G
xthA	GgTAAGC	—17 bp—	cCaTCC	—4 bp—	A
bolA	GTTAAGC	—19 bp—	gCGgCt	—7 bp—	A
ftsQ	GTcAAta	—18 bp—	ACcTtC	—7 bp—	A
mcbA	aTTAtca	—20 bp—	ACGgCa	—7 bp—	A
	–35		–10		+1

FIGURE 3 Alignment of segments of the promoter regions of *katE*, *xthA*, *bolA*, *ftsQ*, and *mcbA* to provide maximum similarity in the –10 and –35 regions and to provide optimum spacing between the two regions. (Data modified from Connell et al. [1987]; Aldea et al. [1989, 1990]; Genilloud et al. [1989]; Lange and Hengge-Aronis [1991a,b].)

and −10 sequences that are very similar, giving rise to consensus sequences of ACGTCC (−10) and GTTAAGC (−35). The promoter from *mcbA* has least similarity, particularly in the −35 region, consistent with its being controlled by *katF*, only weakly, if at all (Lange and Hengge-Aronis 1991b). The promoter of *ftsQ* has not yet been shown to be controlled by *katF*, but the sequence similarity suggests that it is a strong candidate to be a member of this regulon. Additional similarity that would lengthen the −10 sequence to AACGTCCAGT and the −35 sequence to AnTTGnnGTTAAGC exists but is not included in the figure. There is an alternative alignment for the *xthA* −10 region that is also not shown which would increase the spacing to the start site to a more favorable 8 bp, but which would reduce the spacing between the −10 and −35 regions to a less favorable 13 bp. The significance of these sequences will become clearer as more promoters affected by KatF are sequenced.

It is now evident that *katF* is involved in regulating the expression of a group of genes with very diverse functions, including oxidative protection (*katE*), DNA repair (*xthA*), cell morphology changes (*bolA*), phosphate metabolism (*appA*), starvation protection (*pex*), and possibly cell division (*ftsQ*). Maximal expression of *katF* during stationary phase and starvation (Mulvey et al. 1990; Lange and Hengge-Aronis 1991a) suggests that the primary role of the *katF* regulon is to produce gene products necessary for the adaptation of the cell to conditions of nutrient limitation and survival during periods of dormancy. Indeed, strains lacking KatF protein have been found to die off much more rapidly during incubation in stationary phase (Mulvey et al. 1990) and during starvation (McCann et al. 1991). In view of this role and the apparent sigma-like structure of the protein, it has been proposed that KatF protein be named σ^S, where the S denotes starvation, and that *katF* be renamed *rpoS* (Lange and Hengge-Aronis 1991b). The *katF* terminology has been retained for the remainder of this chapter to avoid confusion.

The lack of an easily assayable activity for KatF protein has made the plasmid-borne *katF::lacZ* fusion an invaluable tool in the study of factors affecting the turn-on of transcription of the *katF* gene (Mulvey et al. 1990; P.C. Loewen et al., in prep.).

Generally, factors that induce expression of *katE* also induce expression of *katF*. However, *katF* expression increased gradually throughout exponential phase to a maximum level in stationary phase, whereas *katE* expression remained very low throughout exponential phase and increased substantially only in stationary phase. Consequently, although KatF protein is required for *katE* expression, its presence in the cell is not sufficient to cause the turn-on of *katE* transcription, suggesting that another factor is involved in controlling *katE* expression. As already noted in the case of *xthA* expression, the exonuclease III levels increased throughout exponential phase, seemingly in parallel with the gradual increase in *katF* expression, but decreased in stationary phase, when *katF* levels increase to their highest, implying that yet another protein is required to mediate the expression of *xthA*. KatF may therefore resemble σ^{54} in its requirement for accessory proteins to modulate its influence on gene expression.

The conditions stimulating *katF* expression, namely growth into stationary phase and starvation, are very closely related phenomena in which nutrient depletion is the probable common factor limiting growth. For *katF* expression to occur under starvation conditions in a simple salts medium, a carbon or energy source and an amino acid supplement, sufficient to support protein synthesis but not growth, were required (Mulvey et al. 1990). Expression under these artificial starvation conditions could be prevented by uncouplers, by inhibitors of electron transport (Fig. 4), and by a reduction in pH of the medium, indicating that the proton motive force had a role in mediating the turn-on of *katF* transcription (P.C. Loewen et al., in prep.). In cells growing in rich medium, the response of *katF* expression was quite different, being turned on when the electron transport process was inhibited (Fig. 4) or the proton motive force was reduced. This has given rise to the hypothesis that an optimally charged membrane as would exist in cells growing in rich medium prevents the expression of *katF*, but as the membrane potential drops because of nutrient depletion or slower metabolism, *katF* expression is turned on. However, if the membrane potential drops too low, as when electron flow is inhibited in starved cells or when a very high concentration of KCN is added to growing cells, the cell is no longer capable

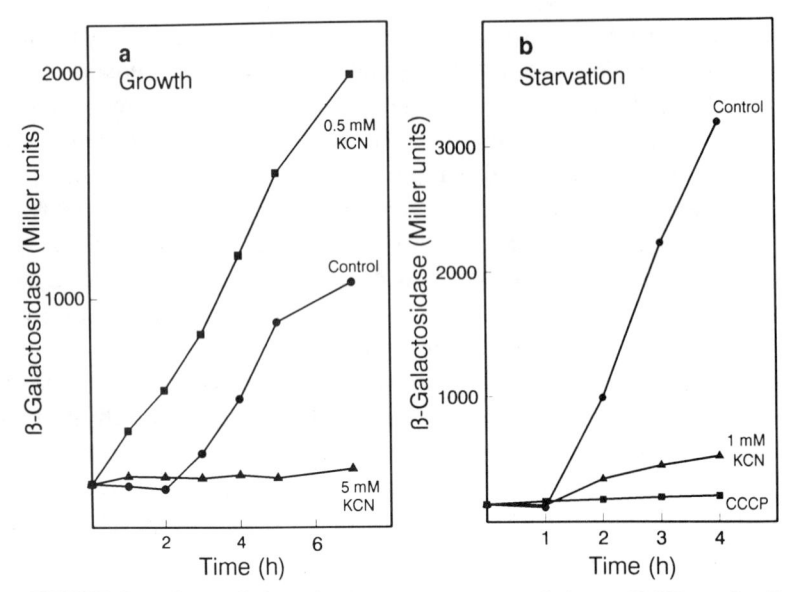

FIGURE 4 Effect of the electron transport inhibitor KCN on *katF* expression in cells growing in LB medium containing 5 g of yeast extract, 10 g of tryptone, and 10 g of NaCl per liter (*a*) and of KCN or the protonophore CCCP on *katF* expression in cells starved for nitrogen and phosphate. Strain NM522 containing plasmid pRSkatF5 (*katF::lacZ* fusion) was used in both experiments. In *a*, growth in the presence of both concentrations of KCN was very slow, whereas the control grew normally. In *b*, no growth occurred in the medium containing 25 mM Tris (pH 8.0), 0.1 M NaCl, 1 mM MgSO$_4$, 16 mM glucose, and 0.006% (w/v) casamino acids.

of supporting the RNA and protein synthesis necessary for expression.

How might changes in the proton motive force be sensed and translated into changes in transcription of *katF*? Several possibilities can be proposed. Because *katF* expression in starved cells is unaffected by valinomycin, it might be the ΔpH component of the membrane potential or the change in activity of proteins involved in the process of pumping protons that is sensed. Alternatively, conformational changes in membrane-associated proteins brought on by changes in membrane potential or proton pumping might be the trigger. Yet another possibility is that changes in the levels of electron carriers such as NAD$^+$, reflecting changes in electron flow, might serve as the signal. Regardless of what mechanism is functioning, the involvement of another protein to sense the metabolic changes and affect *katF* transcription is implied by these results. A mutant with a phenotype consistent with the existence of a locus that influences *katF* expression has recently been isolated and is being characterized.

Changes in *katE* expression largely mimic the changes in *katF* expression under all the metabolic conditions so far studied. This is consistent with KatF protein's being required for the initiation of transcription in the *katE* promoter, but it does not explain the fact that *katE* expression may be only partially induced despite the apparent maximal expression from *katF*. Maximum induction of *katF* supports only one third of maximum induction of *katE* (Mulvey et al. 1990), making it very likely that an as yet unidentified accessory protein is required for maximal *katE* expression. As already noted, an additional protein would seem to be required to mediate *xthA* expression, because the synthesis of exonuclease III is not correlated either with the pattern of HPII synthesis or with the pattern of *katF* expression. A few discrepancies have arisen in various studies about what environmental factors do or do not affect the expression of *katE* and the synthesis of HPII. For example, cAMP may (Lange and Hengge-Aronis 1991a) or may not (P.C. Loewen et al., in prep.) inhibit *katF* expression, and anaerobiosis may (Schellhorn and Hassan 1988; Meir and Yagil 1990) or may not (Mulvey et al. 1990) reduce *katE* expression. The possible involvement of cAMP is probably related to the as yet unexplained phenomenon of *katF* mutations acting as *crp* suppressors (Touati et al. 1991). Glucose has been reported to reduce HPII synthesis (Meir and Yagil 1990), which would be inconsistent with cAMP inhibiting *katF* expression, and this was shown to be a result of the high glucose levels delaying the transition to stationary phase with its associated increase in *katF* expression (Mulvey et al. 1990). Ongoing studies of the mechanism controlling *katE* and *katF* expression will address these questions.

The system controlling the synthesis of HPII is part of a regulon that responds to nutrient depletion but which is intricately involved with a number of other regulons, including those imparting thermotolerance and osmotic shock protection. Another sigma factor, σ^{32}, is involved in the heat shock response; it will be interesting to see how two different sigma factors might affect transcription of the same classes of genes.

SUMMARY

The catalases of *E. coli* are controlled by two different regulons responding to different environmental stimuli. HPI synthesis is controlled by the *oxyR* regulon that responds to oxidative stress by changing the oxidation state of OxyR protein, thereby changing its ability to activate transcription. In this instance, the catalase is serving as one part of a system that protects the cell from reactive oxygen species, including H_2O_2 (Fig. 5). HPII synthesis is controlled by the *katF* regulon (Fig. 5) that responds to nutrient depletion by activating the synthesis of starvation stress proteins, morphological changes, and DNA repair processes, as well as catalase, to prepare the cell for dormancy and to protect it during the dormant state. As just

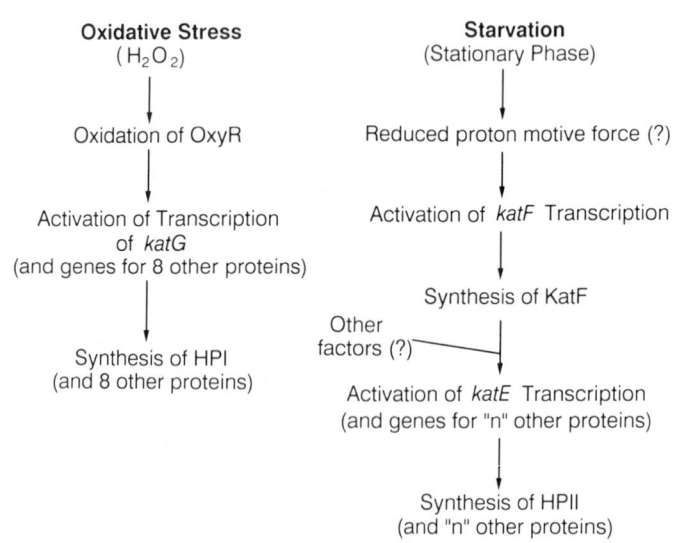

```
         Oxidative Stress                    Starvation
            ( H₂O₂ )                     (Stationary Phase)
               │                                │
               ▼                                ▼
         Oxidation of OxyR         Reduced proton motive force (?)
               │                                │
               ▼                                ▼
       Activation of Transcription    Activation of katF Transcription
              of katG                          │
      (and genes for 8 other proteins)         ▼
               │                       Synthesis of KatF
               ▼                  Other             │
         Synthesis of HPI       factors (?)─────────┤
        (and 8 other proteins)                      ▼
                                  Activation of katE Transcription
                                 (and genes for "n" other proteins)
                                                │
                                                ▼
                                        Synthesis of HPII
                                      (and "n" other proteins)
```

FIGURE 5 Schematic summary of the steps involved in the responses of the two catalases to either oxidative stress or starvation. The designation of "n" for the number of genes and proteins controlled by *katF* was used because the actual number remains uncertain, although it would seem to be in excess of 32.

one enzyme in this very diverse group of proteins, HPII presumably has the role of removing the very reactive H_2O_2 before it can cause cellular damage.

Among other bacterial species, only the catalases from *B. subtilis* have been characterized to provide some limited evidence about factors that affect their synthesis, and the control picture is somewhat similar to that in *E. coli*. Catalase-1 of *B. subtilis* resembles HPI in that it responds to H_2O_2 (Loewen and Switala 1987), but whether or not this is part of an *oxyR*-like regulon response has not been demonstrated. Catalase-2 resembles HPII in several respects: They are structurally and spectrally similar; they both accumulate during nutrient limitation; and the syntheses of both are under the control of an alternate sigma factor (catalase-2 appears only in stage V of the sporulation process [Loewen 1989]). Catalases in other bacterial species have not yet been characterized with respect to their relative levels of expression, preventing any further comparisons.

ACKNOWLEDGMENTS

I thank Drs. R. Hengge-Aronis and A. Matin for providing me with manuscripts prior to their publication. Preparation of this manuscript was supported by a grant (OPG-0009600) from the Natural Sciences and Engineering Council of Canada.

REFERENCES

Aldea, M., T. Garrido, J. Pla, and M. Vicente. 1990. Division genes in *Escherichia coli* are expressed coordinately to cell septum requirements by gearbox promoters. *EMBO J.* **9:** 3787.

Aldea, M., T. Garrido, C. Hernandez-Chico, M. Vicente, and S.R. Kushner. 1989. Induction of a growth-phase-dependent promoter triggers transcription of *bolA*, an *Escherichia coli* morphogene. *EMBO J.* **8:** 3923.

Allgood, G.S. and J.J. Perry. 1986. Characterization of a manganese-containing catalase from the obligate thermophile *Thermoleophilum album. J. Bacteriol.* **168:** 563.

Carlsson, J. and V.S. Carpenter 1980. The *recA*[+] gene product is more important than catalase and superoxide dismutase in protecting *Escherichia coli* against hydrogen peroxide toxicity. *J. Bacteriol.* **142:** 319.

Chiu, J.T., P.C. Loewen, J. Switala, R.B. Gennis, and R. Timkovich. 1989. Proposed structure for the prosthetic group of the catalase HPII for *Escherichia coli*. *J. Am. Chem. Soc.* **111:** 7046.

Christman, M.F., G. Storz, and B.N. Ames. 1989. OxyR, a positive regulator of hydrogen-peroxide inducible genes in *Escherichia coli* and *Salmonella typhimurium*, is homologous to a family of bacterial regulatory proteins. *Proc. Natl. Acad. Sci.* **86:** 3484.

Christman, M.F., R.W. Morgan, F.S. Jacobson, and B.N. Ames. 1985. Positive control of a regulon for defenses against oxidative stress and some heat-shock proteins in *Salmonella typhimurium*. *Cell* **41:** 753.

Claiborne, A. and I. Fridovich. 1979. Purification of the *o*-dianisidine peroxidase from *Escherichia coli* B. *J. Biol. Chem.* **254:** 4245.

Claiborne, A., D.P. Malinowski, and I. Fridovich. 1979. Purification and characterization of hydroperoxidase II of *Escherichia coli* B. *J. Biol. Chem.* **254:** 11664.

Connell, N., Z. Han, F. Moreno, and R. Kolter. 1987. An *E. coli* promoter induced by the cessation of growth. *Mol. Microbiol.* **1:** 195.

Demple, B., J. Halbrook, and S. Linn. 1983. *Escherichia coli xthA* mutants are hypersensitive to hydrogen peroxide. *J. Bacteriol.* **153:** 1079.

Finn, G.J. and S. Condon. 1975. Regulation of catalase synthesis in *Salmonella typhimurium*. *J. Bacteriol.* **123:** 570.

Genilloud, O., F. Moreno, and R. Kolter. 1989. DNA sequence, products, and transcriptional pattern of the genes involved in production of the DNA replication inhibitor microcin B17. *J. Bacteriol.* **171:** 1126.

Goldberg, I. and A. Hochman. 1989a. Purification and characterization of a novel type of catalase from the bacterium *Klebsiella pneumoniae*. *Biochim. Biophys. Acta* **991:** 330.

―――. 1989b. Three different types of catalases in *Klebsiella pneumoniae*. *Arch. Biochem. Biophys.* **268:** 124.

Greenberg, J.T. and B. Demple. 1988. Overproduction of peroxide-scavenging enzymes in *Escherichia coli* suppresses spontaneous mutagenesis and sensitivity to redox-cycling agents in *oxyR*- mutants. *EMBO J.* **7:** 2611.

―――. 1989. A global response induced in *Escherichia coli* by redox-cycling agents overlaps with that induced by peroxide stress. *J. Bacteriol.* **171:** 3933.

Hassan, H.M. and I. Fridovich. 1978. Regulation of the synthesis of catalase and peroxidase in *Escherichia coli*. *J. Biol. Chem.* **253:** 6445.

Herbert, D. and J. Pinsent. 1948. Crystalline bacterial catalase. *Biochem. J.* **43:** 193.

Hochman, A. and A. Shemesh. 1987. Purification and characterization of a catalase-peroxidase from the photosynthetic bacterium *Rhodopseudomonas capsulata. J. Biol. Chem.* **262:** 6871.

Jouve, H.M., J. Gaillard, and J. Pelmont. 1984. Characterization and spectral properties of *Proteus mirabilis* PR catalase. *Can. J. Biochem. Cell Biol.* **62:** 935.

Kono, Y. and I. Fridovich. 1983. Isolation and characterization of the pseudocatalase of *Lactobacillus plantarum. J. Biol. Chem.* **258:** 6015.

Lange, R. and R. Hengge-Aronis. 1991a. Identification of a central regulator of stationary-phase gene expression in *Escherichia coli. Mol. Microbiol.* **5:** 49.

———. 1991b. Growth phase-regulated expression of *bolA* and morphology of stationary phase *Escherichia coli* cells is controlled by the novel sigma factor σ^S. *J. Bacteriol.* **173:** 4474.

Loewen, P.C. 1984. Isolation of catalase-deficient *Escherichia coli* mutants and genetic mapping of *katE*, a locus that affects catalase activity. *J. Bacteriol.* **157:** 622.

———. 1989. Genetic mapping of *katB*, a locus that affects catalase 2 levels in *Bacillus subtilis. Can. J. Microbiol.* **35:** 807.

Loewen, P.C. and J. Switala. 1986. Purification and characterization of catalase HPII from *Escherichia coli* K12. *Biochem. Cell Biol.* **64:** 638.

———. 1987. Multiple catalases in *Bacillus subtilis. J. Bacteriol.* **169:** 3601.

———. 1988. Purification and characterization of spore-specific catalase-2 from *Bacillus subtilis. Biochem. Cell Biol.* **66:** 707.

Loewen, P.C. and B.L. Triggs. 1984. Genetic mapping of *katF*, a locus that with *katE* affects the synthesis of a second catalase species in *Escherichia coli. J. Bacteriol.* **160:** 668.

Loewen, P.C., J. Switala, and B.L. Triggs-Raine. 1985a. Catalases HPI and HPII in *Escherichia coli* are induced independently. *Arch. Biochem. Biophys.* **243:** 144.

Loewen, P.C., B.L. Triggs, C.S. George, and B.E. Hrabarchuk. 1985b. Genetic mapping of *katG*, a locus that affects synthesis of the bifunctional catalase-peroxidase I in *Escherichia coli. J. Bacteriol.* **162:** 661.

Loewen, P.C., B.L. Triggs, G.R. Klassen, and J.H. Weiner. 1983. Identification and physical characterization of a Col E1 hybrid plasmid containing a catalase gene in *Escherichia coli. Can. J. Biochem. Cell Biol.* **61:** 1315.

Loprasert, S., S. Negoro, and H. Okada. 1989. Cloning, nucleotide sequence, and expression in *Escherichia coli* of the *Bacillus stearothermophilus* peroxidase gene (*perA*). *J. Bacteriol.* **171:** 4871.

Matin, A. 1990. Molecular analysis of the starvation stress in *Escherichia coli. FEMS Microbiol. Ecol.* **74:** 185.

McCann, M.P., J.P. Kidwell, and A. Matin. 1991. The putative σ factor KatF has a central role in the development of starvation-mediated general resistance in *Escherichia coli. J. Bacteriol.* **173:** 4188.

Meir, E. and E. Yagil. 1990. Regulation of *Escherichia coli* catalases by anaerobiosis and catabolite repression. *Curr. Microbiol.* **20:** 139.

Morgan, R.W., M.F. Christman, F.S. Jacobson, G. Storz, and B.N. Ames. 1986. Hydrogen peroxide-inducible proteins in *Salmonella typhimurium* overlap with heat shock and other stress proteins. *Proc. Natl. Acad. Sci.* **83:** 8059.

Mulvey, M.R. and P.C. Loewen. 1989. Nucleotide sequence of *katF* of *Escherichia coli* suggests KatF protein is a novel σ transcription factor. *Nucleic Acids Res.* **17:** 9979.

Mulvey, M.R., P.A. Sorby, B.L. Triggs-Raine, and P.C. Loewen. 1988. Cloning and physical characterization of *katE* and *katF* required for catalase HPII expression in *Escherichia coli. Gene* **73:** 337.

Mulvey, M.R., J. Switala, A. Borys, and P.C. Loewen. 1990. Regulation of transcription of *katE* and *katF* in *Escherichia coli. J. Bacteriol.* **172:** 6713.

Richter, H.E. and P.C. Loewen. 1981. Induction of catalase in *Escherichia coli* by ascorbic acid involves hydrogen peroxide. *Biochem. Biophys. Res. Commun.* **100:** 1039.

Sak, B.D., A. Eisenstark, and D. Touati. 1989. Exonuclease III and the catalase hydroperoxidase II in *Escherichia coli* are both regulated by the *katF* gene product. *Proc. Natl. Acad. Sci.* **86:** 3271.

Sammartano, L.J., R.W. Tuveson, and R. Davenport. 1986. Control of sensitivity to inactivation by H_2O_2 and broad-spectrum near-UV radiation by the *Escherichia coli katF* locus. *J. Bacteriol.* **168:** 13.

Saporito, S.M., B.J. Smith-White, and R.P. Cunningham. 1988. Nucleotide sequence of the *xthA* gene of *Escherichia coli* K12. *J. Bacteriol.* **170:** 4542.

Schellhorn, H.E. and H.M. Hassan. 1988. Transcriptional regulation of *katE* in *Escherichia coli* K12. *J. Bacteriol.* **170:** 4286.

Storz, G., L.A. Tartaglia, and B.N. Ames. 1990. Transcriptional regulator of oxidative stress-inducible genes: Direct activation by oxidation. *Science* **248:** 189.

Tartaglia, L.A., G. Storz, and B.N. Ames. 1989. Identification and molecular analysis of *oxyR*-regulated promoters important for the bacterial adaptation to oxidative stress. *J. Mol. Biol.* **210:** 709.

Touati, E., E. Dassa, J. Dassa, P.-L. Boquet, and D. Touati. 1991. Are *appR* and *katF* the same *Escherichia coli* gene encoding a new sigma transcription initiation factor? *Res. Microbiol.* **142:** 29.

Triggs-Raine, B.L. and P.C. Loewen. 1987. Physical characterization of *katG*, encoding catalase HPII of *Escherichia coli. Gene* **52:** 121.

Triggs-Raine, B.L., B.W. Doble, M.R. Mulvey, P.A. Sorby, and P.C. Loewen. 1988. Nucleotide sequence of *katG* encoding catalase HPI of *Escherichia coli. J. Bacteriol.* **170:** 4415.

von Ossowski, I., M.R. Mulvey, P.A. Leco, A. Borys, and P.C. Loewen. 1991. Nucleotide sequence of *Escherichia coli katE*, which encodes catalase HPII. *J. Bacteriol.* **173:** 514.

Yoshpe-Purer, Y., Y. Henis, and J. Yashphe. 1977. Regulation of catalase level in *Escherichia coli* K12. *Can. J. Microbiol.* **23:** 84.

Regulation of the Antioxidant Defense Genes Cat and Sod of Maize

J.G. Scandalios
Department of Genetics, North Carolina State University
Raleigh, North Carolina 27695-7614

The absence of motility among higher plants has resulted in their ability to acquire unique sets of responses to environments from which they cannot escape. In the course of evolution, plants have incorporated a variety of environmental signals into their developmental pathways that have provided for their wide range of adaptive capacities over time. For example, light is such an environmental signal that, in addition to driving photosynthesis, it serves as a trigger and modulator of complex regulatory and developmental mechanisms.

Herein, I discuss some of the mechanisms plants and other aerobes developed for protection against the toxic effects of active oxygen species. Emphasis is placed on the molecular dissection of the two antioxidant gene families, catalase and superoxide dismutase, in an attempt to understand the underlying mechanisms by which the genome perceives and responds to oxidative stress signals.

In plants, the superoxide radical and singlet oxygen are commonly produced in illuminated chloroplasts by the occasional transfer of an electron from an excited chlorophyll molecule to molecular oxygen, or from photosystem I components under conditions of high NADPH/NADP ratios. In addition, such enzymes as xanthine oxidase, aldehyde oxidase, and other flavin dehydrogenases are capable of generating superoxide as a catalytic by-product (Fridovich 1978). The dismutation of two superoxide anions produces hydrogen peroxide, which is also a product of the microbody-associated β-oxidation of fatty acids and glyoxylate cycle, and peroxisomal photorespiration reactions (Beevers 1979; Tolbert 1982). The most reactive and destructive oxygen free radical, the hydroxyl radical (•OH), is generated by the transfer of an electron from the superoxide anion (•O_2^-) to hydrogen peroxide (H_2O_2), or less frequently in a Fenton-type reaction of H_2O_2 with Fe^{++} or reduced ferredoxin (Fridovich 1978; Elstner 1982). These highly reactive species can react with unsaturated fatty acids to cause peroxidation of essential membrane lipids in the plasmalemma or intracellular organelles. Peroxidation damage to the plasmalemma leads to leakage of cellular contents, rapid desiccation, and cell death. Intracellular membrane damage can affect respiratory activity in mitochondria or cause pigment breakdown and loss of carbon-fixing ability in chloroplasts. Several Calvin-cycle enzymes within plant chloroplasts are very sensitive to H_2O_2, and high levels of H_2O_2 can directly inhibit carbon dioxide fixation (Kaiser 1979; Charles and Halliwell 1981). Additionally, H_2O_2 has been shown to be active with mixed function oxidases in marking proteins, particularly several types of enzymes, for proteolytic degradation (Levine et al. 1981; Fucci et al. 1983).

In addition to normal metabolic processes, reactive oxygen species can result from cellular exposure to a variety of external stimuli. Such factors as air pollutants, ultraviolet light and other forms of radiation, herbicides, certain injuries, and various pathogens (e.g., *Cercospora*) are known to induce free radical formation in various organisms.

PROTECTIVE MECHANISMS

As a challenge to the toxic and potentially lethal effects of active oxygen, aerobic organisms evolved protective scavenging or antioxidant defense systems, both nonenzymatic and enzymatic (Halliwell and Gutteridge 1985). Among the former are carotenoids, which occur in great abundance in higher plants and may function to protect plants from solar radiation damage, ascorbic acid, vitamin E or α-tocopherol; and ferredoxin, which scavenges free radicals in chloroplasts via a redox reaction (Allen 1975). Carotenoids protect excited chlorophyll molecules from photooxidation by direct energy transfer at diffusion-controlled rates (Foote 1976).

Enzymatic antioxidant defenses include enzymes capable of removing, neutralizing, or scavenging oxy-intermediates. Examples include ascorbate peroxidase and glutathione reductase, which are believed to scavenge hydrogen peroxide in chloroplasts and mitochondria, respectively (Foyer and Halliwell 1976); catalases (CATs) and other peroxidases that remove H_2O_2; and superoxide dismutases (SODs) that scavenge the superoxide anion. Of these, the CATs and SODs are the most efficient antioxidant enzymes. CATs remove hydrogen peroxide very efficiently according to the overall reaction

$$H_2O_2 + H_2O_2 \rightarrow 2H_2O + O_2 \; (K_1 = 1.7 \times 10^7 \; M^{-1}sec^{-1})$$

SODs comprise a class of metal-combining proteins that are very efficient at scavenging the superoxide radical ($\cdot O_2^-$). The enzyme catalyzes a disproportionation reaction at a rate very near that of diffusion. To accomplish this reaction, the mechanism employs an alternating reduction/oxidation of the respective metal associated with the enzyme

$\cdot O_2^- + M^{++} \rightarrow O_2 + M^+$
$\cdot O_2^- + M^+ + 2H^+ \rightarrow H_2O_2 + M^{++}$

Overall: $2 \cdot O_2^- + 2H^+ \xrightarrow{SOD} H_2O_2 + O_2 \; (K_2 = 2.4 \times 10^9 \; M^{-1}sec^{-1})$

The combined action of CAT and SOD converts the superoxide radical and hydrogen peroxide to water and molecular oxygen, thus abating the formation of the most toxic and highly reactive oxidant, the hydroxyl radical ($\cdot OH$), which can react indiscriminately with all macromolecules. Perhaps this series of cooperative interactions plays a more crucial role in oxygen detoxification than the elimination of $\cdot O_2^-$ or H_2O_2 per se. Although there are no known direct scavengers of singlet oxygen (O_2^1) or the hydroxyl radical ($\cdot OH$), SOD is believed to function in their elimination by chemical reaction (Matheson et al. 1975).

A broad range of stresses affecting plant productivity is directly or indirectly related to oxygen toxicity. Such stressors as air pollutants (e.g., O_3, SO_2), UV radiation, herbicides, hyper- and hypothermia, drought, and others lead to common biochemical and physiological lesions. In addition, plants, because of their growth under very high intensities of sunlight and a high cellular concentration of dioxygen, are subjected to the most severe oxidative stresses relative to other organisms. Thus, the effective scavenging of toxic oxygen species is indispensable to photosynthetic energy conversion, plant growth, and productivity.

Although defenses exist for all aerobes to cope with oxidative stress, the underlying mechanisms are poorly understood. Aside from numerous correlative responses (i.e., increases in oxy-stress leading to increased levels of some antioxidant defenses), there is currently little information or understanding of the underlying molecular mechanisms by which the genome perceives oxidative insult and mobilizes a response to it. To understand these mechanisms, it is essential to identify the responsive genes and to understand their structure, regulation, and expression. Such information is essential in any future attempts to increase tolerance to environmental oxidative stress in organisms and to reduce cellular damage by active oxygen. As mentioned above, two of the most effective enzymatic defenses against oxidative damage are provided by the enzymes CAT and SOD. Among higher plants, these two enzymes have been investigated most thoroughly in the agronomically important monocot *Zea mays* L. (maize). Herein, the CAT and SOD gene-enzyme systems of maize are discussed within the context of the above introduction.

THE CAT GENE-ENZYME SYSTEM

In maize, CAT is encoded by a small gene family. Three unlinked structural genes (*Cat1, Cat2,* and *Cat3*) encode the three biochemically distinct isozymes (CAT-1, CAT-2, and CAT-3) (Scandalios 1965, 1968, 1979; Scandalios et al. 1980a). Each of the *Cat* genes exhibits temporal and spatial specificity in its expression (Scandalios et al. 1984), and each responds variably to different environmental signals (Matters and Scandalios 1986a,b; Scandalios 1987; Skadsen and Scandalios 1987). In addition, the CAT isozymes exhibit cell (Tsaftaris et al. 1983) and organelle (Scandalios 1974;

Scandalios et al. 1980a) specificities. Both overexpression and null mutants have been identified and characterized (Scandalios et al. 1980b; Tsaftaris and Scandalios 1981; Chandlee and Scandalios 1984a; Bethards and Scandalios 1988; Wadsworth and Scandalios 1990) and have been of great utility in the resolution of this system in maize.

The M_r values of all three maize CAT isozyme subunits are approximately 60,000, and the subunits are structurally similar to CATs found in other organisms (Chandlee et al. 1983). The differential spatial and temporal expression of these genes has been characterized in maize, and two temporal regulatory loci, *Car1* and *Car2*, have been genetically defined (Scandalios et al. 1980b; Chandlee and Scandalios 1984b). CAT-1 is the only CAT isozyme expressed in mature pollen, the milky endosperm, aleurone, and the scutellum during early kernel development (Scandalios 1983; Wadsworth and Scandalios 1989; Acevedo and Scandalios 1990). During early sporophytic development, levels of CAT-1 in the scutellum decline, whereas levels of CAT-2 increase, with the CAT-2 developmental profile paralleling that of the glyoxysomes (peroxisomes), the primary intracellular location of these two isozymes (Scandalios 1974). CAT-1 and CAT-3 are the only CAT isozymes present in etiolated leaves and in the coleoptile of the germinating maize seedling. Upon exposure to light, there is rapid accumulation of CAT-2 in leaves, due to both increased transcript accumulation (Redinbaugh et al. 1990b) and increased translation of the *Cat2* message (Skadsen and Scandalios 1987). In mature green leaves of maize, CAT-2 is localized in the peroxisomes of bundle-sheath cells, whereas CAT-1 and CAT-3 are found in mesophyll cells (Tsaftaris et al. 1983).

The CAT-3 isozyme is quite different from the other CATs biochemically. In maize, tobacco, and barley, the CAT-3 (i.e., the isozyme specifically immunoreactive with the maize anti-CAT-3 monospecific antibody) isozyme has enhanced peroxidatic activity (70-, 30-, and 28-fold over "typical" CAT, i.e., CAT-2). CAT can catalyze either the direct dismutation of H_2O_2 into H_2O and O_2 (catalatic mode), or it can use H_2O_2 to oxidize substrates such as methanol, ethanol, formaldehyde, formate, or nitrite (peroxidatic mode) (Havir and McHale 1989).

The ratio of these is usually calculated as $R_{p/c}$ = (mU peroxidatic/U catalatic) × 10. In maize bundle-sheath, the peroxisomal catalase (CAT-2) has high catalatic but low peroxidatic activity ($R_{p/c}$ = 0.25). In contrast, CAT-3, which is expressed in leaf mesophyll and is not peroxisomal (coisolates with mitochondria), has high peroxidatic but low catalatic activity ($R_{p/c}$ = 17.6) (Scandalios et al. 1984; Havir and McHale 1989). Finally, each CAT isozyme exhibits varying degrees of sensitivity to inhibitors such as cyanide (KCN), azide (NaN$_3$), and aminotriazol (AT). CAT-3 proved the least sensitive of the three maize CATs to all inhibitors tested (Scandalios 1990). This again suggests that CAT-3 might have evolved to function under specific conditions or in a specific metabolic role distinct from that of the other CATs. For example, because CAT-3 is cyanide insensitive, one might hypothesize a role or function under conditions favoring cyanide-resistant respiration (i.e., alternate oxidase respiration; Elthon and McIntosh 1987).

CAT PROCESSING

In vitro translation of *Cat2* mRNA using the rabbit reticulocyte lysate system revealed that the M_r of the in-vitro-synthesized CAT-2 protein is 56,000 compared to 54,000 for the purified CAT-2 protein (Skadsen and Scandalios 1986). Two-dimensional polyacrylamide gel electrophoresis of the in-vitro- and in-vivo-labeled CAT-2 protein and Western gel analysis indicated that the CAT-2 protein is processed from a precursor to a lower M_r form in the scutellum. The size reduction of 2 kD corresponds to approximately 19 amino acids, being within the range estimated for signal peptides (Watson 1984). However, CAT-1, which is also associated with glyoxysomes, is synthesized in the same size both in vivo and in vitro, as is CAT-3. Thus, the processing of CAT-2 may not be directly related to its microbody importation.

SIGNAL TRANSDUCTION AND CAT2 ACTIVATION IN THE SCUTELLUM

CAT expression in the scutellum is apparently regulated by "signals" from tissues that spatially interact with the scutellum. Because the scutellum of maize is a fully differentiated, nondividing, diploid embryonic tissue, it was important to determine whether all scutellar cells may be genetically programmed to synchronously activate expression of the *Cat2* gene. Utilizing immunofluorescence microscopy and anti-CAT-2 IgG, it was found that a gradient of *Cat2* gene activation occurs within the scutellar cell mass during postgerminative development (Tsaftaris and Scandalios 1986). The gradient of *Cat2* gene activation occurs from the outer perimeter of the tissue inward toward the embryonic axis. In an effort to determine a potential site of origin for any putative triggering signal for *Cat2* activation, we have demonstrated that the *Cat2* gene is expressed in the single layer of aleurone cells prior to its expression in any other tissue during kernel development. It is conceivable, although it has not yet been proved, that the gradient-type expression of *Cat2* observed in the scutellum possibly involves a molecular signal in the aleurone which diffuses into the scutellum to activate the *Cat2* gene.

It was also demonstrated (Skadsen and Scandalios 1989) that the embryonic axis exerts a specific effect on the accumulation of glyoxysomal proteins, including CAT. Upon excising the embryonic axis from the scutellum prior to imbibition, the developmental accumulation of all glyoxysomal proteins is drastically reduced, whereas the developmental patterns of nonglyoxysomal proteins are unaffected; this suggested that an axis-specific factor modulates the level of expression of the glyoxysomal proteins, including CAT. The exact nature of this "factor" is still unclear. However, it is apparent that possibly two molecular signals may be involved in regulating the expression of *Cat2* in the scutellum during early sporophytic development; one emanating from the aleurone may act to "turn on" the gene, whereas the other emanating from the embryonic axis may modulate the level of *Cat2* expression.

ISOLATION AND CHARACTERIZATION OF THE CAT1, CAT2, AND CAT3 CDNA CLONES

To understand better the molecular mechanisms underlying the temporal and spatial specificities of the three CATs in maize, and to understand more fully the biochemistry and specific physiological function of these isozymes, we isolated and characterized full-length cDNA clones for each of the maize CAT transcripts (Bethards et al. 1987; Redinbaugh et al. 1988; Guan et al. 1991). DNA sequence analyses confirmed that each cDNA encodes a unique CAT protein, and RNA-blot analyses using gene-specific probes confirmed a clear correlation between the *Cat1*, *Cat2*, and *Cat3* transcripts and the tissue-specific expression of CAT-1, CAT-2, and CAT-3 isozymes. The gene-specific probes hybridized with maize genomic DNA blots (Fig. 1) in simple, unique patterns, indicating that there is one copy, or very few copies, of each *Cat* gene in different locations in the maize genome (Redinbaugh et al. 1988); this is in accordance with the earlier genetic and mapping data (Roupakias et al. 1980).

The coding region of the *Cat3* cDNA comprises 66% G+C, which led to a strong codon usage bias in this gene. This codon bias was also observed with the *Cat2* transcripts but

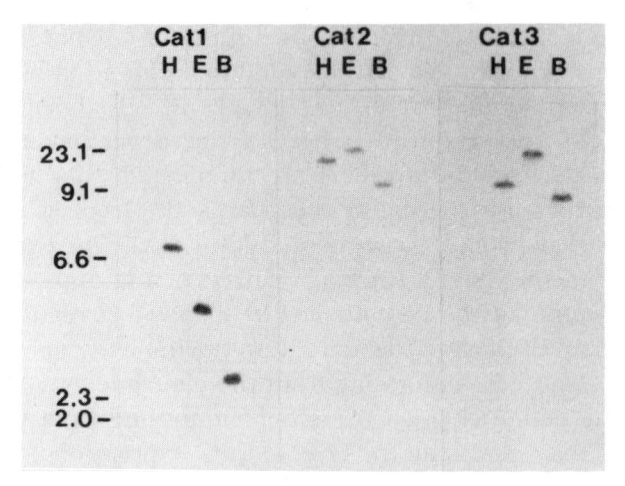

FIGURE 1 Genomic DNA blot analysis. Maize genomic DNA was digested with *Hin*dIII (H), *Eco*RI (E), or *Bam*HI (B) and electrophoresed on a 0.7% agarose gel. After transfer to nitrocellulose, the genomic DNA was hybridized with the indicated gene-specific probe under the conditions appropriate for homologous probes. The migration of the molecular weight standards (in kb) is indicated at the left of the figure. (Reprinted, with permission, from Redinbaugh et al. 1988.)

not with those for *Cat1*. This codon bias might be involved in regulating the expression of these genes. A high degree of similarity was found between the maize catalase nucleic acid and deduced amino acid sequences and those of sweet potato and mammalian catalase (Table 1).

THE SOD GENE-ENZYME SYSTEM

In maize, depending on the inbred lines examined, four or five distinct SOD isozymes are resolved by conventional separation procedures; the four-isozyme phenotype is more common (Baum and Scandalios 1979, 1981). The isozymes SOD-1, SOD-2, SOD-4, and SOD-5 are dimeric, Cu/Zn-containing proteins with subunit M_r ranging from approximately 14,000 to 17,000. SOD-3 is a Mn-containing tetrameric protein with a subunit M_r of 24,000. The maize SODs have similar characteristics to Cu/Zn- and MnSODs from other organisms (Kitagawa et al. 1986; Salin 1988). Among the maize SODs, SOD-3 is unique in its properties in that it is insensitive to cyanide (1 mM), hydrogen peroxide (5 mM), diethyldithiocarbamate (1 mM), and temperature (55°C). The Cu/ZnSOD isozymes exhibit tenfold greater activities at pH 10 than at pH 7.5, whereas the MnSOD-3 isozyme is relatively unaffected by pH. The Cu/Zn-SODs have very low UV absorbance, presumably due to their

TABLE 1 PERCENT SIMILARITY OF THE NUCLEOTIDE AND AMINO ACID SEQUENCES OF CATALASES

	NUCLEIC ACID SIMILARITY				
	Cat1	*Cat2*	*Cat3*	SPOT	RAT
Cat1	–	67	63	66	47
Cat2	77	–	72	65	55
Cat3	67	68	–	62	50
SPOT	74	73	61	–	48
RAT	41	41	40	41	–
	AMINO ACID SIMILARITY				

Similarities were derived from pairwise comparisons of the nucleotide sequences of aligned catalase cDNA. (SPOT) Sweet potato; (RAT) representative mammalian catalase.

low abundance of tryptophan; SOD-3 exhibits an absorbance maximum at 260–270 nm (Baum and Scandalios 1981; Baum et al. 1983).

Turnover studies of SOD during seedling growth demonstrated that SOD is synthesized de novo after seed imbibition and accumulates to roughly 1% of the total soluble protein in scutella between days 1 and 8 postimbibition (Baum and Scandalios 1982a). All the maize SOD isozymes are encoded by nuclear genes, synthesized on cytosolic ribosomes, and translocated to various intracellular compartments.

Monospecific polyclonal antibodies were generated for each of the purified SOD isozymes (Baum and Scandalios 1981) and used to quantitate the levels of SOD proteins synthesized during seedling development (Baum and Scandalios 1982a) and after exposure of plants to various environmental stresses (Matters and Scandalios 1986b).

Intracellular Localization

Unlike CAT, SOD is not highly regulated temporally or spatially, but it does exhibit a great degree of intracellular compartmentalization. Cu/ZnSOD-1 is associated with chloroplasts and etioplasts, and MnSOD-3 is associated with mitochondria (Baum and Scandalios 1979). The other Cu/Zn isozymes (SOD-2, SOD-4, and SOD-5) are located in the cytosolic fraction of maize cells. Although SOD-1 is structurally similar to the other Cu/ZnSODs, antibodies to SOD-1 will not cross-react with the other Cu/Zn isozymes.

The intracellular compartmentalization of SOD is believed to ensure a critical defense against oxygen toxicity in organelles where $\cdot O_2^-$ is generated during electron transport, photorespiration, and other metabolic processes. Because $\cdot O_2^-$ is a charged molecule, it must be eliminated in situ, since it cannot traverse membranes. Understanding the processes by which nuclear-encoded gene products find their way into specific compartments to perform specialized functions is crucial if one wishes eventually to be able to engineer cells for greater capacities to resist the effects of oxidative stress. To this end, the importation of SOD-3 into maize mitochondria was examined, and it was the first such study to be reported in higher plants (White and Scandalios 1989). Deletion mutagenesis, coupled with in vitro transcription, translation, and importation into isolated mitochondria, provided information about the requirements for higher plant mitochondrial import. The data showed that a transit peptide, 31 amino acids in length, was required for efficient import. The relative import efficiency was dependent on the extent of the deletion within the transit peptide region (Fig. 2). Significant findings in the study were that in-vitro-synthesized SOD-3 proteins were not only imported and processed, but also assembled into tetrameric SOD-3 holoenzymes in the mitochondrial matrix (White and Scandalios 1987, 1989).

HISTOGRAPH OF DELETION POSITION VS. PERCENT IMPORTATION

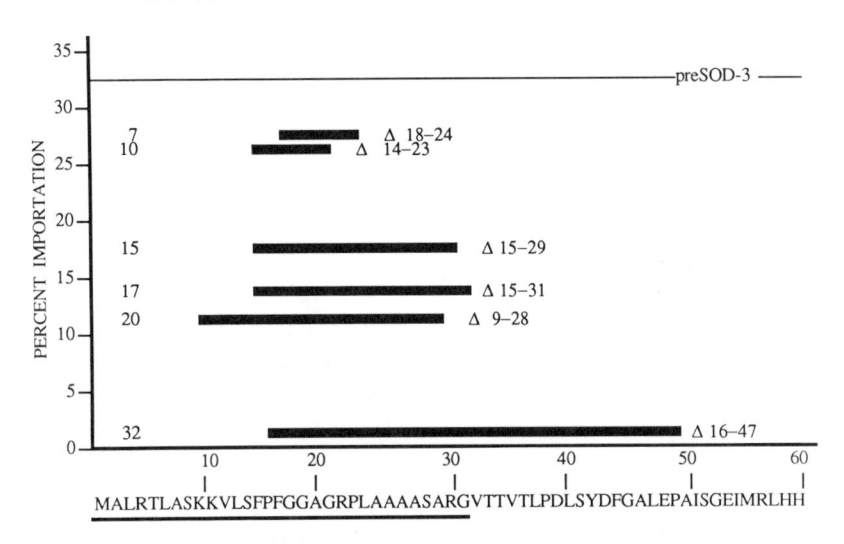

presOD-3 AMINO-TERMINAL SEQUENCE

FIGURE 2 Histographic representation of SOD-3 mitochondrial importation efficiency. Percent importation is plotted against the preSOD-3 amino-terminal sequence. Each bar represents the percent importation of, and the size of the deletion in, the indicated precursor. Numbers to the left of each bar are the number of amino acids deleted from that precursor. The horizontal line indicates the percent importation of unmodified preSOD-3. The transit peptide sequence is underlined. The values for percent importation are the means (±7%) of two experiments. (Reprinted, with permission, from White and Scandalios 1989.)

Genetics

Genetic analysis of electrophoretic SOD variants demonstrated that the maize isozymes SOD-1, SOD-2, SOD-3, SOD-4, and SOD-5, respectively, are encoded by the nuclear, unlinked structural genes *Sod1*, *Sod2*, *Sod3*, *Sod4*, and *Sod5* (Baum and Scandalios 1982b). Furthermore, in the process of isolating cDNAs for the various *Sod* genes, we identified yet another cytosolic SOD isozyme similar to SOD-4. Two cDNAs encoding two proteins differing by only 3 of 153 amino acids were isolated. One of the differing amino acids was located near the amino terminus at residue 12 where GAG (glutamic acid) was changed to GAT (aspartic acid). Upon deciphering this, we purified the SOD-4 protein, amino-terminally sequenced it, and found that at residue 12, both aspartic acid and glutamic acid were present. Genomic DNA and RNA blots confirmed the existence and expression of two genes (*Sod4* and *Sod4A*) that encode two virtually identical SOD-4-like proteins (Cannon and Scandalios 1989).

CLONING AND CHARACTERIZATION OF THE SOD CDNAS

Full-length cDNAs for the maize Cu/Zn*Sod2*, *Sod4*, and *Sod4A*, and for the Mn*Sod3* have been isolated, cloned, sequenced, and characterized (Cannon et al. 1987; White and Scandalios 1988; Cannon and Scandalios 1989). Nucleic acid and amino acid sequences show significant homologies with the respective Cu/ZnSODs and MnSODs characterized and reported for other organisms (Perl-Treves et al. 1988; Scioli and Zilinskas 1988). The Cu/Zn*Sods* are highly conserved in the coding regions, leading to cross-hybridization problems when full-length cDNAs are used as probes. However, this can be alleviated by using gene-specific-probes utilizing the 3'-untranslated region of each Cu/Zn*Sod*. Genomic DNA blots indicate that, like the CATs, the *Sod* genes exist in single copy, or very few copies, and RNA blots indicate that the tissue and temporal distribution of the *Sod* transcripts parallels the earlier isozyme profiles of the various maize tissues.

ISOLATION AND CHARACTERIZATION OF CAT AND SOD GENOMIC CLONES

To be able to identify stress-response elements in the *Cat* and *Sod* genes of maize, the respective cDNAs were used to isolate and characterize *Cat* and *Sod* genomic clones. To date, we have isolated genomic clones for *Cat1*, *Cat3*, *Sod4*, and *Sod4A*. The *Cat1* genomic DNA (7.8 kb) was restriction-mapped, and intron-exon boundaries were determined on comparison with the *Cat1* cDNA sequence. The *Cat1* gene comprises 6 introns and 7 exons, a 2.5-kb promoter, and a 1.5-kb 3'-untranscribed region (Fig. 3). The transcription start site was mapped by primer extension. The promoter region was sequenced, and computer searches identified an abscisic acid (ABA)-response element. Characterization of genomic *Cat3* clones is virtually completed. The *Cat3* gene promoter (2.6 kb) has been sequenced, and the transcription site has been mapped 68 bases from the start of translation by primer extension. Approximately 29 bases 5' of the transcriptional start site is a putative TATA box, and three possible CAAT boxes are located about 150 bases upstream of the start of transcription.

Genomic cytosolic *Sod* DNA blots indicate that the *Sod2*, *Sod4*, and *Sod4A* genes are located on different restriction fragments in the maize genome. Clones of the *Sod4* and *Sod4A* genes were restriction-mapped, DNA was sequenced, and the structure (intron/exon) was determined. The most striking feature is the strong conservation of sequence and organization between the two genes. They contain the same size, number, and location of coding and noncoding DNA regions. The +1 start of transcription has been mapped via reverse-transcriptase-dependent primer extension experiments on the *Sod4A* gene. The upstream region of *Sod4A* contains no consensus TATA, but has a CAAT box. Mapping the 5'end of the *Sod4* gene is presently under way and will provide direct DNA sequence comparisons of putative *cis* regulatory regions of the *Sod4* and *Sod4A* genes. Another interesting feature of the *Sod4* and *Sod4A* genes concerns the location and size of the first intervening sequence (Fig. 4). The intron, located 12 nucleotides upstream of the start of translation in both *Sod4* and *Sod4A*, and having a length of approximately 1300–1500 nucleotides,

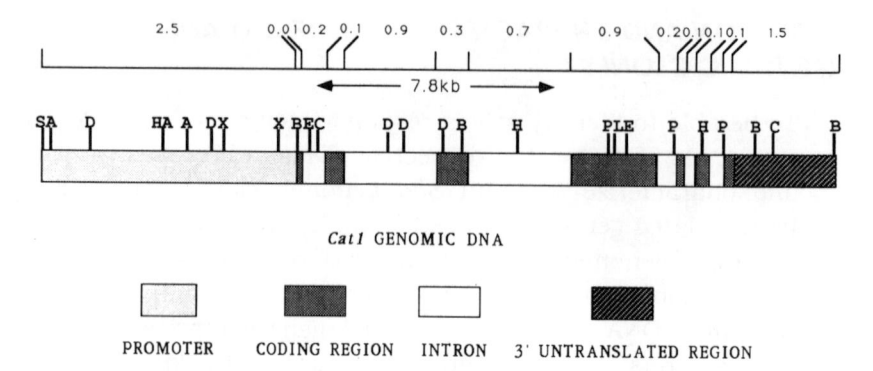

Cat1 GENOMIC DNA

PROMOTER   CODING REGION   INTRON   3' UNTRANSLATED REGION

A=AvaI, B=BamHI, C=ClaI, D=HindIII, E=EcoRI, H=HindII, L=SalI, P=PstI, S=SacI, X=XbaI

FIGURE 3 *Cat1* genomic map showing introns, exons, and 5' and 3' noncoding regions. The 2.5-kb promoter region is to the left.

is quite striking. A rare large intervening sequence, coupled with 5' leader location, suggests a possible regulatory function.

CAT AND SOD GENE EXPRESSION IN RESPONSE TO ENVIRONMENTAL STRESS

There have been numerous reports that, in both prokaryotes and eukaryotes, oxidative stress enhances or induces the activity of SOD and CAT. The formation of superoxide and other active oxygen species can be accelerated as a consequence of various stress conditions, including UV radiation, high light intensity, low CO_2 concentration, and treatment with herbicides that serve as preferred terminal electron acceptors at the reducing site of Photosystem I (e.g., diquat, paraquat) or that are known to block electron transport (e.g., atrazine, diuron). Increases in antioxidant enzyme activities have been reported in response to heat and light conditions that cause sunscald in vegetables, fruits, and flowers (Rabinowitch and Sklan 1980). The fungal toxin cercosporin, produced by the pathogenic fungi *Cercospora* that cause damaging leaf spot diseases on various economically important crops, acts by generating increased levels of singlet oxygen (Daub and Han-

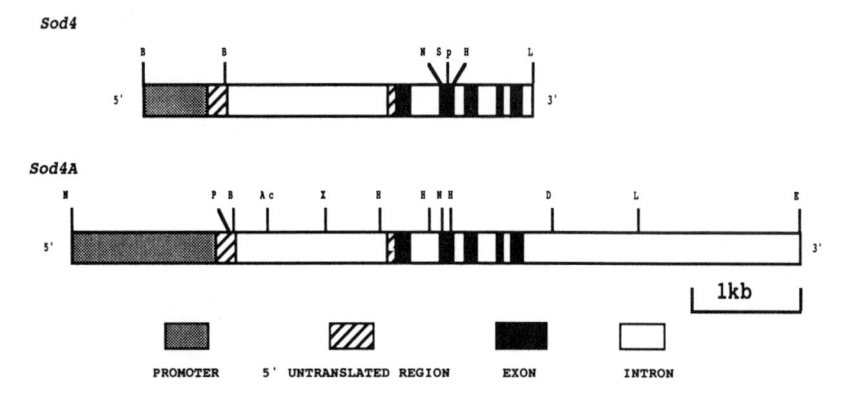

FIGURE 4 Structures of the *Sod4* and *Sod4A* genomic clones. Restriction site abbreviations are listed beneath the maps. Shaded areas represent the putative promoters, black areas represent exons, white boxes represent introns, and the hatched region represents the 5'-untranslated region (5' leader) of the gene. Note the large intervening sequence (intron) in the 5'-untranslated portion of the genes.

garter 1983). Increases in SOD and CAT have been observed in response to ozone (Lee and Bennett 1982) and SO_2 (Tanaka and Sugahara 1980; Alscher et al. 1987) levels in the environment. However, the mechanisms for the observed increases in antioxidant enzymes in response to oxidative stress have yet to be resolved. Changes in individual *Sod* and/or *Cat* genes in response to environmental stresses have not been previously examined in detail, nor have the responses to different stress factors within a single SOD or CAT multienzyme system been studied. The maize systems described above have provided an opportunity to study the response of specific *Sod* and *Cat* genes to imposed environmental stresses in an effort to unravel the mechanisms regulating such responses. To these ends, I describe below some of our studies aimed at deciphering the underlying mechanisms involved in regulating the expression of these important genes in maize. Some broader aspects of responses of these genes to various environmental signals have recently been discussed elsewhere (Scandalios 1990).

THE PHOTORESPONSE OF THE CAT2 AND CAT3 GENES IN MAIZE LEAVES

One of the most fundamental and important processes in nature is the utilization of light by plant cells as the basic energy source to drive biological reactions. Photodynamic reactions are known to occur in green plant tissues in the presence of light and oxygen and are capable of cell injury and tissue death (Rabinowitch et al. 1982).

Green plants have more opportunities to produce activated oxygen species than do animal cells. Consequently, the challenge to overcome oxidative stress is greater in plants than in other eukaryotes because plants both consume O_2 during respiration and generate O_2 during photosynthesis. The question then arises as to the role(s) light may play in modulating, directly or indirectly, the antioxidant defense systems of green plants.

In addition to exhibiting differential temporal and spatial patterns of expression, the three maize CAT genes respond differentially to light in developing maize leaves (Scandalios 1979; Skadsen and Scandalios 1987; Redinbaugh et al. 1990a; Acevedo et al. 1991). CAT-1 protein and *Cat1* message accumulate at low levels in the mesophyll of maize leaves in a light-independent manner throughout development (Redinbaugh et al. 1990a,b). In contrast, both *Cat2* and *Cat3* respond to light, although in quite different ways. The *Cat2* gene is positively regulated by light. CAT-2 protein and *Cat2* mRNA accumulate in the bundle-sheath cells of shoots grown either in a light/dark regime or in constant light (Tsaftaris et al. 1983) but are not detected in young leaves grown in constant dark. The CAT-2 isozyme is dramatically induced (due to de novo synthesis) in leaves when dark-grown seedlings are exposed to light (Fig. 5) (Scandalios 1979). The positive light response of the *Cat2* gene is not phytochrome-mediated (Skadsen and Scandalios 1987). When total poly(A)[+] RNA (mRNA), polysomes, or isolated polysomal mRNA from light- and dark-grown leaves was translated in vitro, CAT-2 protein was detected only among the light-grown leaf products (Skadsen and Scandalios 1987). RNA-blot analyses using a gene-specific *Cat2* cDNA probe showed that *Cat2* mRNA was present in ap-

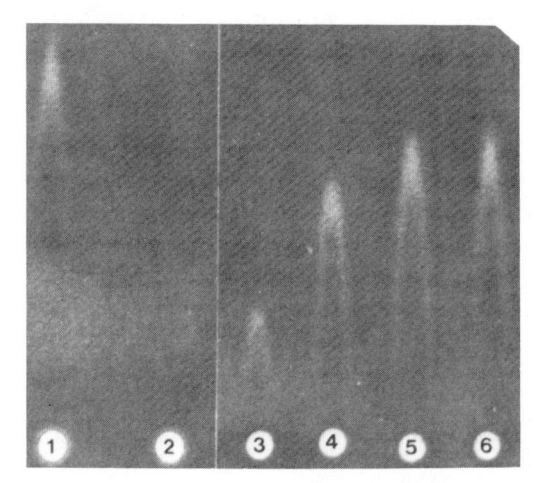

FIGURE 5 Induction of catalase (CAT-2) protein in leaves after exposure to light. Seedlings were grown under a 12-hr photoperiod with 3900 footcandles (1 footcandle = 10.76 lux) of light. Induction of CAT-2 protein was detected by rocket immunoelectrophoresis using anti-CAT-2 monospecific antibodies and staining the immunoelectropherogram for catalase activity (indicated by rockets). Lanes: (1) 0.53 μg of purified maize CAT-2, used as control standard; (2–6) homogenates of leaves after 0, 8, 16, 24, and 32 hr of light exposure, respectively.

proximately equal quantities in total mRNA and polysomal mRNA derived from both light- and dark-grown maize leaves. Furthermore, Cat2 mRNA was equally distributed in identical high-molecular-weight fractions in polysomes from light- and dark-grown leaves (Fig. 6), indicating that the Cat2 mRNA is not sequestered in ribonucleoprotein particles in dark-grown leaves (Skadsen and Scandalios 1987). Thus, the control of Cat2 expression in maize leaves in response to light appears to involve a unique form of translational inhibition in dark-grown leaves, preventing translation of the isolated Cat2 mRNA. The mRNA is rendered translatable only after the leaves are exposed to white light. There is evidence that, like maize Cat2, the mRNA for chlorophyll a/b binding protein is present in dark-grown peas (Giles et al. 1977) and in *Lemma gibba* (Slovin and Tobin 1982) and may be activated by a translational control mechanism in the light.

It is tempting to speculate that a physiological role should

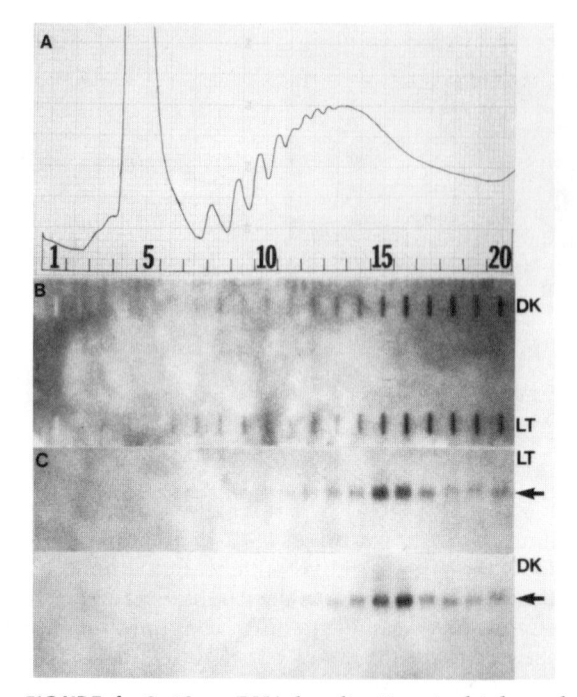

FIGURE 6 Cat2 mRNA localization in high molecular mass polysome fractions. (A) Ten A_{260} units of polysomes from light-grown leaves were separated on a sucrose gradient as described. The dark-grown leaf polysome profile (not shown) was identical. (B) Slot blots of RNA from each polysome fraction. Light (LT)- and dark (DK)-grown leaf RNAs are indicated. (C) RNA blot analysis of light and dark-grown leaf polysomal poly(A)⁺ mRNA from polysome fractions. Duplicate sucrose gradients were formed and fractionated as above. Polysomal poly(A)⁺ mRNA isolation. RNA blot analysis and probing were carried out as described previously. (Reprinted, with permission, from Skadsen and Scandalios 1987.)

attend this CAT-2-specific light response. A plausible explanation is that the cell reserves CAT-2 synthesis until it is needed to destroy H_2O_2 generated during photorespiration (Zelitch 1971). Photorespiration in maize, a C_4 plant, occurs to a lesser degree as compared to C_3 plants. C_3 plants often form large CAT crystals in their peroxisomes (Huang et al. 1983), which may reflect the degree to which CAT is vital. This point is underscored in a CAT underexpression mutant of barley that reportedly does not survive under photorespiratory conditions

(Kendall et al. 1983). However, the existence of normally growing *Cat2* and *Cat3* null mutants in maize (Wadsworth and Scandalios 1990; Abler and Scandalios 1991) suggests that other mechanisms may accommodate the C_4 level of photorespiration and that each of the maize catalases may play a unique metabolic role.

In contrast, the *Cat3* gene appears to be regulated in a much more complex manner. Accumulation of *Cat3* gene products is mesophyll-specific in the leaves of light- and dark-grown maize seedlings. We recently described a novel circadian regulation of *Cat3* gene transcription in developing maize leaves (Fig. 7A). It was also established that this phenomenon is not phytochrome-mediated and that chlorophyll- and carotenoid-deficient mutants and their wild-type siblings exhibit the normal diurnal pattern of *Cat3* RNA accumulation (Redinbaugh et al. 1990b; Acevedo et al. 1991). This indicates that photosynthetic pigments, allelic variation, and genetic background do not directly affect the temporal pattern of *Cat3* accumulation in leaves. Most circadian phenomena previously described in plants at the molecular level involve the regulation of chloroplast-associated gene products directly involved in photosynthesis (Giuliano et al. 1988; Nagy et al. 1988; Taylor 1989). The *Cat3* gene product is not associated with chloroplasts, nor is it regulated by phytochrome (Redinbaugh et al. 1990b; Acevedo et al. 1991). The transcriptionally mediated accumulation of *Cat3* message late in the photoperiod, together with its localization in the leaf mesophyll mitochondrial fraction, suggested that the CAT-3 protein might have a role in nonphotosynthetic reactions in the dark. We observed, however, that when normally pigmented plants are grown in either continuous light or continuous dark, the *Cat3* transcript is present at uniformly high levels throughout the 24-hour sampling period (Fig. 7B). Because the *Cat3* gene is continually transcribed in the absence of a cyclic light regime, the normally observed diurnal variation of *Cat3* gene expression is presumably, therefore, the result of a circadian-regulated transcriptional repressor (Acevedo et al. 1991). This novel type of repressor-mediated circadian regulation has, to our knowledge, not been previously reported. Most previously reported light-induced circadian

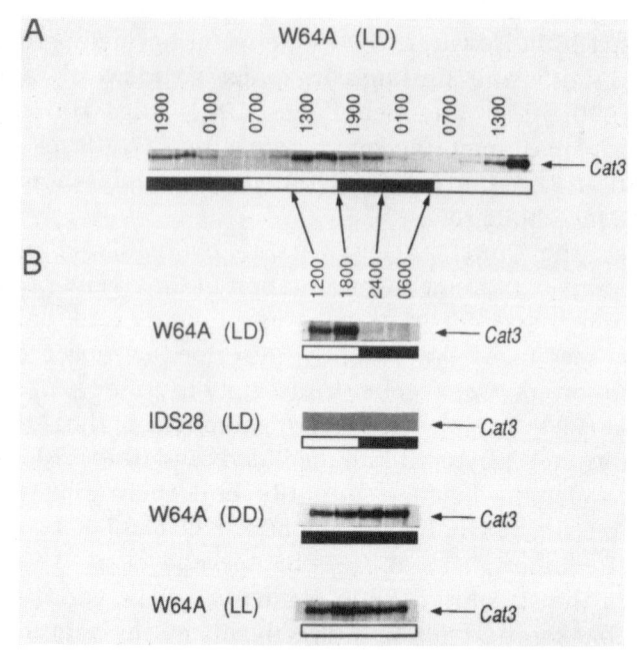

FIGURE 7 Effect of light regime on the accumulation of the *Cat3* catalase transcript in maize leaves. The steady-state levels of *Cat3* transcript in the total RNA population of leaves of the normal inbred maize line W64A and the *Cat3* null line IDS-28 grown under three different light regimes were determined by S1 nuclease protection analysis. For one set of analyses (A), 6-day postimbibition W64A seedlings were grown under a 24-hr photoperiod (12 hr light/12 hr dark), and leaves were collected at 3-hr intervals over a 48-hr period. Samples were collected beginning at 7 P.M. (1900 hours), or 1 hr into the dark period. For all other analyses (B), samples were collected at four representative time points, spaced at 6-hr intervals over the 24-hr period. For these analyses, plants were grown from the time of imbibition in constant light (LL), constant dark (DD), or under a 24-hr light/dark photoperiod (LD). Leaves of W64A grown under a 24-hr light/dark regime exhibit the expected large diurnal variation in *Cat3* RNA levels. In contrast, plants grown in either constant light or constant dark from the time of imbibition accumulate constant, high steady-state levels of the *Cat3* transcript throughout the sampling period. As expected, *Cat3* RNA was not detected in leaves of the *Cat3* null line IDS-28.

rhythms in plants are phytochrome-mediated, involve genes directly related to photosynthesis, and are superimposed on the simple positive light induction of gene transcription; i.e., genes are on in constant light and off in constant dark. Our results, in contrast, suggest that expression of the *Cat3* gene has a "default" mode of expression in leaves where, in the absence of an alternating light/dark cycle, the *Cat3* gene is maximally expressed in leaf mesophyll throughout the 24-hour sampling period. Under a normal diurnal light regime, however, a cyclic repression superimposed on this quasi-constitutive mode of expression in leaves yields the observed variation in *Cat3* gene transcription. Present models of circadian regulation in plants hypothesize the existence of a transcriptional enhancer that is the product of a cyclic regulatory gene or "clock" gene, as has been described for *Neurospora* and *Drosophila* (Hall and Rosbash 1988; McClung et al. 1989). In our case, this clock gene product could be a specific repressor of *Cat3* gene expression or might regulate the cyclic expression of a specific *Cat3* gene repressor. Therefore, the light regulation pathway of *Cat3* represents a completely different set of light regulatory mechanisms and so represents an opportunity to study a completely new type of light signal transduction pathway.

The circadian rhythm in *Cat3* transcript levels is unique among the three maize CATs. In addition, light affects the accumulation of the three CAT transcripts in different ways (Skadsen and Scandalios 1987; Redinbaugh et al. 1990a), suggesting that each isozyme may fulfill a different physiological role. The CAT-3 isozyme that in leaves is expressed only in mesophyll cells (Tsaftaris et al. 1983) has considerably more peroxidatic than catalatic activity (Havir and McHale 1989) as compared to CAT-2, which is localized in the bundle-sheath cells of maize leaves. Because in maize there is a differential localization of the O_2-evolving and CO_2-fixing reactions of photosynthesis in the mesophyll and bundle-sheath cells, respectively, the different CATs may indeed perform distinct metabolic functions.

Studies with other plant genes have fortified the assumption that light-responsive sequences are located in the 5′-flanking regions of structural genes (Nagy et al. 1988; Herrera-Estrella and Simpson 1990). In fact, light-grown CAT-2 null plants (due to a deletion of the 3′ half of the *Cat2* transcript; Abler and Scandalios 1991) produce polyadenylated *Cat2* mRNA that is recruited onto leaf polysomes in vivo. This provides preliminary evidence that the light control of *Cat2* gene transcription lies at the 5'-flanking region of *Cat2* and *Cat3*. If so, *Cat2*-promoter constructs should not express CAT-2 protein in transformed dark-grown callus, whereas they will express CAT-2 in green callus; the opposite should hold for CAT-3. Such results would allow the manipulation of these antioxidant genes under different light regimens and should lead to a better understanding of their response to light and perhaps other environmental signals.

RESPONSE TO PARAQUAT, OZONE, SO_2, AND HYPEROXIA

Treating 10-day-old maize leaves with 10^{-5} M paraquat for 12 hours results in a 40% increase in SOD activity and a smaller (20%) increase in CAT activity (Matters and Scandalios 1986b). The increase in total SOD activity correlated with higher levels of specific SOD isozymes. SOD-1, SOD-2, and SOD-4 were increased significantly, whereas SOD-3 increased only slightly. Higher levels of SOD-4 and SOD-3 activity after paraquat treatment were the result of increased synthesis of these proteins, as determined by in vivo labeling with [^{35}S]methionine (Fig. 8). Polysomal mRNA that codes for SOD-4 and SOD-3 also increased after 10^{-5} M paraquat treatment. These results suggested that the response of SOD to paraquat treatment might be due to enhanced transcription of these genes.

SOD and CAT activities were also determined in maize leaves after treatment with O_3 or SO_2 for 8 hours. Neither O_3 nor SO_2 significantly increased the levels of SOD or CAT. However, after 72 hours of continuous 90% oxygen treatment, total SOD activity was significantly increased (60% over control), and the activity remained high throughout 96 hours of treatment. Immunological analysis indicated that higher levels of the cytosolic SOD-2 and SOD-4 isozymes were present in tissues after the high oxygen treatment. However, no changes were found in the chloroplast (SOD-1) or mitochondrial (SOD-3) isozymes. According to data from immunoprecipitation analysis, the increase of SOD-2 and SOD-4 isozymes was due to

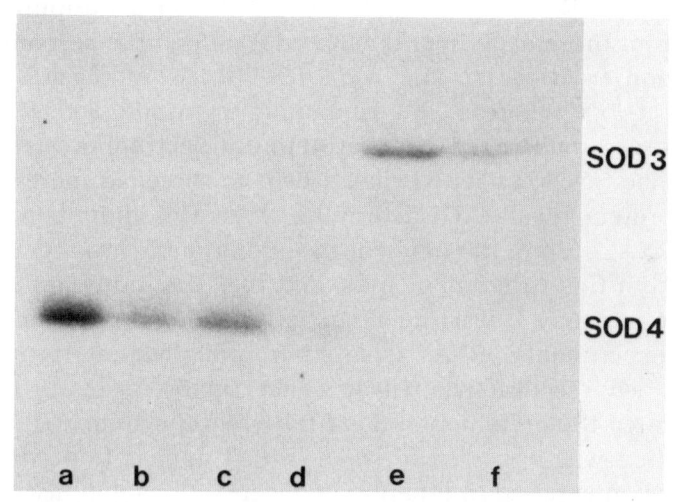

FIGURE 8 Fluorogram of immunoprecipitated SOD-3 and SOD-4 proteins labeled in vivo with [^{35}S]methionine in 10-day-old leaves. Leaves were treated with 10^{-5} M paraquat or buffer for 12 hr and incubated in the presence of radiolabel during the final 4 hr of treatment. To ensure complete recovery of labeled proteins, each immunoprecipitation was done twice. *(a,c)* SOD-4 immunoprecipitated from leaves treated with 10^{-5} M paraquat; *(b,d)* SOD-4 immunoprecipitated from leaves treated with buffer alone; *(e)* SOD-3 immunoprecipitated from leaves treated with 10^{-5} M paraquat; *(f)* SOD-3 immunoprecipitated from leaves treated with buffer alone. (Reprinted, with permission, from Matters and Scandalios 1986b.)

increased amounts of polysome-bound mRNA encoding these proteins (Matters and Scandalios 1987).

RESPONSE OF CU/ZNSOD TO ETHEPHON

Under oxidative stress, the phytohormone ethylene is known to be endogenously increased in plants. The chemical ethephon (2-chloroethylphosphoric acid) is metabolized by plants to ethylene and phosphoric acid. Maize seedlings (12 days postgermination) were treated hydroponically for 17 hours with various concentrations of ethephon. After the treatments, maize leaf tissue samples were collected, the respective RNAs and proteins were isolated, and their quantitative levels were determined by RNA blots and rocket immunoelectrophoresis, respectively (R.E. Cannon and J.G. Scandalios, unpubl.). Results from these experiments showed a differential response to ethephon treatments. The *Sod4* RNA levels increase significantly (~7x), whereas the remaining cytosolic *Sod* RNA levels remain unchanged. Rocket immunoelectrophoresis of the cytosolic SODs indicated that a two- to threefold increase in SOD protein results. To determine when this induction of *Sod4* mRNA occurs, a time-course experiment was done. Results from this experiment indicated that the induction occurs approximately 7–8 hours after treatment begins. In more detailed experiments, RNAs isolated from individual tissues were probed separately. These data more dramatically demonstrated the induction of *Sod4* mRNA in the stem (Fig. 9). No induction was apparent in roots, which appear to become necrotic when exposed to ethephon concentrations ≥50 mM, as evidenced by the absence of RNA. Leaf tissue exhibits a slight induction of *Sod4* mRNA after treatment with 25–100 mM ethephon. This stem-specific *Sod4* mRNA induction is a novel phenomenon and provides a significant starting point to help unravel the differential biological roles for the different cytosolic Cu/ZnSOD forms in maize. We find the expression of *Sod4* and not *Sod4A* quite intriguing, especially considering the strong sequence and structural conservation of the two genes. We believe that this organ-specific differential response may give the first insight into the need for multiple cytosolic forms of SOD in higher eukaryotes. Thus, the localization of

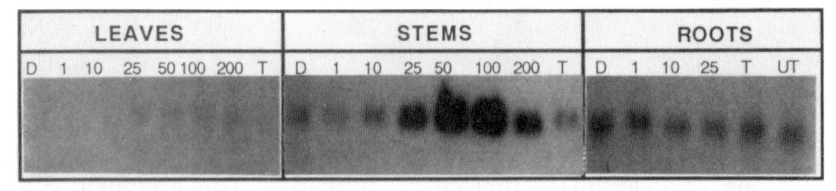

FIGURE 9 Organ-specific *Sod4* mRNA induction. Each lane of the RNA blot contains 15 μg total RNA from either leaves, stems, or roots of 12-day-old seedlings treated with distilled (D) or tap (T) water or 1, 10, 25, 50, 100, or 200 mM ethephon. Treatments were performed via hydroponic uptake for 15 hr. All blots were hybridized with ^{32}P-labeled gene-specific *Sod4* probes. Note absence of induction in roots, slight induction in leaves, and significant (~10-fold) increase in stems. (UT) Untreated control.

pertinent regions of DNA responsible for providing the *Sod4/Sod4A* organ-specific differential response to ethephon may provide valuable, biologically relevant information toward understanding both the differential regulation of and the need for multiple cytosolic Cu/ZnSODs in maize.

RESPONSE OF THE *CAT1* GENE TO ABA

In further efforts to examine the manner by which the antioxidant genes respond to various signals (environmental and/or developmental), we investigated the effects of the phytohormone ABA (J.D. Williamson and J.G. Scandalios, unpubl.). Under a variety of stresses (including oxidative stress), the endogenous levels of ABA are altered significantly. Therefore, it seems a reasonable assumption that responses to oxidative stress may involve ABA, and experiments were done to that end. Maize embryos were removed from the kernel and cultured in vitro. If 10^{-4} M ABA was presented in the culture medium, the *Cat1* mRNA and protein were rapidly accumulated. Sequencing analysis of the 5' end of *Cat1* genomic DNA (L. Guan and J.G. Scandalios, unpubl.) reveals that *Cat1* has the ABA-response consensus sequence (CACGTGGA) reported for wheat, rice, and cotton (Guiltinan et al. 1990). This evidence suggests the possible transcriptional regulation of the *Cat1* gene by ABA. To understand the transcriptional control mechanism of ABA induction and to identify the *cis*-acting regulatory sequence, chimeric genes were made by fusing the 5' end of *Cat1* promoter DNA with the reporter gene *Gus*. This DNA will be introduced into rice protoplasts. Analysis of stable chimeric genes transformed into plants has been the primary approach to characterize *cis*-acting sequences involved in gene regulation. In the case of most cereals, however, the routine use of transgenic plants for analysis of cereal genes in homologous systems is not yet possible. Alternatively, protoplasts have been extensively used to introduce and characterize cereal genes by using transient assays. The transient assay system has been used to identify ABA response elements in wheat genomic clones (Marcotte et al. 1988). We are using this system to functionally test the ABA response elements in the *Cat1* promoter region.

RESPONSE OF THE CAT AND SOD GENES TO THE PHOTOACTIVATED FUNGAL TOXIN CERCOSPORIN

Fungi of the genus *Cercospora* produce a light-induced photoactivated polyketide toxin, cercosporin. By itself, cercosporin does no damage to the host plant. However, in the presence of light, an excited triplet form of this molecule reacts directly with molecular oxygen to produce (depending on the redox potential of the environment) singlet oxygen and/or superoxide radicals (Daub and Hangarter 1983). The effect of cercosporin-generated oxygen radicals on the expression of individual CAT and SOD isozymes has not been previously determined. Using fungal extracts and purified cercosporin, it was demonstrated that total CAT activity, protein, and RNA steady-state levels changed in parallel in response to applied toxin (Williamson and Scandalios 1992). Responses were not identical for all CAT isozymes and appeared to vary with tissue and developmental stage. In contrast, neither total SOD activity nor individual isozyme protein levels changed in response to cercosporin treatment. However, steady-state RNA levels for several of the SODs changed dramatically in response to the toxin. This suggests that protein turnover may be an important aspect of the response of SOD to activated oxygen species.

TRANSFER AND EXPRESSION OF MAIZE CAT AND SOD GENES IN ALIEN GENOMES: PLANT GENE TRANSFORMATION SYSTEM

The analysis of gene expression in many plants has been greatly facilitated by the *Agrobacterium*-mediated transformation system (Willmitzer 1988). *Agrobacterium tumefaciens* is a naturally occurring soil bacterium. It harbors a tumor-inducing (Ti) plasmid of more than 100 kb that stably transfers to the plant nuclear genome (Chilton et al. 1980). A portion of the Ti plasmid DNA (T-DNA, especially the border sequences) is essential for transfer (Shaw et al. 1984). Most of the essential functions for the T-DNA transfer are provided by the Ti plasmid virulence (*vir*) genes. *vir* gene products are involved in initial recognition of the plant inducers and induction of other *vir* genes. Taking advantage of this naturally occurring process, vectors for plant transformation have been developed. The binary plasmid vector pBin19 can replicate in *Escherichia coli* and *Agrobacterium*, allowing the insertion of foreign DNA into the vector followed by screening in *E. coli* prior to transfer into *Agrobacterium*. This vector contains a kanamycin-resistance marker for direct selection in bacteria as well as T-DNA border sequences flanking a dominant selectable marker for use in plant cells. The *Agrobacterium*-mediated transformation systems have provided an opportunity to study the in vivo effects of mutations and deletions created on any cloned sequence. Bacterial reporter genes are used for easy detection of the effects. These gene products are not present in plant cells, thus eliminating any background from endogenous gene expression. The β-glucuronidase gene (*Gus*) from bacteria has been developed as a reporter gene system for the transformation of plants (Jefferson et al. 1987). Several useful features of the *Gus* gene make it a superior reporter gene for plant studies. Many plants assayed to date lack detectable GUS activity, providing a null background in which to assay chimeric gene expression, and the *Gus* gene is an easy, sensitive assay both in vitro and in vivo. The histochemical method can provide a tool for resolving differences in gene expression in individual cells and cell types within tissues.

CHIMERIC CAT AND SOD GENE CONSTRUCTS USING THE GUS REPORTER GENE

To functionally define the promoter regions of the *Cat* and *Sod* genes, we have placed the putative *cis* regions in the pBI101 *Gus* vector and have begun introducing various constructs into tobacco via *Agrobacterium*-mediated transformation in an attempt to express *Gus* under the control of the respective *Sod* and *Cat* promoters. Various constructs are currently being tested. Some of the fusions contain intervening sequences (i.e.,

Sod4 and Sod4A), whereas others do not. This should provide some information relative to the role(s) introns play in the expression of these genes.

Chimeric *Cat1* promoter-*Gus* constructs have been used to transform tobacco plants using the binary vector pBI101.2 that was derived from *A. tumefaciens*. Preliminary data indicate that the *Cat1* promoter can drive the expression of *Gus* in the transgenic tobacco plants. Two transgenic tobacco lines have been established that contain 2.5 kb and 0.8 kb of the maize *Cat1* promoter sequence and the coding sequence of the reporter gene *Gus*. Preliminary results have shown that GUS activity can be detected in young transgenic tobacco seedlings and seeds. No GUS activity was found in tobacco plants containing "no promoter" or the *Cat1* promoter in antisense orientation. Deletion analysis is under way to determine the putative essential *cis*-acting elements for *Cat1* gene expression.

EXPRESSION OF MAIZE SOD3 IN YEAST

Maize MnSOD (SOD-3) is synthesized as a precursor with a cleavable amino-terminal extension of 31 amino acids and is posttranslationally imported into the mitochondrial matrix (White and Scandalios 1987). In vitro import results indicated that preSOD-3 is imported and processed in isolated maize mitochondria (White and Scandalios 1989). We have recently introduced the maize *Sod3* cDNA and its deletion mutants into a yeast MnSOD-deficient mutant strain (Sod2d), which is hypersensitive to oxygen (van Loon et al. 1986). Results indicate that the maize transgene is properly expressed and that its product, the SOD-3 protein, is effectively synthesized in the yeast cells. SOD-3 is properly targeted, imported, and processed in the yeast mitochondria in vivo, and sufficiently complements the MnSOD deficiency in the mutant yeast cells. In addition, the complemented yeast containing active maize SOD-3 in its mitochondrial matrix regains its resistance to oxidative stress imposed by treatment with paraquat, whereas the mutant yeasts not transformed with maize *Sod3* are still susceptible to paraquat (Fig. 10). These results indicate that

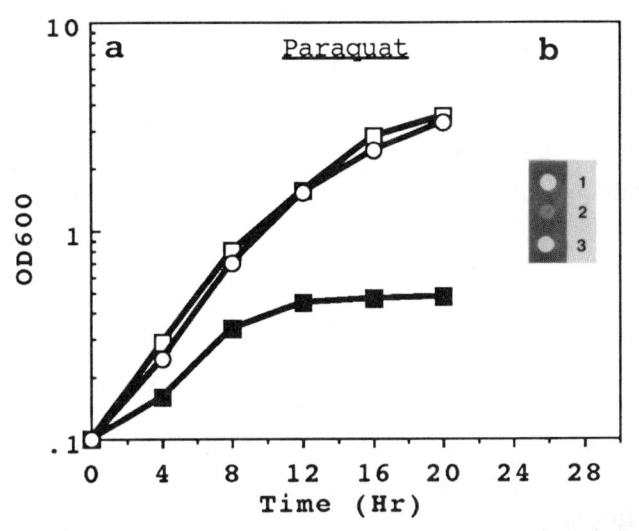

FIGURE 10 Growth of the yeast strains treated with 0.2 mM paraquat. Sensitivity of the transformed yeast strains of paraquat was tested by growing the yeast cells in MM medium containing 0.2 mM paraquat. In a, the overnight cultures were diluted to an OD_{600} about 0.025 with 50 ml of fresh MM. Paraquat was added to the cultures of 0.2 mM when the OD_{600} was almost 0.1. Growth was measured at 600 nm for 20 hr. (Filled boxes) MnSOD-deficient yeast strain Sod2d; (open boxes) wild-type yeast strain DL-1wt; (open circles) mutant yeast Sod2d transformed with plasmid pYZ0 bearing full-length maize *Sod3* cDNA. In b, the sensitivity of the yeast cells to paraquat on MM plates containing 0.2 mM paraquat. The transformants were grown in 5 ml of MM liquid culture overnight, and equal numbers of cells were spotted onto the MM plates containing 0.2 mM paraquat and incubated at 30°C for 72 hr. The white (opaque) color indicates that the yeast cells grew, and dark (transparent) color indicates that the yeast cells did not grow. (1) Wild-type yeast strain DL-1wt; (2) mutant yeast strain Sod2d; (3) mutant yeast Sod2d transformed with plasmid pYZ0 bearing full-length maize *Sod3* cDNA.

the ability of the yeast mutant cells to become resistant to paraquat is dependent on targeting the maize preSOD-3 into the yeast mitochondria. These findings further support the notion that MnSOD functions to protect cells from the lethal effects of active oxygen and that this functional role is highly conserved. The results further imply that maize SOD proteins will likely be properly targeted into intracellular compartments in cells from other species (e.g., tobacco cells) and may play an important role in protecting transgenic plants from environmentally imposed oxidative stress. Given the conservative nature of the *Sod* genes (both Mn- and Cu/ZnSODs) and the results from our recent yeast transformation experiments, it is likely that SOD-deficient mutants of yeast and other readily transformable organisms may serve as useful bioassays for uncovering and differentiating the functional roles for each of the SOD isozymes in maize and other organisms.

CONCLUDING REMARKS

It seems appropriate to conclude, from available data, that both CATs and SODs play significant roles in protecting living cells against the toxic and mutagenic effects of active oxygen species. Whether CATs and SODs play another biological role(s), in addition to their capacity to scavenge active oxygen, is not known and is fertile ground for future research.

The CAT and SOD gene-enzyme systems of maize discussed above provide excellent models to study the regulation and expression of these structurally and functionally conserved genes characteristic of virtually all aerobes. They provide an opportunity to examine how higher plant (and other) genomes may perceive oxidative insult and mobilize a response toward cell survival. The *Cat* and *Sod* genes of maize are some of the most highly regulated genes described with some rigor in plants. As new molecular technologies are applied to help us unravel the underlying mechanisms by which these genes are regulated to respond to various signals (e.g., developmental, environmental, or physiological), it is inevitable that their biological role(s) will begin to unfold. The differential responses observed for the different isozymes to given stressors or signals may help to define the reason for the existence of multiple forms of these enzymes. Furthermore, the use of null mutants, as well as the ability to transform alien genomes with these genes (transgenics), is beginning to shed new light on the functional role of each *Sod* and *Cat* gene in response to given stressors. The use of transgenic plants will help determine how the *Cat* and *Sod* genes of maize are regulated and how they respond to specific signals. The identification and characterization of *cis*-acting elements and *trans*-acting factors involved in *Cat* and *Sod* gene expression will help us to understand mechanisms of the entire signal transduction pathway during oxidative stress and to enhance our efforts toward engineering organisms to better cope with oxidative stress.

ACKNOWLEDGMENTS

This work has been supported, in part, by research grants from the United States Environmental Protection Agency, the National Institutes of Health, and the National Science Foundation to J.G.S. I thank Stephanie Ruzsa and Sheri Plant Kernodle for expert technical assistance, and all present and former associates who have contributed to these studies. I thank Suzanne Quick for expert typing of the manuscript.

REFERENCES

Abler, M.L. and J.G. Scandalios. 1991. The CAT-2 null phenotype in maize is likely due to a DNA insertion into the *Cat2* gene. *Theor. Appl. Genet.* **81:** 635.

Acevedo, A. and J.G. Scandalios. 1990. Expression of the catalase and superoxide dismutase genes in mature pollen in maize. *Theor. Appl. Genet.* **80:** 705.

Acevedo, A., J.D. Williamson, and J.G. Scandalios. 1991. Photoregulation of the *Cat2* and *Cat3* catalase genes in pigmented and pigment-deficient maize: The circadian regulation of *Cat3* is superimposed on its quasi-constitutive expression in maize leaves. *Genetics* **127:** 601.

Allen, J.F. 1975. A two-step mechanism for the photosynthetic reduction of oxygen by ferredoxin. *Biochem. Biophys. Res. Commun.* **66:** 36.

Alscher, R., M. Franz, and C. Jeske. 1987. Sulfur dioxide and chloroplast metabolism. In *Phytochemical effects of environmental compounds* (ed. J. Saunders et al.), p. 1. Plenum Press, New York.

Baum, J.A. and J.G. Scandalios. 1979. Developmental expression and intracellular localization of superoxide dismutases in maize. *Differentiation* **13:** 133.

———. 1981. Isolation and characterization of the cytosolic and mitochondrial superoxide dismutases of maize. *Arch. Biochem. Biophys.* **206:** 249.

———. 1982a. Expression of genetically distinct superoxide dismutases in the maize seedling during development. *Dev. Genet.* **3:** 7.

———. 1982b. Multiple genes controlling superoxide dismutase expression in maize. *J. Hered.* **73:** 95.

Baum, J.A., J.M. Chandlee, and J.G. Scandalios. 1983. Purification and partial characterization of a genetically defined superoxide dismutase (SOD-1) associated with maize chloroplasts. *Plant Physiol.* **73:** 31.

Beevers, H. 1979. Microbodies in higher plants. *Annu. Rev. Plant Physiol.* **30:** 159.

Bethards, L.A. and J.G. Scandalios. 1988. Molecular basis for the CAT-2 null phenotype in maize. *Genetics* **118:** 149.

Bethards, L.A., R.W. Skadsen, and J.G. Scandalios. 1987. Isolation and characterization of a cDNA clone for the *Cat2* gene in maize and its homology with other catalases. *Proc. Natl. Acad. Sci.* **84:** 6830.

Cannon, R.E. and J.G. Scandalios. 1989. Two cDNAs encode two nearly identical Cu/Zn superoxide dismutase proteins in maize. *Mol. Gen. Genet.* **219:** 1.

Cannon, R.E., J.A. White, and J.G. Scandalios. 1987. Cloning cDNA for maize superoxide dismutase 2 (SOD-2). *Proc. Natl. Acad. Sci.* **84:** 179.

Chandlee, J.M. and J.G. Scandalios. 1984a. Analysis of variants affecting the catalase developmental program in maize scutellum. *Theor. Appl. Genet.* **69:** 71.

———. 1984b. Regulation of *Cat1* gene expression in the scutellum of maize during early sporophytic development. *Proc. Natl. Acad. Sci.* **81:** 4903.

Chandlee, J.M., A.S. Tsaftaris, and J.G. Scandalios. 1983. Purification and partial characterization of three genetically defined catalases of maize. *Plant Sci. Lett.* **29:** 117.

Charles, S. and B. Halliwell. 1981. Light activation of fructose bisphosphatase in isolated spinach chloroplasts and deactivation by hydrogen peroxide—A physiological role for the thioredoxin system. *Planta* **151:** 242.

Chilton, M.-D., R.K. Saiki, N. Yadav, M.P. Gordon, and F. Quetier. 1980. T-DNA from *Agrobacterium* Ti plasmid is in the nuclear DNA fraction of crown gall tumor cells. *Proc. Natl. Acad. Sci.* **77:** 4060.

Daub, M.E. and R.P. Hangarter. 1983. Production of singlet oxygen and superoxide by the fungal toxin, cercosporin. *Plant Physiol.* **73:** 855.

Elstner, E.F. 1982. Oxygen activation and oxygen toxicity. *Annu. Rev. Plant Physiol.* **33:** 73.

Elthon, T.E. and L. McIntosh. 1987. Identification of the alternate terminal oxidase of higher plant mitochondria. *Proc. Natl. Acad. Sci.* **82:** 8399.

Foote, C.S. 1976. Photosensitized oxidation and singlet oxygen: Consequences in biological systems. In *Free radicals in biology* (ed. W.A. Pryor), vol. 2, p. 85. Academic Press, New York.

Foyer, C.H. and B. Halliwell. 1976. The presence of glutathione and glutathione reductase in chloroplasts: A proposed role in ascorbic acid metabolism. *Planta* **133:** 21.

Fridovich, I. 1978. Superoxide dismutases. *Annu. Rev. Biochem.* **44:** 147.

Fucci, L., C. Oliver, M. Coon, and E. Stadtman. 1983. Inactivation of key metabolic enzymes by mixed-function oxidation reactions: Possible implication in protein turnover and aging. *Proc. Natl. Acad. Sci.* **80:** 1521.

Giles, A.B., D. Grierson, and H. Smith. 1977. *In vitro* translation of mRNA from developing bean leaves. Evidence for the existence of stored mRNA and its light-induced mobilization onto polyribosomes. *Planta* **136:** 31.

Giuliano, G., N.E. Hoffman, K. Ko, P.A. Scolnik, and A.R. Cashmore. 1988. A light-entrained circadian clock controls transcription of several plant genes. *EMBO J.* **7:** 3635.

Guan, L., S. Ruzsa, R.W. Skadsen, and J.G. Scandalios. 1991. Comparison of the *Cat2* complementary DNA sequences of a normal catalase activity line (W64A) and a high catalase activity line (R6-67) of maize. *Plant Physiol.* **96:** 1379.

Guiltinan, M.J., W.R. Marcotte, and R.S. Quatrano. 1990. A plant leucine zipper protein that recognizes an abscisic acid response element. *Science* **250:** 267.

Hall, J.C. and M. Rosbash. 1988. Mutations and molecules influencing biological rhythms. *Annu Rev. Neurosci.* **11:** 373.

Halliwell, B. and J.M.C. Gutteridge. 1985. The importance of free radicals and catalytic metal ions in human diseases. *Mol. Aspects Med.* **8:** 89.

Havir, A.E. and N.A. McHale. 1989. Enhanced-peroxidatic activity in specific catalase isozymes of tobacco, barley and maize. *Plant Physiol.* **91:** 812.

Herrera-Estrella, L. and J. Simpson. 1990. Influence of environmental factors on photosynthetic genes. *Adv. Genet.* **28:** 133.

Huang, A.H.C., R.N. Trelease, and T.S. Moore. 1983. *Plant peroxisomes*. Academic Press, New York.

Jefferson, R.A., T.A. Kavanagh, and M.W. Bevan. 1987. GUS fusions: β-Glucuronidase as a sensitive and versatile gene fusion marker in higher plants. *EMBO J.* **6:** 3901.

Kaiser, W. 1979. Carbon metabolism of chloroplasts in the dark. *Planta* **144:** 193.

Kendall, A.C., A.J. Keys, J.C. Turner, P.J. Lea, and B.J. Miflin. 1983. The isolation and characterization of a catalase-deficient mutant of barley. *Planta* **159:** 505.

Kitagawa, Y., S. Tsunasawa, N. Tanaka, Y. Katsube, F. Sakiyama, and K. Asada. 1986. Amino acid sequence of Cu/Zn-superoxide

dismutase from spinach leaves. *J. Biochem.* **99:** 1289.
Lee, E. and J. Bennett. 1982. Superoxide dismutase: A possible protective enzyme against ozone injury in snap beans (*Phaseolus vulgaris*). *Plant Physiol.* **69:** 1444.
Levine, R., C. Oliver, R. Fulks, and E. Stadtman. 1981. Turnover of bacterial glutamine synthetase oxidative inactivation precedes proteolysis. *Proc. Natl. Acad. Sci.* **78:** 2120.
Marcotte, W., C. Bayley, and R. Quatrano. 1988. Regulation of a wheat promoter by abscisic acid in rice protoplasts. *Nature* **335:** 454.
Matheson, I.B.C., R.D. Etheridge, N.R. Kratowich, and J. Lee. 1975. The quenching of singlet oxygen by amino acids and proteins. *Photochem. Photobiol.* **21:** 165.
Matters, G.L. and J.G. Scandalios. 1986a. Changes in plant gene expression during stress. *Dev. Genet.* **7:** 167.
———. 1986b. Effect of the free radical-generating herbicide paraquat on the expression of the superoxide dismutase (*Sod*) genes in maize. *Biochim. Biophys. Acta* **882:** 29.
———. 1987. Synthesis of isozymes of superoxide dismutase in maize leaves in response to O_3, SO_2 and elevated O_2. *J. Exp. Bot.* **38:** 842.
McClung, C.R., B.A. Fox, and J.C. Dunlap. 1989. The *Neurospora* clock gene *frequency* shares a sequence element with the *Drosophila* clock gene *period*. *Nature* **339:** 558.
Nagy, F., S.A. Kay, and N.-H. Chua. 1988. A circadian clock regulates transcription of the wheat *Cab-1* gene. *Genes Dev.* **2:** 376.
Perl-Treves, R., B. Nachmias, D. Aviv, E.P. Zeelon, and E. Galun. 1988. Isolation of two cDNA clones from tomato containing two different superoxide dismutase sequences. *Plant Mol. Biol.* **11:** 609.
Rabinowitch, H.D. and D. Sklan. 1980. Superoxide dismutase: A possible protective agent against sunscald in tomatoes. *Planta* **148:** 162.
Rabinowitch, H.D., D. Sklan, and P. Budowski. 1982. Photo-oxidative damage in the ripening tomato fruit: Protective role of superoxide dismutase. *Physiol. Plant.* **54:** 369.
Redinbaugh, M.G., M. Sabre, and J.G. Scandalios. 1990a. The distribution of catalase activity, isozyme protein, and transcript in the tissues of the developing maize seedling. *Plant Physiol.* **92:** 375.
———. 1990b. Expression of the maize *Cat3* catalase gene is under the influence of a circadian rhythm. *Proc. Natl. Acad. Sci.* **87:** 6853.
Redinbaugh, M.G., G.J. Wadsworth, and J.G. Scandalios. 1988. Characterization of catalase transcripts and their differential expression in maize. *Biochim. Biophys. Acta* **951:** 104.
Roupakias, D.G., D.E. McMillin, and J.G. Scandalios. 1980. Chromosomal location of the catalase structural genes in *Zea*

mays, using B-A translocations. *Theor. Appl. Genet.* **58:** 211.

Salin, M.L. 1988. Plant superoxide dismutase: A means of coping with oxygen radicals. *Curr. Top. Plant Biochem. Physiol.* **7:** 188.

Scandalios, J.G. 1965. Subunit dissociation and recombination of catalase isozymes. *Proc. Natl. Acad. Sci.* **53:** 1035.

───. 1968. Genetic control of multiple molecular forms of catalase in maize. *Ann. N.Y. Acad. Sci.* **151:** 274.

───. 1974. Subcellular localization of catalase variants coded by two genetic loci during maize development. *J. Hered.* **65:** 28.

───. 1979. Control of gene expression and enzyme differentiation. In *Physiological genetics* (ed. J.G. Scandalios), p. 63. Academic Press, New York.

───. 1983. Multiple varieties of isozymes and their role in studies of gene regulation and expression during eukaryote development. *Isozymes Curr. Top. Biol. Med. Res.* **9:** 1.

───. 1987. The antioxidant enzyme genes *Cat* and *Sod* of maize: Regulation, functional significance, and molecular biology. *Isozymes: Curr. Top. Biol. Med. Res.* **14:** 19.

───. 1990. Response of plant antioxidant defense genes to environmental stress. *Adv. Genet.* **28:** 1.

Scandalios, J.G., W.F. Tong, and D.G. Roupakias. 1980a. *Cat3*, a third gene locus coding for a tissue-specific catalase in maize: Genetics, intracellular location, and some biochemical properties. *Mol. Gen. Genet.* **179:** 33.

Scandalios, J.G., A.S. Tsaftaris, J.M. Chandlee, and R.W. Skadsen. 1984. Expression of the developmentally regulated catalase (*Cat*) genes in maize. *Dev. Genet.* **4:** 281.

Scandalios, J.G., D.Y. Chang, D.E. McMillin, A.S. Tsaftaris, and R. Moll. 1980b. Genetic regulation of the catalase developmental program in maize scutellum: Identification of a temporal regulatory gene. *Proc. Natl. Acad. Sci.* **77:** 5360.

Scioli, J.R. and B.A. Zilinskas. 1988. Cloning and characterization of a cDNA encoding chloroplastic Cu-Zn-superoxide dismutase. *Proc. Natl. Acad. Sci.* **85:** 7661.

Shaw, C.H., M.D. Watson, G.H. Carter, and C.H. Show. 1984. The right hand copy of the nopaline Ti-plasmid 25 bp repeat is required for tumour formation. *Nucleic Acids Res.* **12:** 6031.

Skadsen, R.W. and J.G. Scandalios. 1986. Evidence for processing of maize catalase 2 and purification of its messenger RNA aided by translation of antibody-bound polysomes. *Biochemistry* **25:** 2027.

───. 1987. Translational control of photo-induced expression of the *Cat2* catalase gene during leaf development in maize. *Proc. Natl. Acad. Sci.* **84:** 2785.

───. 1989. Pretranslational control of the levels of glyoxysomal protein gene expression by the embryonic axis in maize. *Dev. Genet.* **10:** 1.

Slovin, J.P. and E.M. Tobin. 1982. Synthesis and turnover of the light-harvesting chlorophyll a/b-protein in *Lemma gibba* grown with intermittent red light: Possible translational control. *Planta* **154:** 465.

Tanaka, K. and K. Sugahara. 1980. Role of superoxide dismutase in defense against SO_2 toxicity and an increase in superoxide dismutase activity with SO_2 fumigation. *Plant Cell Physiol.* **21:** 601.

Taylor, W.C. 1989. Transcriptional regulation by a circadian rhythm. *Plant Cell* **1:** 259.

Tolbert, N.E. 1982. Leaf peroxisomes. *Ann. N.Y. Acad. Sci.* **386:** 254.

Tsaftaris, A.S. and J.G. Scandalios. 1981. Genetic and biochemical characterization of a *Cat2* catalase null mutant of *Zea mays. Mol. Gen. Genet.* **181:** 158.

———. 1986. Spatial pattern of catalase (*Cat2*) gene activation in scutella during postgerminative development in maize. *Proc. Natl. Acad. Sci.* **83:** 5549.

Tsaftaris, A.S., A.M. Bosabalidis, and J.G. Scandalios. 1983. Cell-type-specific gene expression and acatalasemic peroxisomes in a null *Cat2* catalase mutant of maize. *Proc. Natl. Acad. Sci.* **80:** 4455.

van Loon, A.P., B. Hurt, and G. Schatz. 1986. A yeast mutant lacking mitochondrial MnSOD is hypersensitive to oxygen. *Proc. Natl. Acad. Sci.* **83:** 3820.

Wadsworth, G.J. and J.G. Scandalios. 1989. Differential expression of the maize catalase genes during kernel development: The role of steady-state mRNA levels. *Dev. Genet.* **10:** 304.

———. 1990. Molecular characterization of a catalase null allele at the *Cat3* locus in maize. *Genetics* **125:** 867.

Watson, M.E. 1984. Compilation of published signal sequences. *Nucleic Acids Res.* **12:** 5145.

White, J.A. and J.G. Scandalios. 1987. *In vitro* synthesis, importation and processing of Mn-superoxide dismutase (SOD-3) into maize mitochondria. *Biochim. Biophys. Acta* **926:** 16.

———. 1988. Isolation and characterization of a cDNA for mitochondrial manganese superoxide dismutase (SOD-3) of maize and its relation to other manganese superoxide dismutases. *Biochim. Biophys. Acta* **951:** 61.

———. 1989. Deletion analysis of the maize mitochondrial superoxide dismutase transit peptide. *Proc. Natl. Acad. Sci.* **86:** 3534.

Williamson, J.D. and J.G. Scandalios. 1992. Differential response of maize catalases and superoxide dismutases to the photoactivated fungal toxin cercosporin. *Plant J.* (in press).

Willmitzer, L. 1988. The use of transgenic plants to study plant gene expression. *Trends Genet.* **4:** 13.

Zelitch, I. 1971. *Photosynthesis, photorespiration, and plant productivity.* Academic Press, New York.

# Regulation of Yeast Catalase Genes

H. Ruis and B. Hamilton
Institut für Allgemeine Biochemie der Universität Wien
Ludwig Boltzmann-Forschungsstelle für Biochemie
A-1090 Vienna, Austria

Catalases are enzymes present in virtually all aerobic organisms. In eukaryotes, they are characteristic constituents of peroxisomes, but evidence for the existence of extraperoxisomal catalases has been presented in a number of cases. In vitro these enzymes, which are mostly tetrameric hemoproteins consisting of four identical subunits, catalyze the decomposition of hydrogen peroxide and the peroxidation of a number of substrates. Concerning the in vivo function of catalases, there is no doubt that they protect cells together with other antioxidant enzymes (superoxide dismutases, peroxidases) against toxic and mutagenic effects of oxygen metabolites. However, it is impossible from the information presently available to give a realistic estimate of the quantitative importance of catalases within this complex protective system or to describe the relevant details of their mode of action in vivo.

Although there is obviously most interest in understanding the functions and the mechanism of action of catalase in mammals and in humans, unicellular microorganisms like *Escherichia coli* or *Saccharomyces cerevisiae* offer tremendous technical advantages for the investigation of antioxidant enzymes, especially because of their suitability for genetic studies. Genetic methods have been particularly valuable in the investigation of control of expression of genes encoding antioxidant enzymes. Studies focusing on the regulation of catalase genes in microorganisms have turned out to be valuable in two ways: (1) Since it can be assumed that, at least in general, levels of catalase are controlled according to functional needs, regulatory studies provide important information concerning the functions of this enzyme in vivo. (2) Catalase genes of unicellular organisms have turned out to be useful model systems, which can be studied to obtain more general insights into mechanisms of control of gene expression.

The yeast S. cerevisiae produces two catalase proteins, the peroxisomal catalase A (Seah et al. 1973; Skoneczny et al. 1988), encoded by the CTA1 gene (Cohen et al. 1985), and the cytosolic catalase T (Seah and Kaplan 1973), encoded by the CTT1 gene (Spevak et al. 1983). Little if any molecular information is available concerning regulation of catalase genes in other yeasts. In this review, we therefore concentrate on the CTA1 and CTT1 genes of S. cerevisiae, which have been demonstrated to be differentially regulated in a rather complex way. It seems likely from the insights gained from these investigations that the peroxisomal catalase A and the cytosolic catalase T have different, but overlapping, functions.

THE CTA1 GENE ENCODING THE PEROXISOMAL CATALASE A IS CONTROLLED BY OXYGEN, CARBON SOURCE, AND FATTY ACIDS

Since hydrogen peroxide, which is formed from molecular oxygen in various ways, is the substrate of catalase, it is not surprising that catalase formation is induced by oxygen in an organism like S. cerevisiae, which is able to live fermentatively under anaerobic or oxygen-limiting conditions or can derive its energy from respiration when sufficient oxygen is available. The absence of catalase A and catalase T from anaerobic yeast, and catalase induction by oxygen, have been demonstrated (Zimniak et al. 1976). Although other mechanisms of oxygen control cannot be entirely eliminated by the evidence presently available, it is clear that in yeast, heme plays a prominent role in signaling availability of oxygen to catalase genes and other genes controlled by oxygen (Winkler et al. 1988; Forsburg and Guarente 1989). Little if any heme is present in yeast cells grown under strict anaerobic conditions, since molecular oxygen is required as a cosubstrate in two steps of heme biosynthesis, which are catalyzed by coproporphyrinogen III oxidase and protoporphyrinogen IX oxidase, respectively. No catalase T or A apoproteins are detectable in heme-deficient yeast mutants grown in the presence of oxygen (Woloszczuk et al. 1980), and catalase T and catalase A mRNA accumulation was demonstrated to require heme (Richter et al. 1980; Spevak et al. 1983: Cohen et al. 1985). Detailed information is available concerning heme control regions of the *CTT1* gene and the role of transcription activator HAP1 (Pfeifer et al. 1987) as a mediator of the effect of heme on transcription (see below). There is much less evidence concerning this point in the case of *CTA1*, although a HAP1-binding site has been detected in the upstream region of this gene (Ruis 1989).

Like some other genes encoding individual peroxisomal proteins and like peroxisomes of *S. cerevisiae* (Veenhuis et al. 1987; Skoneczny et al. 1988), the *CTA1* gene is strongly repressed by glucose and derepressed on nonfermentable carbon sources (Cross and Ruis 1978; Rytka et al. 1978). Glucose repression affects *CTA1* expression at the transcript level (Hörtner et al. 1982). Recently, the transcription activator ADR1 (Denis et al. 1981), originally characterized as a regulator of the *ADH2* gene, which encodes the glucose-repressible alcohol dehydrogenase II of *S. cerevisiae*, was demonstrated to mediate derepression of *CTA1* and to be a main target of its glucose repression (Simon et al. 1991). In addition to the gene encoding the peroxisomal catalase, ADR1 affects several other genes encoding proteins involved in peroxisomal fatty acid β-oxidation or in peroxisome assembly. A *CTA1* gene upstream element with limited sequence similarity to the ADR1-binding site of the *ADH2* gene was shown to bind ADR1 protein in vitro and to function in derepression of transcription in vivo.

Peroxisomes of *S. cerevisiae* are dramatically induced by exogenous fatty acids (Veenhuis et al. 1987). This mode of regulation reflects a central role of these organelles in the metabolism of fatty acids and lipids. As a main constituent of peroxisomes, catalase A is induced by fatty acids (Skoneczny et al. 1988). Coordinate induction of genes encoding peroxisomal proteins by fatty acids may be triggered by a signal mechanism also causing peroxisome proliferation in mammalian liver, where these organelles are most dramatically induced by a chemically heterogeneous class of compounds known to have hypolipidemic effects. These peroxisome proliferators are of medical interest because of their effect on blood lipid levels, but also because of their carcinogenic effects, which may be linked to stimulation of peroxisomal metabolism (Lock et al. 1989). The recent observation that at least one of these peroxisome proliferators, nafenopin, induces transcription of yeast genes encoding peroxisomal proteins (M. Simon and W. Rapatz, unpubl.) greatly increases the appeal of the hypothesis that peroxisome proliferation in mammalian liver and in yeast is controlled by similar basic mechanisms. To elucidate such mechanisms, work is in progress in the authors' laboratory to identify control elements of the *CTA1* gene mediating induction by fatty acids and by peroxisome proliferators. From the evidence presently available, it appears possible that ADR1 itself or some other factor interacting with the ADR1-binding region of *CTA1* is a target of fatty acid induction. However, results of a deletion analysis of the *CTA1* upstream region also suggest that this gene possesses at least one further control element important for fatty acid induction. It seems possible that only this second element is a direct target for peroxisome-specific signals, whereas the ADR1 element mediates the more general carbon source control of the gene and enhances its fatty acid induction in a synergistic manner. A scheme summarizing the control of *CTA1* gene expression according to the present status of information is presented in Figure 1.

LIKE CTA1, *THE* CTT1 *GENE ENCODING THE CYTOSOLIC CATALASE T IS UNDER* O_2 *CONTROL; IN CONTRAST TO* CTA1, *IT IS UNDER NUTRIENT AND HEAT STRESS CONTROL*

As summarized above, formation of the cytosolic catalase T is positively controlled by molecular oxygen via heme. A region involved in heme control of *CTT1* transcription was identified by deletion analysis (Spevak et al. 1986). This region, which centers around bp –460, was demonstrated to bind the transcription activator HAP1, which is a positive regulator of *CTT1* expression according to genetic criteria (Winkler et al. 1988). A

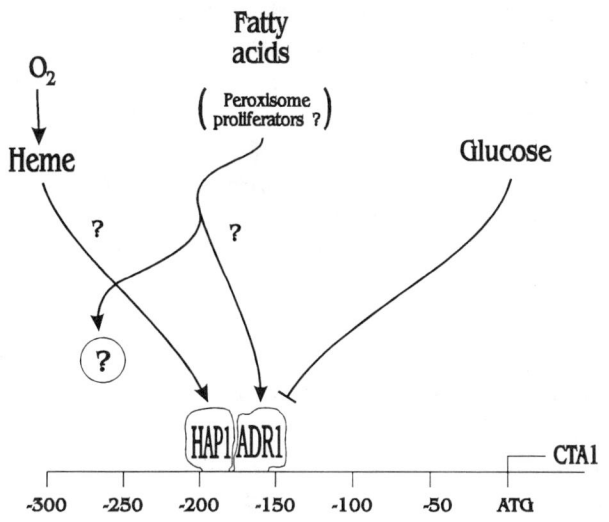

FIGURE 1 Control of transcription of the catalase A (CTA1) gene.

second HAP1-binding site of the CTT1 promoter was detected further downstream (-210) (Ruis 1989), and the function of this binding site as a positive control element was demonstrated by deletion analysis (Belazzi et al. 1991). Binding of HAP1 to the control regions of CTT1 (Winkler et al. 1988) and of CYC1 (Pfeifer et al. 1987) in vitro is stimulated by hemin. It is possible that hemin is a HAP1 ligand and that binding of hemin triggers activation of transcription, but final evidence for the correctness of this hypothesis has not yet been obtained.

Heme acts as a regulator of yeast catalase gene expression not only at the transcriptional level, but also at the translational level. Evidence for posttranscriptional heme control of CTT1 was initially obtained by genetic studies. It was demonstrated that the regulatory mutations cgr4 and cas1 allow transcription of CTT1 in the absence of heme, but no accumulation of apocatalase T was observed under these conditions (Sledziewski et al. 1981). Although it might be possible to explain this observation by instability of catalase apoprotein in the absence of its prosthetic group, more direct evidence for translational control was obtained by in vitro studies (Hamilton et al. 1982). It was demonstrated that a cell-free translation system isolated from a heme-deficient mutant of *S. cerevisiae* is inefficient in the translation of catalase T and catalase A mRNAs, although its capacity to translate most yeast mRNAs is quite comparable to that of a system isolated from heme-containing cells. Addition of hemin to the system from heme-deficient cells stimulates catalase mRNA translation significantly. No further information has yet been obtained concerning the mechanism of this specific translational control by hemin or its more general relevance.

Early investigations on the carbon source control of catalase T formation had provided evidence for glucose repression of *CTT1* expression (Cross and Ruis 1978; Rytka et al. 1978; Hörtner et al. 1982). However, more recent studies have demonstrated that this gene is under a more general form of nutrient control, which mediates repression of transcription when essential nutrients are available to cells in high quality and sufficient quantity. *CTT1* transcription is derepressed when any one of a number of essential nutrients (e.g., a nitrogen source) becomes limiting or is available only in lower quality (e.g., shift from glucose to ethanol medium) (Bissinger et al. 1989). This derepression of the catalase T gene by nutrient starvation is part of a pleiotropic response of yeast cells, which is characterized by accumulation of storage carbohydrates, acquisition of thermotolerance, early G_1 arrest, and sporulation (in the case of **a**/α diploid cells). It is known from a number of investigations that this pleiotropic response is mainly mediated by a signal pathway, which involves the action of the *S. cerevisiae* CDC25 protein; of RAS proteins, which act as positive regulators of adenylate cyclase in this organism; and of cAMP-dependent protein kinases. In *S. cerevisiae*, cAMP-dependent protein phosphorylation triggers cell growth and contributes to the entry of cells into mitotic division. Simultaneously, high protein kinase A activity acts as a negative regulator of events characteristic for starvation conditions (see, e.g., Cameron et al. 1988).

Studies by Bissinger et al. (1989) have provided genetic and biochemical evidence that repression of *CTT1* transcription by nutrients is mediated by cAMP-dependent protein kinases. Mutations in the pathway like *ras2* causing decreased adenylate cyclase activity lead to increased *CTT1* expression. In contrast, *bcy1* mutants with a defective regulatory subunit of cAMP-dependent protein kinase, and therefore high constitutive protein kinase A activity, exhibit very low catalase T levels and a dramatically reduced response to nitrogen starvation. Under proper conditions, exogenous cAMP significantly reduces derepression of *CTT1* expression by nitrogen limitation. Derepression of *CTT1* mRNA accumulation by nitrogen starvation occurs rapidly and is insensitive to cycloheximide (Belazzi et al. 1991). It can be concluded from these results that the response of *CTT1* expression to nutrients is fairly direct and is not a consequence of entry of cells into stationary phase.

Studies by Cameron et al. (1988) have demonstrated that nutrient control of cellular events in *S. cerevisiae* is not exclusively mediated by the RAS-cAMP pathway. Mutants lacking a functional protein kinase A regulatory subunit, and therefore unable to respond to cAMP, are still able to respond to changes in nutrient conditions (e.g., by glycogen accumulation, G_0 arrest, sporulation) if their constitutive protein kinase A activity is drastically lowered by mutations in the *TPK* genes encoding catalytic protein kinase A subunits.

Experiments by Belazzi et al. (1991) have demonstrated that nutrient control of transcription of *CTT1* is also partly mediated by this cAMP-independent signaling mechanism. Further studies from the authors' laboratory (G. Adam, unpubl.) have demonstrated that, in addition to protein kinase A, two further protein kinases encoded by the *YAK1* and *SCH9* genes are involved in *CTT1* control. Recessive *yak1* mutations have been isolated as suppressors of the growth defect of yeast *ras1 ras2* double mutants (Garrett and Broach 1989). The *YAK1* gene was cloned and was demonstrated to exhibit significant homology with protein kinase genes. The results obtained by Garrett and Broach (1989) place the product of the *YAK1* gene downstream from protein kinase A or in a parallel nutrient signaling pathway. Results obtained in the authors' laboratory demonstrate that disruption of *YAK1* has a dramatic negative effect on derepression of *CTT1* by nutrient starvation. They further show that this effect is partially epistatic over the effect of the *ras2* mutation, and that multiple copies of *YAK1* partially suppress the negative effect of the *bcy1* mutation on *CTT1*. Whereas all these results are also consistent with a function of the *YAK1* protein kinase downstream from protein kinase A, further results, which are discussed below, place it in a parallel, cAMP-independent nutrient signaling pathway. In contrast to the *YAK1* kinase, which acts as a functional antagonist of protein kinase A, the product of the *SCH9* gene (Toda et al. 1988) appears to be functionally related to cAMP-dependent protein kinases. The *SCH9* gene was isolated by its ability to complement the growth defect of a *cdc25* mutation and was shown to be homologous to yeast and mammalian protein kinase A catalytic subunits. It was proposed that the *SCH9* protein kinase functions in a growth control pathway that is at least partially redundant with the cAMP pathway. Experiments in the authors' laboratory have demonstrated that disruption of *SCH9* enhances expression of *CTT1* on complete medium and that there is a synergistic effect of low protein kinase A (*ras2* mutant) and absence of *SCH9* protein kinase (*sch9* disruption) activity on *CTT1* expression. Further experiments (R. Wieser, unpubl.) have demonstrated that the *sch9* effect on *CTT1* occurs independent of the nutrient status (carbon or nitrogen source) of yeast cells. These results, and further results described below, place the *SCH9* protein kinase in a cAMP-independent signaling pathway. It remains to be clarified, however, which signal controls *SCH9* kinase activity.

Transcription of *CTT1*, but not of *CTA1*, was further demonstrated to be induced by heat shock (Wieser et al. 1991). When cells are shifted from 23°C to 37°C, maximal induction of transcription takes place within 30 minutes. Considerable induction of the gene occurs at temperatures above 27°C. Therefore, expression of *CTT1* at standard laboratory growth temperatures (30°C) is significantly induced by heat stress, whereas the gene is expressed only at a low basal level at 23°C.

A dramatic synergism of the effects of the three positive signals influencing *CTT1* transcription (O_2 via heme, nutrient starvation via the RAS-cAMP pathway, heat) was observed (Belazzi et al. 1991). In the presence of heme, low protein kinase A activity (*ras2*) has no effect on catalase T levels at 23°C, but it enhances heat shock induction. High constitutive protein kinase A (*bcy1*) prevents heat shock induction virtually completely. No heat induction is observed in heme-deficient cells grown on complete medium (high cAMP), but the heme requirement for transcription can be bypassed by simultaneous heat shock and low cAMP. Heat stress seems to be absolutely required to raise *CTT1* transcription above a low basal level, but it is efficient only in combination with at least a second positive factor. Obviously, two interesting questions have to be raised in connection with these observations: (1) What is the mechanistic basis of these synergistic effects? (2) What is their functional significance? Attempts to give at least partial answers to these questions are presented below.

A NOVEL TYPE OF STRESS CONTROL ELEMENT IN THE CTT1 PROMOTER

Standard techniques of molecular genetics were used to localize elements of the *CTT1* promoter, mediating its control by nutrients and by heat stress. Deletion analysis of the upstream region of a *CTT1-lacZ* fusion gene (Belazzi et al. 1991) provided evidence that a positive promoter element located between bp −330 and bp −380 was responding to cAMP-mediated nutrient control. The results obtained further demonstrate that sequences mediating cAMP-independent nutrient control fall at least partly outside this region. Further experiments demonstrated that an oligonucleotide corresponding to bp −325 to bp −382 confers cAMP-mediated nutrient control to a yeast *LEU2* promoter (Belazzi et al. 1991). When the heat shock element of *CTT1* was localized with the help of the deletion mutations and *LEU2* promoter constructs used for the identification of the cAMP-responsive element, it was found that the same DNA sequences were necessary and sufficient for heat shock and cAMP control (Wieser et al. 1991).

To test which sequences within the DNA region shown to be sufficient for heat and nutrient control were actually important for regulation, this region was analyzed further by introducing blocks of point mutations and by testing smaller regions in combination with the *LEU2* promoter (Wieser et al. 1991; G. Marchler and C. Schüller, unpubl.). These studies showed that mutation of a sequence with limited similarity to a canonical

cAMP-responsive
heat stress element   SCH9-responsive element

```
                -362                                -336
CTT1     GGTAAGGGGC            CTT1     GCGTATTGTTTCCT

                -176
         AATTAGGGAT                      -385
SSA3         -110              UBI4     CAGTATTGTTTCTA
         GTTAAGGGAT
                                         -201
                -170           DDRA2    TTCCCCTGTTTCCA
         CTTATGGGGA
DDRA2        -175
         CATAAGGGGT  *)

                -434
HSP12    GTTAAGGGGA
```

FIGURE 2 Promoter elements of the catalase T (*CTT1*) gene involved in control by heat and nutrient stress. (*) Indicates reverted orientation.

heat shock element (Bienz 1985) (bp −351 to bp −358) had no effect on expression. This result is in line with in vitro DNA-protein-binding studies, which provided no evidence for binding of purified yeast heat shock transcription factor to the region mediating heat shock control. In summary, these results make a role of heat shock transcription factor in heat shock control of *CTT1* transcription very unlikely. Mutagenesis experiments demonstrated that mutation of two regions centering around −362 and around −336, respectively (see Fig. 2), causes at least partial loss of nutrient and heat shock control. If tested separately in front of the *LEU2* promoter, both elements have UAS activity, but only the element around −362 responds to cAMP (*ras2*) and to heat shock. The second element (−336) was demonstrated to respond to carbon source control and to a disruption of the *SCH9* gene. However, no evidence was obtained for an involvement of *SCH9* in cAMP-independent carbon source control, since the up effect of the *sch9* mutation was similar on different carbon sources. Whereas a larger promoter fragment (−522 to −142) was demonstrated to confer *YAK1* control to the *LEU2* promoter, this was not the case with the cAMP-responsive fragment (−382 to −325) (R. Wieser, unpubl.). This finding places the

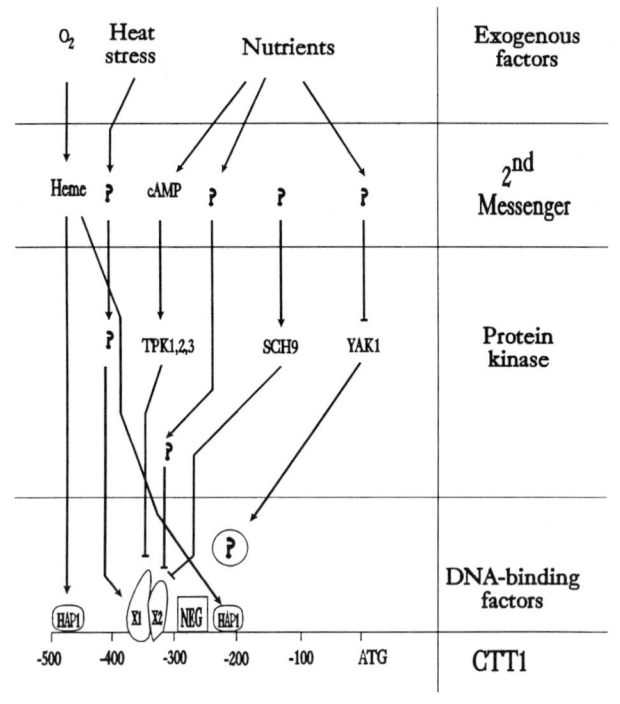

FIGURE 3 Control of transcription of the catalase T *(CTT1)* gene. (NEG) Negative control element.

YAK1 protein kinase in a separate, cAMP-independent nutrient control pathway. A summary of regulatory pathways and control elements of the *CTT1* gene is presented in Figure 3.

No regulatory signal has yet been identified that affects the activity of the negative region of the promoter localized by deletion experiments (Spevak et al. 1986). However, the results obtained indicate that this region contributes significantly to the synergism of the different regulatory signals observed with the wild-type *CTT1* promoter, since this synergism is partially lost with constructs lacking this negative region.

When the two DNA regions of the *CTT1* promoter contributing directly or indirectly to cAMP and heat control are compared with promoter regions of other yeast genes reported to be regulated by cAMP (the heat shock genes *UBI4* [polyubiquitin] [Tanaka et al. 1988], *SSA3* [an HSP70 gene] [Boorstein and Craig 1990], and *HSP12* [Praekelt and Meacock 1990]) or by heat shock transcription factor-independent heat shock control (see Fig. 2) (the DNA damage responsive gene *DDRA2*; Kobayashi and McEntee 1990), it appears very likely that at least one of these elements, the cAMP-responsive heat stress element, is of fairly general importance. Not all the DNA regions depicted in Figure 2 have been unequivocally shown to have regulatory functions. The *SSA3* element is undoubtedly involved in negative cAMP control (Boorstein and Craig 1990). The two *DDRA2* sequences are part of a 49-bp region shown to mediate heat shock factor-independent heat shock control (Kobayashi and McEntee 1990), but no information is available concerning cAMP control of *DDRA2*. The *HSP12* gene has been demonstrated to be under both heat shock and cAMP control (Praekelt and Meacock 1990), but no functional analysis of control elements of this gene has been published. At our present state of knowledge, it can be hypothesized that this element has a fairly broad function in control of genes by stress. Apparently its specificity concerning the response to various types of stress signals (responsiveness to heat and nutrient stress) is overlapping, but not identical with that of the canonical heat shock element, which is not controlled by cAMP levels. Further studies are necessary to analyze whether the novel element responds to other stress signals, which protein factors interact with it, and which mechanisms are involved in transduction of the nutrient signal from the level of protein kinase A and of the heat stress signal. From the evidence presently available, it is obvious that there are three classes of yeast genes induced by heat shock: (1) those controlled by heat shock transcription factor via the canonical heat shock element (heat shock genes which are not under cAMP control); (2) cAMP-controlled genes regulated via the newly identified stress control element (e.g., *CTT1*); and (3) heat shock genes possessing both types of elements (e.g., *SSA3*). The general importance of the second element detected in the *CTT1* promoter (Fig. 2), which is controlled by the *SCH9* kinase and contributes to derepression by nonfermentable carbon sources, is less clear, although the striking 10-bp identity of this region with an upstream sequence of the *UBI4* gene located next to a canonical heat shock element suggests that it will also be found to function in a number of other yeast promoters.

DIFFERENTIAL FUNCTIONS OF THE TWO YEAST CATALASES?

Although regulatory patterns cannot provide reliable proof for the specific functions of gene products, they should at least allow the formulation of working hypotheses that can then be tested by further experiments. According to their mode of regulation, both catalases are required only under aerobic conditions, which is consistent with the assumption that they are involved in detoxification of hydrogen peroxide. However, beyond this general point, the considerable differences in control of expression of *CTA1* and *CTT1* suggest that the cytosolic and the peroxisomal catalase differ in their functions. Coordinate induction of catalase A with peroxisomal structures and with the peroxisomal fatty acid β-oxidation system suggests a predominant function of catalase A in peroxisomal metabolism, particularly in fatty acid β-oxidation. This is reasonable, since H_2O_2 is a product of this metabolic pathway, but direct proof for the correctness of this assumption is lacking, and catalase-deficient *S. cerevisiae* strains have been reported to grow on oleic acid (van der Klei et al. 1990). Limited functional redundancy among antioxidant enzymes of *S. cerevisiae* may explain this finding.

Induction of catalase T under aerobic conditions by heat and nutrient stress and the synergistic effect of the regulatory signals discussed above suggest a predominant function of the cytosolic catalase under conditions where yeast cells are exposed to oxidative stress in combination with other types of stress. Although such a function is presently not understood in detail, some experimental observations are consistent with this hypothesis. It was demonstrated with the help of isogenic wild-type and *ctt1* null mutant strains that the wild-type strains are more resistant to a "lethal heat shock" regime (exposure to 50°C for 20 min under aerobic conditions) than are catalase-T-deficient mutants (Wieser et al. 1991). Since the protective effect provided by catalase T is much smaller than differences observed, e.g., between heat shock resistance of logarithmic and stationary phase cells, it is obvious that the cytosolic catalase is only one of a number of factors contributing to heat stress protection. Pretreatment of yeast cells by a mild heat shock (37°C) provides a dramatic protection against the lethal effect of hydrogen peroxide, and this effect is much more pronounced in catalase-T-containing cells than in *ctt1* mutants. This observation indicates that heat stress induction of catalase T in these cells provides fairly effective protection against oxidative stress, which might be more severe and damaging to cells when it occurs in combination with other types of stress, and therefore requires more efficient protection.

CONTROL OF CATALASE GENE EXPRESSION IS COUPLED TO REGULATION OF AEROBIC GENES AND STRESS GENES, TO CONTROL OF NUTRIENT UTILIZATION, AND TO THE DECISION BETWEEN CELL GROWTH AND GROWTH ARREST

Figure 4 presents a simplified version of the integration of control of yeast catalase genes into a regulatory network of *S. cerevisiae*. The complex coupling of the regulation of expression of catalase genes to the control of numerous other yeast genes is probably typical for the mode of control of expression of the majority of eukaryotic genes. An attempt to identify such regulatory networks and to understand their mechanisms in detail is therefore not only important in the context of investigation of oxygen radical stress response, but should also contribute to our general understanding of eukaryotic gene expression.

Both yeast catalases are hemoproteins, which undoubtedly function as part of the system protecting cells against oxygen metabolites. It is not surprising, therefore, that they are controlled by oxygen coordinately with other hemoproteins functioning in aerobic metabolism, particularly in mitochondrial respiration (Forsburg and Guarente 1989). However, control by oxygen and heme is also shared with at least some non-hemoproteins needed in the presence of oxygen, e.g., Mn-superoxide dismutase (Autor 1982) and at least one enzyme involved in the biosynthesis of steroids and other compounds derived from mevalonic acid, 3-hydroxy-3-methylglutaryl coenzyme A reductase (Thorsness et al. 1989). It remains to be clarified whether oxygen or oxygen metabolites make a more general contribution to the control of stress proteins of *S. cerevisiae*.

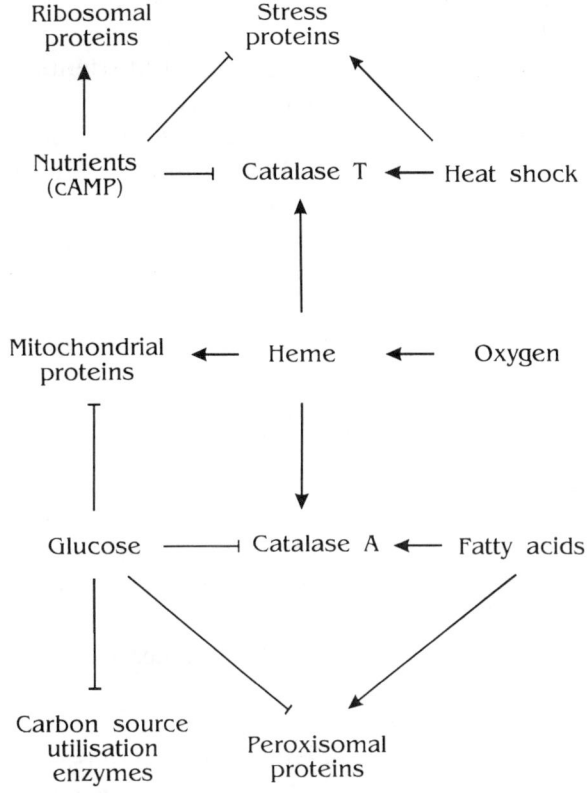

FIGURE 4 Regulatory network involved in control of expression of catalase genes of S. cerevisiae.

Negative control of expression by nutrients is also common to both catalase genes. However, whereas the *CTT1* gene is controlled by general nutrient availability (cAMP), catalase A is under glucose repression, which links its control to that of enzymes involved in utilization of carbon sources, among them a number of mitochondrial proteins, and of peroxisomal proteins. Induction of peroxisomal proteins by fatty acids might be considered as a special case of carbon source control. However, fatty acid utilization allows growth of *S. cerevisiae* only in some strains, whereas induction of peroxisomes in their presence is a more general phenomenon. It appears possible, therefore, that this induction of peroxisome proliferation is another example of a stress response mechanism, which protects yeast cells against toxic effects of fatty acids.

The ability to sense the general availability of nutrients and to make an appropriate decision accordingly either for cell growth and division or for entry into a G_0-like state or into sporulation (Pringle and Hartwell 1981) is of fundamental importance for unicellular microorganisms. Many relevant details of this regulatory process are not yet understood, but it is clear that the RAS-cAMP pathway and at least one parallel cAMP-independent pathway play a role in signaling nutrient availability and in triggering a proper response in S. cerevisiae. The second-messenger cAMP should therefore be a positive regulator of expression of genes whose products are required for cell growth. In S. cerevisiae, little evidence exists concerning this point, but some information is available on positive transcriptional control of a number of ribosomal protein genes by cAMP (J. Broach, pers. comm.). The finding that a number of genes encoding stress proteins, among them the catalase T gene, are turned off when growth conditions are optimal and are derepressed when nutrients are limiting makes sense from a teleological point of view. Under optimal growth conditions and in the absence of significant stress, it is advantageous for unicellular microorganisms to concentrate on increasing the cell population. However, when growth is slow or not possible at all, protection of cells against damage caused by stress factors should have increased survival value.

It is obvious from these considerations that a certain set of stress proteins derepressed by nutrient starvation should also be induced by other types of stress. It is unexpected, however, that two different types of positive promoter elements (the canonical heat shock element and the novel element described above), and perhaps two different signaling pathways, are mediating heat stress control. Since the specificity of the two types of elements with respect to stress signals is overlapping, but not identical, it can be hypothesized that the increased flexibility of response gained by the combination of these two mechanisms is of selective advantage. It is in line with this hypothesis that different stress genes possess only one or both of these elements. It remains to be elucidated whether the factors interacting directly or indirectly with the novel stress control element have any similarity with those functioning in prokaryotic stress protection systems. It has been reported that an overlapping set of proteins is induced by oxidative stress and by heat shock in *Salmonella typhimurium* (Morgan et al. 1986). Furthermore, starvation-induced cross-protection against heat and hydrogen peroxide has been observed in *E. coli* (Jenkins et al. 1988), and one of the catalases of *E. coli*, the *katE* gene product, has been shown to be induced by nutrient starvation (Mulvey et al. 1990). It should further be clarified whether induction of the *DDRA2* gene by DNA damage and by heat occurs via the same DNA element and whether other genes controlled via this type of DNA element are also induced by DNA damage. Control of expression of a catalase gene by DNA damage would not be surprising, since DNA damage has been postulated to be the main cause for toxicity of oxygen radicals (Imlay and Linn 1988). As with other results described in this chapter, a clarification of this question would contribute to our general understanding of the molecular mechanisms enabling cells to minimize damage caused by the negative effects of the availability and utilization of oxygen in the biosphere.

ACKNOWLEDGMENTS

The authors thank Gerhard Adam and Manuel Simon for the design and drawing of the figures; and Gerhard Adam, Pavel Kovarik, Wolfgang Löffelhardt, and Rotraud Wieser for critically reading this manuscript. Work from the authors' laboratory was supported by grants from the Fonds zur Förderung der wissenschaftlichen Forschung and from the Bundesministerium für Wissenschaft und Forschung, Vienna, Austria.

REFERENCES

Autor, A. 1982. Biosynthesis of mitochondrial manganese superoxide dismutase in *Saccharomyces cerevisiae*. *J. Biol. Chem.* **257:** 2713.

Belazzi, T., A. Wagner, R. Wieser, M. Schanz, G. Adam, A. Hartig, and H. Ruis. 1991. Negative regulation of transcription of the *Sac-*

charomyces cerevisiae catalase T (CTT1) gene by cAMP is mediated by a positive control element. *EMBO J.* **10:** 585.

Bienz, M. 1985. Transient and developmental activation of heat-shock genes. *Trends Biochem. Sci.* **10:** 157.

Bissinger, P.H., R. Wieser, B. Hamilton, and H. Ruis. 1989. Control of Saccharomyces cerevisiae catalase T gene (CTT1) expression by nutrient supply via the RAS-cyclic AMP pathway. *Mol. Cell. Biol.* **9:** 1309.

Boorstein, W.R. and E.A. Craig. 1990. Regulation of a yeast HSP70 gene by a cAMP responsive transcriptional control element. *EMBO J.* **9:** 2543.

Cameron, S., L. Levin, M. Zoller, and M. Wigler. 1988. cAMP-independent control of sporulation, glycogen metabolism, and heat shock resistance in S. cerevisiae. *Cell* **53:** 555.

Cohen, G., F. Fessl, A. Traczyk, J. Rytka, and H. Ruis. 1985. Isolation of the catalase A gene of Saccharomyces cerevisiae by complementation of the cta1 mutation. *Mol. Gen. Genet.* **200:** 74.

Cross, H.S. and H. Ruis. 1978. Regulation of catalase synthesis in Saccharomyces cerevisiae by carbon catabolite repression. *Mol. Gen. Genet.* **166:** 37.

Denis, C.L., M. Ciriacy, and E.T. Young. 1981. A positive regulatory gene is required for accumulation of the functional mRNA for the glucose-repressible alcohol dehydrogenase from Saccharomyces cerevisiae. *J. Mol. Biol.* **148:** 355.

Forsburg, S.L. and L. Guarente. 1989. Communication between mitochondria and the nucleus in regulation of cytochrome genes in the yeast Saccharomyces cerevisiae. *Annu. Rev. Cell Biol.* **5:** 153.

Garrett, S. and J. Broach. 1989. Loss of RAS activity in Saccharomyces cerevisiae is suppressed by disruptions of a new kinase gene, YAK1, whose product may act downstream of the cAMP-dependent protein kinase. *Genes Dev.* **3:** 1336.

Hamilton, B., R. Hofbauer, and H. Ruis. 1982. Translational control of catalase synthesis by hemin in the yeast Saccharomyces cerevisiae. *Proc. Natl. Acad. Sci.* **79:** 7609.

Hörtner, H., G. Ammerer, E. Hartter, B. Hamilton, J. Rytka, T. Bilinski, and H. Ruis. 1982. Regulation of synthesis of catalases and iso-1-cytochrome c in Saccharomyces cerevisiae by glucose, oxygen and heme. *Eur. J. Biochem.* **128:** 179.

Imlay, J.S. and S. Linn. 1988. DNA damage and oxygen radical toxicity. *Science* **240:** 1302.

Jenkins, D.E., J.E. Schultz, and A. Matin. 1988. Starvation-induced cross protection against heat or H_2O_2 challenge in Escherichia coli. *J. Bacteriol.* **170:** 3910.

Kobayashi, N. and K. McEntee. 1990. Evidence for a heat shock transcription factor-independent mechanism for heat shock induction of transcription in Saccharomyces cerevisiae. *Proc. Natl. Acad. Sci.* **87:** 6550.

Lock, E.A., A.M. Mitchell, and C.R. Elcombe. 1989. Biochemical mechanisms of induction of hepatic peroxisome proliferation. *Annu. Rev. Pharmacol. Toxicol.* **29:** 145.

Morgan, R.W., M.F. Christman, F.S. Jacobson, G. Storz, and B. Ames. 1986. Hydrogen peroxide-inducible proteins in *Salmonella typhimurium* overlap with heat shock and other stress proteins. *Proc. Natl. Acad. Sci.* **83:** 8059.

Mulvey, M.R., J. Switala, A. Borys, and P.C. Loewen. 1990. Regulation of transcription of *katE* and *katF* in *Escherichia coli. J. Bacteriol.* **172:** 6713.

Pfeifer, K., B. Arcangioli, and L. Guarente. 1987. Yeast HAP1 activator competes with the factor RC2 for binding to the upstream activation site UAS1 of the *CYC1* gene. *Cell* **49:** 9.

Praekelt, U.M. and P.A. Meacock. 1990. *HSP12*, a new small heat shock gene of *Saccharomyces cerevisiae*: Analysis of structure, regulation and function. *Mol. Gen. Genet.* **223:** 97.

Pringle, J.R. and L.H. Hartwell. 1981. The *Saccharomyces cerevisiae* cell cycle. In *The molecular biology of the yeast* Saccharomyces: *Life cycle and inheritance* (ed. J. Strathern et al.), p. 97. Cold Spring Harbor Laboratory, Cold Spring Harbor, New York.

Richter, K., G. Ammerer, E. Hartter, and H. Ruis. 1980. The effect of δ-aminolevulinate on catalase T-messenger RNA levels in δ-aminolevulinate synthase-defective mutants of *Saccharomyces cerevisiae. J. Biol. Chem.* **255:** 8019.

Ruis, H. 1989. Heme as a regulator of catalase biosynthesis in the yeast *Saccharomyces cerevisiae*. In *Highlights of modern biochemistry* (ed. A. Kotyk et al.), vol. 1, p. 339. VSP International Science Publishers, Zeist, The Netherlands.

Rytka, J., A. Sledziewski, J. Lukaszkiewicz, and T. Bilinski. 1978. Haemoprotein formation in yeast. III. The role of carbon catabolite repression in the regulation of catalase A and T formation. *Mol. Gen. Genet.* **160:** 51.

Seah, T.C.M. and J.G. Kaplan. 1973. Purification and properties of the catalase of bakers' yeast. *J. Biol. Chem.* **248:** 2889.

Seah, T.C.M., A.R. Bhatti, and J.G. Kaplan. 1973. Novel catalase proteins of bakers' yeast. I. An atypical catalase. *Can. J. Biochem.* **51:** 1551.

Simon, M., G. Adam, W. Rapatz, W. Spevak, and H. Ruis. 1991. The *Saccharomyces cerevisiae ADR1* gene is a positive regulator of transcription of genes encoding peroxisomal proteins. *Mol. Cell. Biol.* **11:** 699.

Skoneczny, M., A. Chelstowska, and J. Rytka. 1988. Study of the coinduction by fatty acids of catalase A and acyl-CoA oxidase. *Eur. J. Biochem.* **174:** 297.

Sledziewski, A., J. Rytka, T. Bilinski, H. Hörtner, and H. Ruis. 1981. Posttranscriptional heme control of catalase synthesis in the yeast *Saccharomyces cerevisiae. Curr. Genet.* **4:** 19.

Spevak, W., A. Hartig, P. Meindl, and H. Ruis. 1986. Heme control region of the catalase T gene of the yeast *Saccharomyces cerevisiae*. *Mol. Gen. Genet.* **203:** 73.

Spevak, W., F. Fessl, J. Rytka, A. Traczyk, M. Skoneczny, and H. Ruis. 1983. Isolation of the catalase T structural gene of *Saccharomyces cerevisiae* by functional complementation. *Mol. Cell. Biol.* **3:** 1545.

Tanaka, K., K. Matsumoto, and A. Toh-e: 1988. Dual regulation of the expression of the polyubiquitin gene by cyclic AMP and heat shock in yeast. *EMBO J.* **7:** 495.

Thorsness, M., W. Schafer, L. Dari, and J. Rine. 1989. Positive and negative transcriptional control by heme of genes encoding 3-hydroxy-3-methylglutaryl coenzyme-A reductase in *Saccharomyces cerevisiae*. *Mol. Cell. Biol.* **9:** 5702.

Toda, T,, S. Cameron, P. Sass, and M. Wigler. 1988. *SCH9*, a gene of *Saccharomyces cerevisiae* that encodes a protein distinct from, but functionally and structurally related to, cAMP-dependent protein kinase catalytic subunits. *Genes Dev.* **2:** 517.

van der Klei, I.J., J. Rytka, W.H. Kunau, and M. Veenhuis. 1990. Growth of catalase A and catalase T deficient mutant strains of *Saccharomyces cerevisiae* on ethanol and oleic acid. *Arch. Microbiol.* **153:** 513.

Veenhuis, M., M. Mateblowski, W.H. Kunau, and W. Harder. 1987. Proliferation of microbodies in *Saccharomyces cerevisiae*. *Yeast* **3:** 77.

Wieser, R., G. Adam, A. Wagner, C. Schüller, G. Marchler, H. Ruis, Z. Krawiec, and T. Bilinski. 1991. Heat shock-factor-independent heat control of transcription of the *CTT1* gene encoding the cytosolic catalase T of *Saccharomyces cerevisiae*. *J. Biol. Chem.* **266:** 12406.

Winkler, H., G. Adam, E. Mattes, M. Schanz, A. Hartig, and H. Ruis. 1988. Co-ordinate control of synthesis of mitochondrial and non-mitochondrial hemoproteins: A binding site for the HAP1 (CYP1) protein in the UAS region of the yeast catalase T gene (*CTT1*). *EMBO J.* **7:** 1799.

Woloszczuk, W., D.B. Sprinson, and H. Ruis. 1980. The relation of heme to catalase apoprotein synthesis in yeast. *J. Biol. Chem.* **255:** 2624.

Zimniak, P., E. Hartter, W. Woloszczuk, and H. Ruis. 1976. Catalase biosynthesis in yeast: Formation of catalase A and catalase T during oxygen adaptation of *Saccharomyces cerevisiae*. *Eur. J. Biochem.* **71:** 393.

Production and Scavenging of Active Oxygen in Chloroplasts

K. Asada
The Research Institute for Food Science
Kyoto University, Uji, Kyoto 611, Japan

Chloroplasts have the potential to convert photon energy to biochemical energy mediated by NADPH and ATP at a high efficiency. Under specific environmental conditions, photosynthetic organisms are able to fix one molecule of carbon dioxide to form carbohydrate by eight quanta of 680-nm light, that is, 36% efficiency in energy conversion. This efficiency is attained only under conditions of low light intensity with favorable spectra, a higher than atmospheric concentration of CO_2, and a low concentration of O_2; i.e., under conditions of no excess photons compared to the physiological electron acceptor. The maximum potential to convert solar energy is estimated to be 12%, considering its spectrum. The conversion efficiency of solar energy under natural conditions, however, is only 2–4% at maximum, which is mainly due to a wide divergence of the actual global environment from the above conditions.

Natural environments for photosynthesis vary greatly within a day. For example, plants are exposed to a 20°C difference of temperature during a day, and a shortage of water supply under a high intensity of sunlight is not unusual. Under such conditions, the stomata are closed due to water stress, and the supply of CO_2 to chloroplasts is suppressed. In oxygenic photosynthetic plants, the two reaction center complexes, photosystems I and II, convert in series to drive a linear electron flow from water to NADP, producing dioxygen as a result of the oxidation of water. Because of the production of dioxygen in cells under strong sunlight, the oxygenic plants are exposed to the most severe environments among organisms with respect to the toxicity of active oxygen. Under conditions where CO_2 in chloroplasts is depleted, excess photon energy is dissipated in several ways so as not to produce active molecules that decompose the chloroplast components. One important way is the photoreduction of dioxygen in place of CO_2. In this paper, I first present a mechanism by which the chloroplasts suppress the release of superoxide from the thylakoid membranes. Subsequently, the scavenging system of active oxygen in chloroplasts is described, with an emphasis on its microlocalization in chloroplasts and its molecular evolution.

PRODUCTION OF SUPEROXIDE IN CHLOROPLASTS

The major producing site of superoxide in chloroplasts is the reducing side of photosystem I. The electron generated in photosystem II by the oxidation of water is transported through the intersystem carriers to the photosystem I reaction center complex and is able to reduce NADP after its excitation by the reaction center chlorophyll, P700. Superoxide is generated by the autooxidation of the thylakoid membrane-bound primary electron acceptor of photosystem I (Asada et al. 1974) and of peripheral reduced ferredoxin (Furbank and Badger 1983; for review, see Asada and Takahashi 1987). The production of superoxide in photosystem I does not occur on the surface of the thylakoid, but in the aprotic interior of the membrane (Takahashi and Asada 1988). This is also the case in the production of superoxide by NADPH-oxidase in neutrophile, where the producing site is estimated to be 4–5 Å inside from the outer surface of the membrane (Fujii and Kakinuma 1990). In the reaction center of photosystem I, the electron released from P700 after charge separation by the absorption of photon energy is transferred successively to four intrinsic redox centers: A_0 (chlorophyll), A_1 (vitamin K), X ([2Fe-2S] cluster), and A/B (2[4Fe-4S] cluster). P700, A_0, A_1, and X reside in the core subunits, and A/B resides in a 9-kD peptide (Oh-oka et al. 1989). The actual photoreductant of dioxygen would be the centers X and A/B, because of very rapid rates of the reactions to the steps of the center X from P700. Neither of the Fe-S clusters is exposed to the surface of the membranes (Fig. 1).

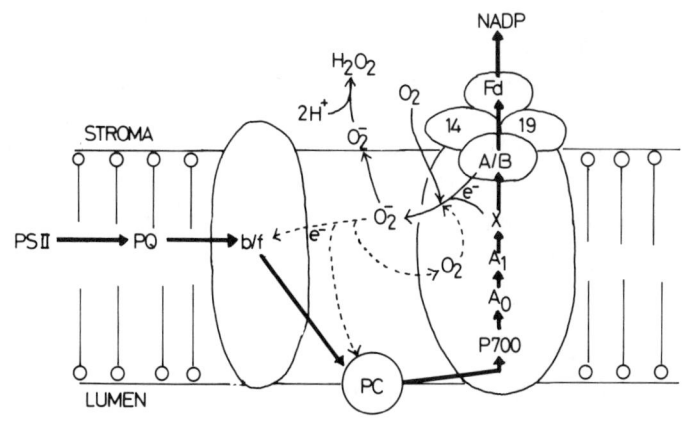

FIGURE 1 Structure of photosystem I complex of thylakoids, and the reduction of dioxygen and fate of the superoxide formed within the membrane. A_0, A_1, X, and A/B represent the respective centers. Center A/B resides on a 9-kD peptide, and the other centers reside on the photosystem I core subunits (PSI-A and PSI-B). (Fd) Peripheral ferredoxin, (14) 14-kD peptide, (19) 19-kD peptide, (PC) plastocyanin, (b/f) cytochrome b/f complex, and (PQ) the pool of plastoquinone. Thick lines show the linear electron flow from photosystem II (PSII) to NADP, thin lines show the photoreduction of dioxygen and the release of the superoxide from the membrane, and dotted lines show a proposed superoxide-mediated cyclic electron flow.

SUPPRESSION OF RELEASE OF SUPEROXIDE FROM THYLAKOID MEMBRANES

When thylakoids are illuminated by a flash of light for a single turnover of P700, one molecule of superoxide is generated per P700 in the reaction center, as determined by the reduction of cytochrome c (Takahashi and Asada 1988). However, the production rate of superoxide observed under continuous, saturating light is about 30 µmole mg Chl^{-1} hr^{-1}, even in the absence of the physiological electron acceptor, as determined from the photoreduction rate of $^{18}O_2$ in chloroplasts (Asada and Badger 1984). This rate is less than 10% of the total electron transport rate as estimated by the reduction rate of either NADP$^+$ or CO_2. These observations indicate that the superoxide anions produced within the thylakoid membranes are reduced to dioxygen before ejection to either the stroma or the lumen. The probable candidates for the oxidant of the superoxide anion in the membrane are plastocyanin and cytochrome f. Plastocyanin is a peripheral electron carrier localized in the lumen side (Takano et al. 1985) and is the immediate electron donor to P700, and cytochrome f is the electron donor to plastocyanin (Fig. 1). Both of them are reduced at an appreciable rate by superoxide (1 x 10^6 M^{-1} sec^{-1} to 6 x 10^6 M^{-1} sec^{-1}; Tanaka et al. 1978; Takahashi et al. 1980). If the superoxide anions in the membrane donate the electron to the electron carriers in the oxidizing side of P700, a cyclic electron flow operates around photosystem I. Direct evidence for the superoxide-mediated cyclic electron flow has not been obtained, but the following observations support its operation in intact thylakoids.

1. The photoproduction rate of superoxide is increased by treatment of thylakoids with Triton X-100, indicating the structural requirement for the operation of superoxide-mediated cyclic flow (Takahashi and Asada 1982).
2. When protons are donated to the thylakoid membranes by ammonium ions with their ionophore nonactin, the photoreduction rate of dioxygen and the production rate of hydrogen peroxide are increased severalfold (Takahashi and Asada 1988). At pH 5, the photoreduction rate of dioxygen is increased compared with that at neutral pH (Schreiber et al. 1991). These observations suggest that only when the concentration of protons either within or outside the membrane is high, the superoxide anions are disproportionated before donation of the electrons to the membrane-bound carriers.
3. Photoreduction of dioxygen is enhanced by the addition of methyl viologen at high light intensities due to a high reactivity of reduced methyl viologen with dioxygen. However, this is not the case at low light intensities (T. Shiraishi et al., unpubl.). In photosystem I, the donation rate of electrons to P700 from plastocyanin ($t_{1/2}$ = ~ 20 msec) is slower than that to the acceptors from P700 ($t_{1/2}$ = 14 psec - 200 nsec). At low light intensities, the donors to P700 would occur in a reduced form for a longer time because of a low turnover rate of P700. Under such conditions, the super-

oxide cannot reduce the donors and is released from the membranes.

Superoxide anions produced within the aprotic membranes would have a longer life compared with those produced in aqueous environments (Takahashi and Asada 1988). The superoxide anions in thylakoid membranes are subjected to (1) diffusion to the surface of the membrane where proton and superoxide dismutase (SOD) are available for the disproportionation, (2) disproportionation to hydrogen peroxide and dioxygen when protons are available within the membranes, or (3) oxidation by the oxidized cytochrome f or plastocyanin to form dioxygen within the membrane. The uptake of O_2 is observed only in the cases of (1) and (2), and the case of (3) does not result in a net change of dioxygen. Thus, the O_2 uptake or the superoxide production as determined by the probes in the medium under strong light at neutral pH seems to represent only a part of the actual photoreduction of dioxygen.

D1 protein of photosystem II is decomposed under conditions of excess photon energy (Kyle 1987). The same photoinhibition has been observed in thylakoids, intact chloroplasts, and cells under anaerobic conditions (Asada and Takahashi 1987; Krause and Cornic 1987). In Figure 2 are shown the results for the case of thylakoids, and this inhibition in photosystem II is found even in the presence of dichloroindophenol, an electron acceptor in photosystem II. Thus, dioxygen protects from photoinhibition by functioning as an electron acceptor even if its apparent reduction rate is low, and the superoxide-mediated cyclic flow of electrons would contribute to the protection.

INHIBITION OF PHOTOSYNTHESIS BY ACTIVE OXYGEN

As described above, the release of superoxide from the thylakoid membranes is suppressed even under saturating light, which seems to be an effective system for protection from oxidative damage under conditions of excess light energy. Even so, superoxide and hydrogen peroxide are released to the stroma at a rate of 240 µM sec^{-1} and 120 µM sec^{-1}, respectively, assuming a chloroplast volume of 35 µl mg Chl^{-1} (Asada

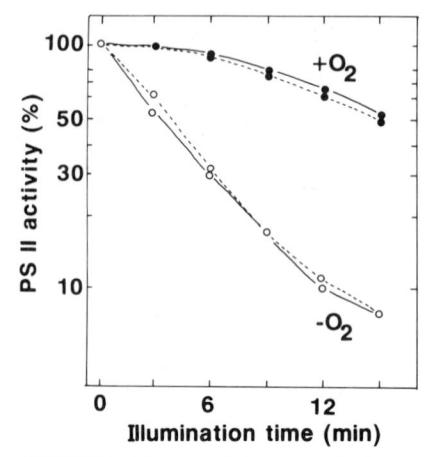

FIGURE 2 Photoinhibition of photosystem II activity of thylakoids under aerobic (1% O_2,[+O_2]) and anaerobic conditions (-O_2) (Y. Matoba and K. Asada, unpubl.). Spinach thylakoids were illuminated by a halogen projector lamp at 860 W min^{-2} for the indicated times. After the light treatment, the photoreducing activity of dichloroindophenol was determined in the absence (solid line) and presence (dashed line) of 0.1 mM diphenylcarbazide. Similar inhibition, irrespective of the presence or absence of the electron donor to photosystem II diphenylcarbazide, indicates that the inhibition site by anaerobic illumination is not the side of the oxygen evolution, but rather the reducing side of photosystem II. Under present conditions, no inhibition of photosystem I was observed as assayed by the photoreduction of dioxygen in the presence of DCMU, ascorbate, and dichloroindophenol.

and Takahashi 1987). This rate is nearly two orders of magnitude higher than that in *Escherichia coli* cells (Imlay and Fridovich 1991). Photosynthetic CO_2 fixation is half inhibited by 10 μM hydrogen peroxide added exogenously (Kaiser 1976), indicating the nearly complete inhibition of photosynthesis within a second if no scavenging system of active oxygen operates.

The primary target of active oxygen has been assigned to be the enzymes participating in the CO_2-fixation cycle: fructose-1, 6-bisphosphatase, glyceraldehyde-3-phosphate dehydrogenase, and ribulose-5-phosphate kinase. All these enzymes are thiol enzymes and are inactivated by hydrogen peroxide through the formation of a disulfide bridge. At high concentrations of hydrogen peroxide, ribulose bisphosphate carboxylase and CuZnSODs also are inhibited, but these enzymes are not the primary targets. Inactivation of these enzymes occurs only after hydrogen peroxide accumulates because of the inactivation of CO_2 fixation and of hydrogen peroxidase-scavenging enzymes. The superoxide-sensitive enzyme in chloroplasts is nitrite reductase, and its molybdenum center is the site of inhibition (for review, see Asada and Takahashi 1987).

SCAVENGING OF SUPEROXIDE IN CHLOROPLASTS

The superoxide anions released from the thylakoid membranes to the stroma and lumen sides are disproportionated by SOD localized in both the compartments at 40–50 µM (Hayakawa et al. 1984). The concentration of SOD in *E. coli* has been estimated to be 2–18 µM (Imlay and Fridovich 1991). The steady-state concentration of superoxide in chloroplasts is lowered by the SOD to 3×10^{-9} M from 2.4×10^{-5} M, as estimated using $v(\cdot O_2^-) = k_{SOD} [\cdot O_2^-][SOD]$ and $v(\cdot O_2^-) = k_{Sp}[\cdot O_2^-]^2$, where $v(\cdot O_2^-)$ is the production rate of superoxide (2.4×10^{-4} M sec^{-1}), k_{Sp} is the rate constant for spontaneous disproportionation of superoxide at pH 7 (4×10^5 M^{-1} sec^{-1}), and k_{SOD} is the reaction rate constant between superoxide and SOD (2×10^9 M^{-1} sec^{-1}). The above estimation, however, should be regarded as a first approximation, because the following assumptions for the estimation are different from the actual conditions in chloroplasts having a size of about 5 µm × 3 µm.

1. Superoxide anions are assumed to be uniformly generated in chloroplasts, but most superoxide is produced in photosystem I of the thylakoid membranes. When the thylakoids are illuminated for 20 seconds in the presence of lactoperoxidase and $^{125}I^-$, only the core subunit of the reaction center of photosystem I (Fig. 1) is iodinated (Takahashi and Asada 1988). These observations indicate that the producing site of superoxide and then of hydrogen peroxide in the membranes is actually limited to the complex of the photosystem I reaction center.

2. The values of k_{SOD} and k_{Sp} determined in aqueous medium are used for the estimation, but the protein concentration in the chloroplast stroma is about 30%. Thus, the diffusion

of either superoxide anion or SOD in the stroma would be suppressed due to its high viscosity. Especially, the reaction of superoxide with SOD is a nearly diffusion-controlled one, and the reaction rate is lowered to 37% with a fivefold increase in viscosity by the addition of glycerol (Rotilio et al. 1972).
3. SOD is assumed to be distributed uniformly in the lumen and stroma of chloroplasts, but this has not been proven. Considering the results in Figure 3, it is likely that SOD is localized in a site of the stroma where superoxide is released from the thylakoid.

SCAVENGING OF HYDROGEN PEROXIDE IN CHLOROPLASTS

The superoxide anions released into the lumen are disproportionated by the SOD there, and the hydrogen peroxide is diffused to the stroma side. This is expected by a rapid diffusion of hydrogen peroxide through the thylakoid membrane, but the permeability of superoxide anions through the membranes is low (Takahashi and Asada 1983).

In place of water, hydrogen peroxide can donate electrons to the oxidizing side of photosystem II accompanied by the evolution of dioxygen under flash lights (Mano et al. 1987), but this reaction is negligibly low under continuous light and cannot account for the scavenging of hydrogen peroxide. The evolution of $^{16}O_2$ from water upon the addition of $H_2^{18}O_2$ to the illuminated chloroplasts (Asada and Badger 1984) and the quenching of chlorophyll fluorescence by the addition of hydrogen peroxide (Neubauer and Schreiber 1989; Miyake et al. 1991) indicate that the hydrogen peroxide is not oxidized in photosystem II, but is reduced by the electrons from water catalyzed with peroxidase.

The participation of ascorbate peroxidase in the scavenging of hydrogen peroxide in chloroplasts, and the regeneration of ascorbate from the primary oxidation product of the peroxidase reaction (monodehydroascorbate) and from its disproportionation product (dehydroascorbate), have been established (Asada and Takahashi 1987). The electron donors (NAD[P]H and GSH) for the reduction of the oxidation products

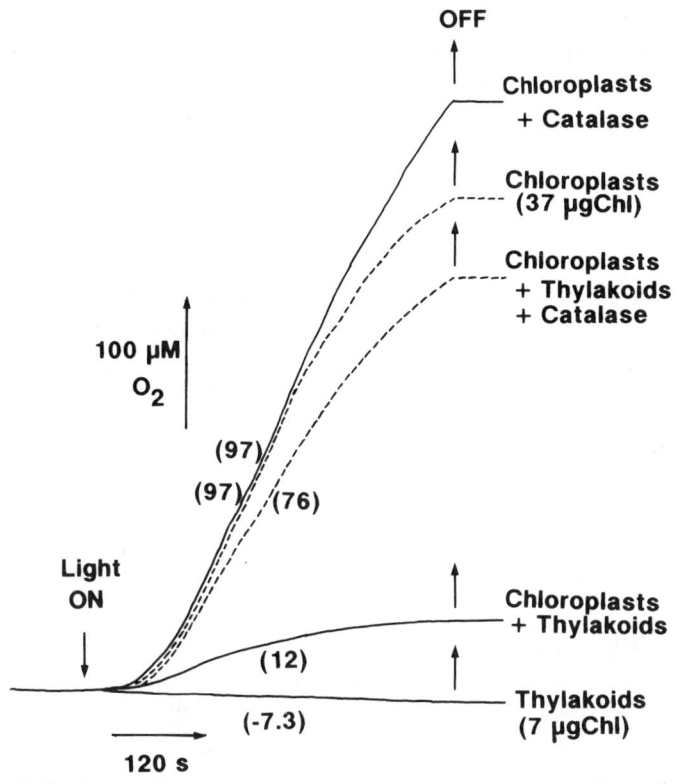

FIGURE 3 Effect of hydrogen peroxide generated by thylakoids on bicarbonate-dependent evolution of dioxygen in spinach chloroplasts (K. Asada, unpubl.). The reaction mixture (1 ml) contained 0.33 M sorbitol, 10 mM NaCl, 1 mM $MgCl_2$, 2 mM EDTA, 0.5 mM KH_2PO_4, 50 mM HEPES-KOH (pH 7.6), 10 mM $NaHCO_3$, and intact chloroplasts (37 µg Chl) and/or thylakoids (7 µg Chl) from spinach. Where indicated, 1000 units of catalase were added. The reaction was started by illumination by a tungsten projector lamp at 2.7 mmole quanta m^{-2} sec^{-1}. The reaction temperature was 25°C. Values in parentheses are the maximum rate of oxygen evolution (chloroplasts) or uptake (thylakoids) in µmole mg Chl^{-1} hr^{-1}.

of the peroxidase reaction are supplied through photosystem I by linear electron flow. The heme protein ascorbate peroxidase (Asada 1992), the FAD enzyme monodehydroascorbate reductase (Hossain and Asada 1985), and the thiol enzyme dehydroascorbate reductase (Hossain and Asada 1984) have been found in chloroplasts at a sufficient concentration to scavenge the hydrogen peroxide generated. Ascorbate and GSH occur at about 10 mM in chloroplasts, which is also high enough for the peroxidase and dehydroascorbate reductase, respectively. NAD(P) (~1 mM) also occurs at the saturating concentration for monodehydroascorbate reductase.

The contents of ascorbate peroxidase in the stroma (e_0, 3.7 x 10^{-5} M), its K_m for hydrogen peroxide (8 x 10^{-5} M for the chloroplast isozyme), and its molecular activity (k_0, 3.1 x 10^2 mole sec^{-1}) (Chen and Asada 1989) allow an estimate of the steady-state concentration of hydrogen peroxide in chloroplasts with the equation [$v(H_2O_2) = k_0 e_0 [H_2O_2]/(K_m + [H_2O_2])$], where $v(H_2O_2)$ is the generation rate of hydrogen peroxide (1.2 x 10^{-4} M sec^{-1}). The estimated concentration of hydrogen peroxide is 8 x 10^{-7} M. For this estimation, it is assumed that ascorbate is photoregenerated so rapidly in chloroplasts that ascorbate is not a limiting factor for the peroxidase reaction. This is actually the case, as observed by the transient stopping of CO_2 fixation during the reduction of exogenous hydrogen peroxide (Nakano and Asada 1980), which indicates the photoreductant is preferentially used for the reduction of hydrogen peroxide. Quenching of chlorophyll fluorescence upon the addition of hydrogen peroxide (Neubauer and Schreiber 1989; Miyake et al. 1991) also sustains the same preferential reduction of hydrogen peroxide. A similar estimation of the steady-state concentration of monodehydroascorbate radical in illuminated chloroplasts gives a value of about 3 x 10^{-7} M (Asada and Takahashi 1987).

These estimations for hydrogen peroxide and monodehydroascorbate using kinetic constants also should be regarded as a first approximation, as in the case of the superoxide in chloroplasts. The production of hydrogen peroxide in a limiting site of the thylakoid from the superoxide would require a modification of the estimation, and the results in Figure 3 indicate that the scavenging system of hydrogen peroxide does not uniformly distribute in whole chloroplasts.

Intact chloroplasts without any contamination of thylakoids show the bicarbonate-dependent evolution of dioxygen for several minutes under light without the addition of catalase. This is due to the operation of the scavenging system mentioned above, even though the hydrogen peroxide is generated at a rate of 1.2×10^{-4} M sec^{-1} in chloroplasts. However, when a small amount of thylakoids is added to such chloroplasts, photosynthesis is dramatically inhibited, as shown by the suppression of oxygen evolution, probably due to the inactivation of enzymes for the CO_2-fixation cycle as discussed in the previous section. This inhibition is recovered by the addition of catalase, indicating that photosynthesis is inhibited by the hydrogen peroxide generated with the thylakoids in the medium. The generation rate of hydrogen peroxide by the thylakoids, however, is only 9×10^{-9} M sec^{-1} in the reaction medium. Then, the hydrogen peroxide accumulated by the thylakoids during a 100-second illumination (9×10^{-7} M) is similar to the steady-state concentration of hydrogen peroxide in chloroplasts estimated above (8×10^{-7} M). A greater inhibition of photosynthesis by a small amount of the hydrogen peroxide penetrated through the envelope of chloroplasts indicates that the actual concentration of hydrogen peroxide in chloroplasts should be lower than the estimated concentration, probably by localization of the scavenging system including SOD in the producing site of superoxide. Thus, the chloroplast has little scavenging ability for the hydrogen peroxide penetrating into the stroma through the envelope, although it has a very high capacity for the hydrogen peroxide generated in the thylakoids.

In accordance with this observation, the thylakoid-bound ascorbate peroxidase has been found in addition to the enzyme in the stroma (Groden and Beck 1979). This membrane-bound enzyme accounts for about half the activity in chloroplasts, and monodehydroascorbate reducing activity is also bound to the membranes (C. Miyake and K. Asada; H. Hormann and U. Schreiber; both unpubl.).

The photoreductions of two molecules of dioxygen to superoxide and of the resulting one molecule of hydrogen peroxide to water by the above system allow the flow of the four electrons generated by one molecule of dioxygen evolved in photosystem II. This electron flow would protect the photoinhibition of chloroplasts under the conditions of a shortage of the electron acceptor, in addition to the postulated superoxide-mediated cyclic flow discussed above. Furthermore, it has been shown that the photoreducing system of dioxygen to water contributes to the membrane energization as assayed by nonphotochemical quenching of chlorophyll fluorescence (Schreiber and Neubauer 1990). Thus, the superoxide-ascorbate system generates ATP in chloroplasts without the generation of NADPH and can adjust the photogenerating ratio of ATP to NADP, depending on the metabolic requirement.

CHLOROPLASTIC AND CYTOSOLIC CUZNSODS IN PLANTS, AND THEIR MOLECULAR EVOLUTION

In plant chloroplasts and cyanobacteria, MnSOD binds to thylakoids (Okada et al. 1979; Hayakawa et al. 1985). The major SOD in angiosperm chloroplasts is the CuZn-enzyme localized in the stroma. The amino acid sequence and three-dimensional structure of the chloroplastic isozyme of CuZnSOD from spinach have been determined (Kitagawa et al. 1986, 1991). In addition to the chloroplastic isozyme of CuZnSOD, plants contain the cytosolic isozyme of CuZnSOD. The cytosolic isozyme is mainly expressed in nonphotosynthetic tissues and in the cell constituents other than chloroplasts in leaf tissues (Baum et al. 1983; Kwiatowski and Kaniuga 1986; Kanematsu and Asada 1989a,b; 1990). Both types of CuZnSOD isozyme have been found in the green alga *Spirogyra*, ferns, and angiosperms. *Spirogyra* is classified in the most evolved group of the green algae with respect to its type of cell division, and only this group of algae contain CuZnSOD (Kanematsu and Asada 1989b). Other eukaryotic algae lack CuZnSOD and contain FeSOD and/or MnSOD (Asada et al. 1980). Therefore, the divergence of CuZnSOD to the chloroplastic and cytosolic isozymes, and their independent molecular evolutions, started immediately after the CuZnSOD was acquired by the algae.

Both isozymes show similar enzymatic properties but can be distinguished from each other by their immunological properties and amino acid sequences, in addition to their cellular locations. Although both isozymes keep the conserved residues found in other CuZnSOD, the chloroplastic isozymes are characterized by two additional residues in the amino terminus and one more residue in the carboxyl terminus. Both isozymes are characterized by the conserved residues for each isozyme (Kanematsu and Asada 1990). Comparison of the amino acid sequences of both isozymes from several plants indicates a greater difference between sequences among cytosolic isozymes than among chloroplastic isozymes. Thus, the mutation rate of the cytosolic CuZnSOD is higher than that of the chloroplastic CuZnSOD.

Why is the mutation rate of the chloroplastic CuZnSOD lower? One possible answer is the occurrence of several isozymes of the cytosolic CuZnSOD in plants. For example, rice contains three isozymes of the cytosolic CuZnSOD (Kanematsu and Asada 1989a); however, only one isozyme of the chloroplastic CuZnSOD has been found in each plant surveyed so far. Therefore, the cytosolic isozymes were able to accept a mutation of one isozyme without a lethal risk, compared to chloroplastic isozyme. Another reason for the low mutation rate of the chloroplastic isozyme may be a high generation rate of superoxide in chloroplasts, as described above, compared with that in other cell constituents, which would result in a lethal effect if the mutation of the chloroplastic isozyme induced a decrease in, or loss of, the activity for the disproportionation of superoxide.

ACQUISITION AND DIVERGENCE OF PEROXIDASE FOR SCAVENGING OF HYDROGEN PEROXIDE

Although an electron donor is required, peroxidase provides an effective scavenging system for hydrogen peroxide by its reduction, as compared with its disproportionation by catalase. Because of a high K_m value for hydrogen peroxide of catalase, a high content of the enzyme is necessary to lower the concentration of hydrogen peroxide for the suppression of oxidative damages. The hydrogen peroxide produced by the two-electron oxidases such as glycolate oxidase and D-amino acid oxidase in peroxisomes is scavenged by catalase. The amount of catalase in peroxisomes is very high; in leaf peroxisome, catalase occurs in a crystalline state, and in mammalian peroxisome, catalase occupies 16% of total protein.

Cyanobacteria are the most primitive oxygenic photosynthetic organisms. With respect to the scavenging of hydrogen peroxide, cyanobacteria are divided into the peroxidase-containing and -lacking species. The peroxidase-containing species produces $^{16}O_2$ upon the addition of $H_2{}^{18}O_2$ to the illuminating cells, because the photoreductant is used as the electron donor for the peroxidase, and the evolution of $^{16}O_2$ is inhibited by the electron transport inhibitor 3-(3,4-dichlorophenyl)1,1-dimethylurea (DCMU) (Fig. 4A). On the contrary, the peroxidase-lacking species evolved $^{18}O_2$ from $H_2{}^{18}O_2$, irrespective of light and the inhibitor (Fig. 4B). These observations indicate that the peroxidase system for the scavenging of hydrogen peroxide generated by illuminated thylakoids was acquired during the evolution of cyanobacteria, 3×10^9 to 2×10^9 years ago (Miyake et al. 1991). At the time cyanobacteria appeared on the earth, the atmospheric concentration of dioxygen is deduced to have been 0.002%. This concentration of dioxygen was at least two orders of magnitude lower than the apparent K_m value for dioxygen in the photoproduction of superoxide by thylakoids (Takahashi and Asada 1982). Therefore, the photoproduction rates of superoxide and hydrogen peroxide in cyanobacteria should have increased accompanied by an accumulation of atmospheric dioxygen with cyanobacteria themselves, and an effective peroxidase system for scavenging of hydrogen peroxide was absolutely necessary to cyanobacteria. This peroxidase system acquired by cyanobacteria has been conserved in eukaryotic algae and in chloroplasts of angiosperms.

The hydrogen peroxide-scavenging peroxidase has diverged to several forms having different specificities of the electron donor and various prosthetic groups: the Se-enzyme glutathione peroxidase in mammalian mitochondria, the heme protein cytochrome *c* peroxidase in yeast mitochondria, the heme protein ascorbate peroxidase in plants, and the flavoenzyme NAD(P)H peroxidase in bacteria (Asada 1992). Plant cells have little glutathione, cytochrome *c*, and NADPH peroxidases, and ascorbate peroxidase plays a role in the scavenging in chloroplasts and other cell constituents. Thus, each hydrogen peroxide-scavenging peroxidase seems to show a characteristic, phylogenetic distribution. As in the case of SOD, the

ACTIVE OXYGEN IN CHLOROPLASTS 187

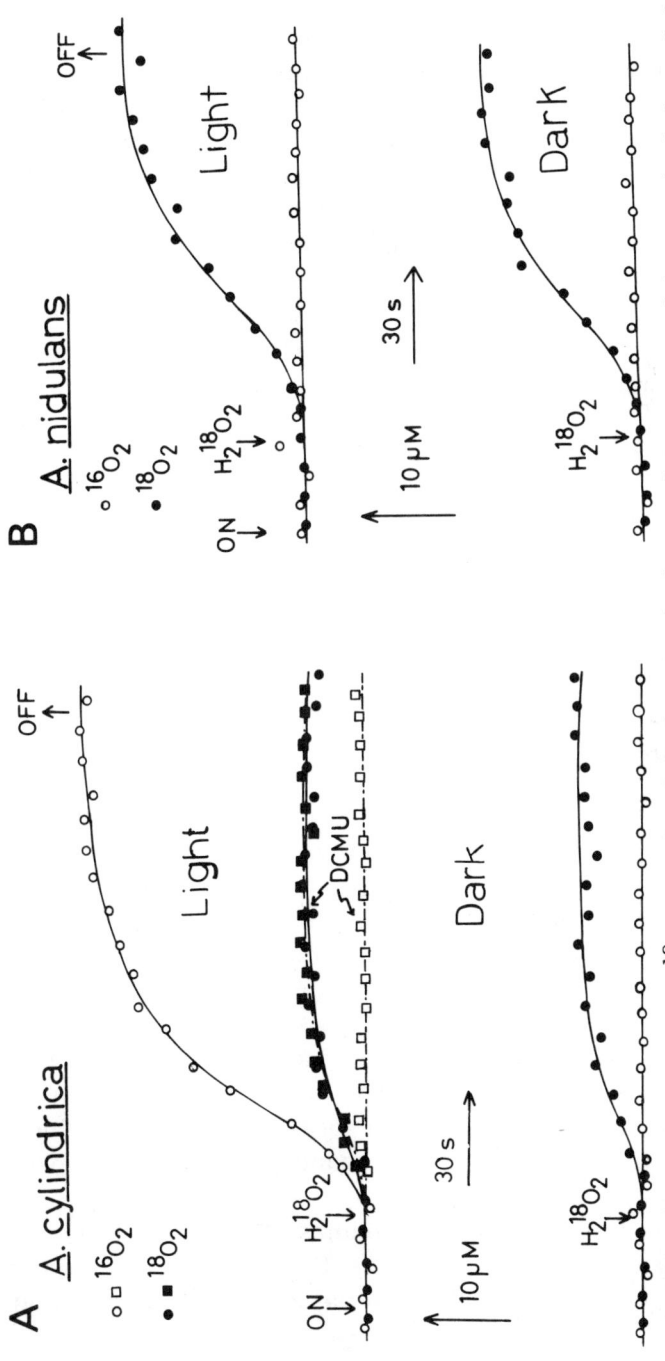

FIGURE 4 Decomposition of $H_2^{18}O_2$ in cells of *Anabaena cylindrica* and *Anacystis nidulans* in the light and dark (Miyake et al. 1991). Upon addition of 80 μM $H_2^{18}O_2$, *A. cylindrica* cells showed the evolution of $^{16}O_2$, indicating the peroxidatic decomposition of H_2O_2 using the photoreductant as the electron donor. The evolution of $^{18}O_2$ indicates the operation of the catalatic decomposition of the H_2O_2 also, which was not affected by the light and DCMU. On the contrary, *A. nidulans* cells showed the same rate of evolution of $^{18}O_2$ in the dark and light, but no evolution of $^{16}O_2$, indicating the operation of only the catalatic decomposition of H_2O_2.

chloroplast and cytosol isozymes have been found also in ascorbate peroxidase (Chen and Asada 1989).

The amino acid sequences around the proximal histidine residue of ascorbate, cytochrome c, and guaiacol peroxidases show a high homology, indicating that they belong to a superfamily of heme peroxidases. However, the homology of the sequences between ascorbate peroxidase and cytochrome c peroxidase from yeast is rather higher than that between ascorbate and guaiacol peroxidases from plants (Chen et al. 1992). Guaiacol peroxidase is represented by the enzyme from horseradish and participates in the biosynthesis of lignin and in a broad spectrum of other reactions, but not in the scavenging of hydrogen peroxide. In accordance with a difference of physiological function, two plant peroxidases show different molecular properties (Asada 1992). For example, guaiacol peroxidase is a glycoprotein, but ascorbate peroxidase is not. Aminophenoxy radical is produced by both peroxidases from p-aminophenol, but only ascorbate peroxidase is inactivated by this radical (Chen and Asada 1990). On the contrary, ascorbate peroxidase and cytochrome c peroxidase share many molecular properties (Chen et al. 1992).

CONCLUDING REMARKS

Univalent reduction of dioxygen is inevitable in illuminated chloroplasts under natural conditions, especially when the physiological electron acceptor CO_2 is not available. However, the release of superoxide from the thylakoid membranes seems to be suppressed in a manner not yet clearly understood, and this is the primary protection system against the photooxidative damages to plants. Even though the suppression system operates in thylakoids, the releasing rate of superoxide is high compared with other organisms, and the scavenging enzymes appear to be localized near the producing site of active oxygen in the cell organella for effective scavenging.

Cyanobacteria were the first photosynthetic organisms supplying dioxygen to the atmosphere. However, SOD had already been acquired by anaerobic bacteria, including photosynthetic bacteria, before the appearance of cyanobacteria (Asada et al. 1980). These observations indicate the importance of SOD for protection from oxidative damages under an ultra-low concentration of dioxygen. Catalase also would be acquired at the same time, but it cannot effectively scavenge hydrogen peroxide, so a large amount of the enzyme is required for protection from oxidative damages. The effective scavenging enzyme of hydrogen peroxide, peroxidase, was acquired during the evolution of cyanobacteria to adapt to an increased production of hydrogen peroxide accompanied by the increase in atmospheric dioxygen by cyanobacteria themselves. At present, the hydrogen peroxide-scavenging peroxidases have diverged to several forms with respect to the electron donor and the prosthetic group, and each peroxidase shows a characteristic distribution among organisms. Plants prefer ascorbate peroxidase, but it is not known why plants have not used glutathione peroxidase and mammals have not used ascorbate peroxidase.

ACKNOWLEDGMENT

This work was supported by a research grant from the Ministry of Education, Science and Culture, Japan.

REFERENCES

Asada, K. 1992. Ascorbate peroxidase: A hydrogen peroxide-scavenging enzyme in plants. *Physiol. Plant.* **83:** (in press).

Asada, K. and M.R. Badger. 1984. Photoreduction of $^{18}O_2$ and $H_2{}^{18}O_2$ with a concomitant evolution of $^{18}O_2$ in intact spinach chloroplasts: Evidence for scavenging of hydrogen peroxide by peroxidase. *Plant Cell Physiol.* **25:** 1169.

Asada, K. and M. Takahashi. 1987. Production and scavenging of active oxygen in photosynthesis. In *Photoinhibition* (ed. D.J. Kyle et al.), p. 227. Elsevier, Amsterdam.

Asada, K., K. Kiso, and K. Yoshikawa. 1974. Univalent reduction of molecular oxygen by spinach chloroplasts on illumination. *J. Biol. Chem.* **249:** 2175.

Asada, K., S. Kanematsu, S. Okada, and T. Hayakawa. 1980. Phylogenic distribution of three types of superoxide dismutase in organisms and cell organelles. In *Chemical and biochemical aspects of superoxide and superoxide dismutase* (ed. J.V. Bannister and H.A.O. Hill), vol. 1, p. 136. Elsevier, Amsterdam.

Baum, J.A., J.M. Chandlee, and J.G. Scandalios. 1983. Purification and partial characterization of a genetically-defined superoxide dismutase (SOD-1) associated with maize chloroplasts. *Plant Physiol.* **73**: 31.

Chen, G.-X. and K. Asada. 1989. Ascorbate peroxidase in tea leaves: Occurrence of two isozymes and their differences in enzymatic and molecular properties. *Plant Cell Physiol.* **30**: 987.

———. 1990. Hydroxyurea and p-aminophenol are the suicide inhibitors of ascorbate peroxidase. *J. Biol. Chem.* **265**: 2775.

Chen, G.-X., S. Sano, and K. Asada. 1992. Amino acid sequence of ascorbate peroxidase has a homology with that of cytochrome c peroxidase from yeast. *Plant Cell Physiol.* **33**: (in press).

Fujii, H. and K. Kakinuma. 1990. Studies on the superoxide releasing site in plasma membranes of neutrophils with ESR spin-labels. *J. Biochem.* **108**: 292.

Furbank, R.T. and M.R. Badger. 1983. Oxygen exchange associated with electron transport and photophosphorylation in spinach chloroplasts. *Biochim. Biophys. Acta* **723**: 400.

Groden, D. and E. Beck. 1979. H_2O_2 destruction by ascorbate-dependent systems from chloroplasts. *Biochim. Biophys. Acta* **546**: 426.

Hayakawa, T., S. Kanematsu, and K. Asada. 1984. Occurrence of CuZn-superoxide dismutase in the intrathylakoid space of spinach chloroplasts. *Plant Cell Physiol.* **25**: 883.

———. 1985. Purification and characterization of thylakoid-bound Mn-superoxide dismutase in spinach chloroplasts. *Planta* **166**: 111.

Hossain, M.A. and K. Asada. 1984. Purification of dehydroascorbate reductase from spinach and its characterization as a thiol enzyme. *Plant Cell Physiol.* **25**: 85.

———. 1985. Monodehydroascorbate reductase from cucumber is a flavin adenine dinucleotide enzyme. *J. Biol. Chem.* **260**: 12920.

Imlay, J.A. and I. Fridovich. 1991. Assay of metabolic superoxide production in *Escherichia coli*. *J. Biol. Chem.* **266**: 6957.

Kaiser, W. 1976. The effect of hydrogen peroxide on CO_2 fixation of isolated intact chloroplasts. *Biochim. Biophys. Acta* **440**: 476.

Kanematsu, S. and K. Asada. 1989a. CuZn-superoxide dismutases in rice: Occurrence of an active, monomeric enzyme and two isozymes in leaf and non-photosynthetic tissues. *Plant Cell Physiol.* **30**: 381.

———. 1989b. CuZn-superoxide dismutases from the fern *Equisetum arvense* and the green alga *Spirogyra* sp.: Occurrence of chloroplast and cytosol types of enzyme. *Plant Cell Physiol.* **30**: 717.

———. 1990. Characteristic amino acid sequences of chloroplast and cytosol isozymes of CuZn-superoxide dismutase in spinach, rice and horsetail. *Plant Cell Physiol.* **31:** 99.

Kitagawa, Y., S. Tsunasawa, N. Tanaka, Y. Katsube, F. Sakiyama, and K. Asada. 1986. Amino acid sequence of copper, zinc-superoxide dismutase from spinach leaves. *J. Biochem.* **99:** 1289.

Kitagawa, Y., N. Tanaka, Y. Hata, M. Kusunoki, G.-P. Lee, Y. Katsube, K. Asada, S. Aibara, and Y. Morita. 1991. Three-dimensional structure of CuZn-superoxide dismutase from spinach at 2.0 Å resolution. *J. Biochem.* **109:** 477.

Krause, G.H. and G. Cornic. 1987. CO_2 and O_2 interactions in photoinhibition. In *Photoinhibition* (ed. D.J. Kyle et al.), p. 169. Elsevier, Amsterdam.

Kwiatowski, J. and Z. Kaniuga. 1986. Isolation and characterization of cytosolic and chloroplast isozymes of CuZn-superoxide dismutase from tomato leaves and their relationship to other CuZn-superoxide dismutase. *Biochim. Biophys. Acta* **874:** 99.

Kyle, D.J., ed. 1987. The biochemical basis for photoinhibition of photosystem II. In *Photoinhibition*, p. 197. Elsevier, Amsterdam.

Mano, J., M. Takahashi, and K. Asada. 1987. Oxygen evolution from hydrogen peroxide in photosystem II: Flash-induced catalatic activity of water-oxidizing photosystem II membranes. *Biochemistry* **26:** 2495.

Miyake, C., F. Michihata, and K. Asada. 1991. Scavenging of hydrogen peroxide in prokaryotic and eukaryotic algae: Acquisition of ascorbate peroxidase during the evolution of cyanobacteria. *Plant Cell Physiol.* **32:** 33.

Nakano, Y. and K. Asada. 1980. Spinach chloroplasts scavenge hydrogen peroxide on illumination. *Plant Cell Physiol.* **21:** 1295.

Neubauer, C. and U. Schreiber. 1989. Photochemical and non-photochemical quenching of chlorophyll fluorescence induced by hydrogen peroxide. *Z. Naturforsch.* **44C:** 262.

Oh-oka, H., Y. Takahashi, and H. Matsubara. 1989. Topological consideration of a 9-kDa polypeptide which contains centers A and B, associated with the 14- and 19-kDa polypeptides in the photosystem I complex of spinach. *Plant Cell Physiol.* **30:** 869.

Okada, S., S. Kanematsu, and K. Asada. 1979. Intracellular distribution of manganic and ferric superoxide dismutases in blue-green algae. *FEBS Lett.* **103:** 106.

Rotilio, G., R.C. Bray, and E.M. Fielden. 1972. A pulse radiolysis study of superoxide dismutase. *Biochim. Biophys. Acta* **268:** 605.

Schreiber, U. and C. Neubauer. 1990. O_2-dependent electron flow, membrane energization and the mechanism of non-photochemical quenching of chlorophyll fluorescence. *Photosynth. Res.* **25:** 279.

Schreiber, U., H. Reising, and C. Neubauer. 1991. Contrasting pH-optima of light-driven O_2- and H_2O_2-reduction in spinach chloroplasts as measured via chlorophyll fluorescence quenching. *Z. Naturforsch.* **46C:** 635.

Takahashi, M. and K. Asada. 1982. Dependence of oxygen affinity for Mehler reaction on photochemical activity of chloroplast thylakoids. *Plant Cell Physiol.* **23:** 1457.

──────. 1983. Superoxide anion permeability of phospholipid membranes and chloroplast thylakoids. *Arch. Biochem. Biophys.* **226:** 558.

──────. 1988. Superoxide production in aprotic interior of chloroplast thylakoids. *Arch. Biochem. Biophys.* **267:** 714.

Takahashi, M., Y. Kono, and K. Asada. 1980. Reduction of plastocyanin with superoxide and superoxide dismutase-dependent oxidation of plastocyanin by hydrogen peroxide. *Plant Cell Physiol.* **21:** 1431.

Tanaka, K., M. Takahashi, and K. Asada. 1978. Isolation of monomeric cytochrome f from Japanese radish and a mechanism of autoreduction. *J. Biol. Chem.* **253:** 7397.

Takano, M., M. Takahashi, and K. Asada. 1985. Molecular orientation of plastocyanin on spinach thylakoid membranes as determined by acetylation of lysine residues. *J. Biochem.* **98:** 1333.

Iron and Manganese Superoxide Dismutases: Catalytic Inferences from the Structures

W.C. Stallings,[1,3] **C. Bull,**[1,4] **J.A. Fee,**[2]
M.S. Lah,[1] **and M.L. Ludwig**[1]

[1]*Biophysics Research Division and Department of Biological Chemistry
University of Michigan, Ann Arbor, Michigan 48109*
[2]*Isotope and Nuclear Chemistry Division, Los Alamos National Laboratory
Los Alamos, New Mexico 87545*

Superoxide dismutases (SODs) are metalloproteins that catalyze the conversion of superoxide anions to peroxide and molecular oxygen. SODs have been classified into three groups according to the metal ions bound at the active site: Cu (bridged to Zn), Fe, or Mn. However, the three-dimensional structures and sequences divide the known SODs into just two structural families. The CuZn enzymes, represented by the SOD from bovine erythrocytes (Tainer et al. 1982) are dimeric molecules whose three-dimensional motif is a β-barrel. On the other hand, the Mn and Fe enzymes from mitochondria, bacteria, and plants are all close relatives (Barra et al. 1987; Chan et al. 1990) with two-domain structures that incorporate high contents of α-helix. The Mn and Fe SODs occur as dimers that sometimes aggregate to tetrameric species.

In all the enzymes, whether they contain CuZn, Fe, or Mn, superoxide dismutation is thought to proceed in two steps,

Present addresses: [3]Monsanto, Mail Zone BB4K, 700 Chesterfield Village Parkway, St. Louis, Missouri 63198; [4]Washington Research Center, W.R. Grace and Co., 7379 Route 32, Columbia, Maryland 21044.

with cyclic reduction and reoxidation of the metal ion (Lavelle et al. 1977; McAdam et al. 1977; Bull and Fee 1985; Bull et al. 1991). Our discussion here focuses on elements of the structures of Fe and Mn SODs that appear to be important in the catalysis of the dismutation reaction.

THE STRUCTURES OF FESODS AND MNSODS ARE HIGHLY CONSERVED

We have determined the crystal structures of MnSOD from *Thermus thermophilus* (Stallings et al. 1984, 1985; Ludwig et al. 1991) and of FeSOD from *Escherichia coli* (Stallings et al. 1983; Carlioz et al. 1988). Table 1 summarizes these and other structural analyses of Mn and Fe SODs. Figure 1 demonstrates the conservation of the three-dimensional structures for the enzymes studied in our laboratory; the folding patterns shown in Figure 1 recur in the SODs from *Pseudomonas ovalis* (Stoddard et al. 1990a) and *Bacillus stearothermophilus* (Parker and Blake 1988), and in recombinant human mitochondrial MnSOD (Wagner et al. 1989; U. Wagner and J. Sussman, pers. comm.). Strong sequence homologies among some 15 species further suggest a common ancestor for the FeSOD and MnSOD families (Harris et al. 1980; Chan et al. 1990). Sequence divergence is most pronounced in the residues following the first helix, in a region that is an important determinant in defining the quaternary structures of FeSODs and MnSODs.

In all the known Fe and Mn dismutase structures, the helices at the carboxyl terminus of the chain are separated from one another by a short extended region that lies inside the domain connector. This topological feature suggests that the domains can come together only at a late stage in folding (Ludwig et al. 1986) and that the mature metal-binding site, part of the domain interface, is formed only when the carboxyl terminus is properly placed with respect to the domain connector. Like linker sequences that are known to be flexible in other proteins (Texter et al. 1988), the domain connectors in SODs are rich in alanine and glycine residues. Curiously, the domain connector sequences are not rigorously conserved, and

TABLE 1 CURRENT STATUS OF CRYSTAL STRUCTURES OF MNSODS AND FESODS

Species	Source	Quaternary structure	Resolution (Å)	Crystallographic residual	References
Mn(III)	T. thermophilus	tetramer	1.8	0.176	Stallings et al. (1985); Ludwig et al. (1991)
Mn(II)	T. thermophilus	tetramer	2.3	0.173	Ludwig et al. (1991)
Mn(III)	B. stearothermophilus	dimer	2.4	0.26	Parker and Blake (1988)
Mn	human recombinant	tetramer	3.2		Wagner et al. (1989); U. Wagner and J. Sussman (pers. comm.)
Fe(III)	E. coli	dimer	1.9	0.196	Stallings et al. (1983); Carlioz et al. (1988); M.S. Lah (unpubl.)
Fe(III)-azide	E. coli	dimer	1.8	0.179	Stallings et al. (1983, 1991)
Fe(II)	E. coli	dimer	1.8	0.187	M.S. Lah (unpubl.)
Fe(III)	P. ovalis	dimer	2.1	0.24	Ringe et al. (1983); Stoddard et al. (1990b)
Fe(III)-azide	P. ovalis	dimer	2.9	0.23	Stoddard et al. (1990a)

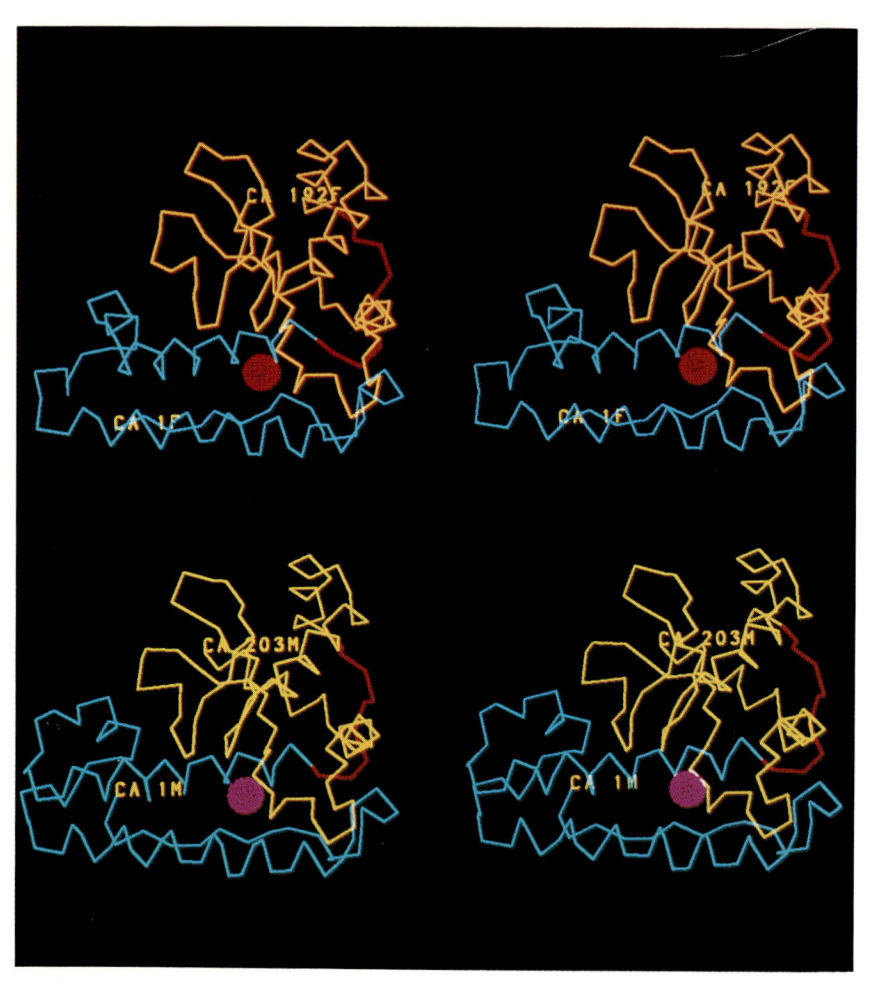

FIGURE 1 Stereo views of the polypeptide folds of FeSOD (*above*) and MnSOD (*below*). Distance plots that display the neighbors of C_α atoms clearly reveal two domains, the first comprising residues 1–90 in MnSOD, and the second, residues 100–203. The domains are differentiated by color coding, blue for domain I and maize for domain II. The domain connector is red. Although the chains begin with an extended strand, two antiparallel helices crossing at an angle of 38° constitute the principal motif of domain I. The linker connecting the two major helices of domain I differs in length in *T. thermophilus* MnSOD and *E. coli* FeSOD. Domain II is a mixed α+β structure incorporating a three-stranded β-sheet with the somewhat unusual topology: −1, 2x (Richardson and Richardson 1989). The metal ions (stippled spheres) are bound by four residues, two from each of the domains.

the conformations of the linker residues are not identical in the known SOD structures (Parker and Blake 1988; Ludwig et al. 1991).

FOUR PROTEIN LIGANDS AND A SOLVENT MOLECULE COORDINATE THE METAL IONS IN E. COLI FE(III)SOD AND T. THERMOPHILUS MN(III)SOD

The electron density maps show that four protein ligands coordinate the active-site metal ions. In *T. thermophilus* MnSOD, the ligands from the amino-terminal domain are His-28 and His-83, whose imidazole side chains are contributed by two long crossing helices (Fig. 1). The Asp-166 and His-170 ligands are from the carboxy-terminal domain; position 166 is at the carboxyl terminus of the β-sheet, and residue 170 is part of a β-turn bordering the sheet. Identical ligands (His-26, His-70, Asp-156, and His-160 in *E. coli* FeSOD) from equivalent regions of secondary structure bind the metal ions in all the known structures. The imidazole ligands all use their NE2 nitrogens for metal coordination; the carboxyl ligand is monodentate and the geometry suggests that the Me-O bond utilizes an oxygen *syn* lone pair of electrons for coordination.

Electron density maps of *E. coli* FeSOD and *T. thermophilus* MnSOD also demonstrate clearly that solvent is a fifth metal ligand. The crystallographic evidence for solvent ligation in the structure of *T. thermophilus* MnSOD is displayed in Figure 2. The five ligands are at the vertices of a trigonal bipyramid (Fig. 3A); solvent and the imidazole side chain from the first crossing helix of the amino-terminal domain are the axial ligands. We propose that the solvent ligand is hydroxide, rather than water, in the oxidized structures. Assignment of the solvent ligand as hydroxide is consistent with the acid-base chemistry of these transition metals and compensates the positive charge on the buried Me(III) ions. The carboxylate ligand is presumed to be negatively charged, but the imidazole ligands are expected to be uncharged. Complexation with OH^- would reduce the formal charge of the Me(III) ligand cluster to +1. According to our interpretation of the structures of reduced SODs, the formal charge of the Me(II) centers would also be +1.

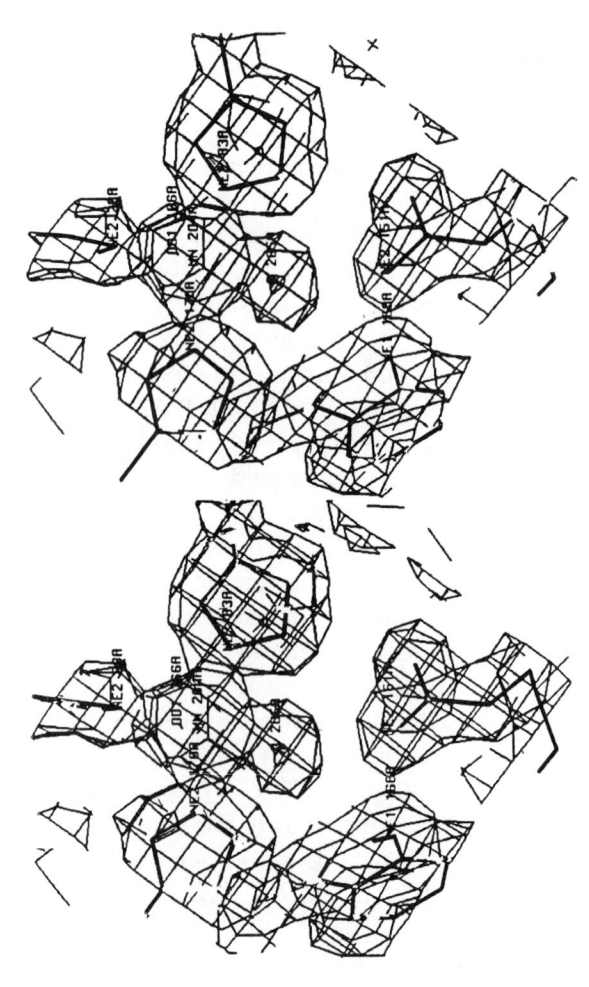

FIGURE 2 Electron density of Mn(III)SOD in a $(2|F_o|-|F_c|)$ map calculated after refinements from which the solvent ligand was omitted. The density at the center of the view corresponds to the position of solvent 205. Corresponding maps with amplitudes $(|F_o|-|F_c|)$ display a positive peak corresponding to solvent at a level of 10σ.

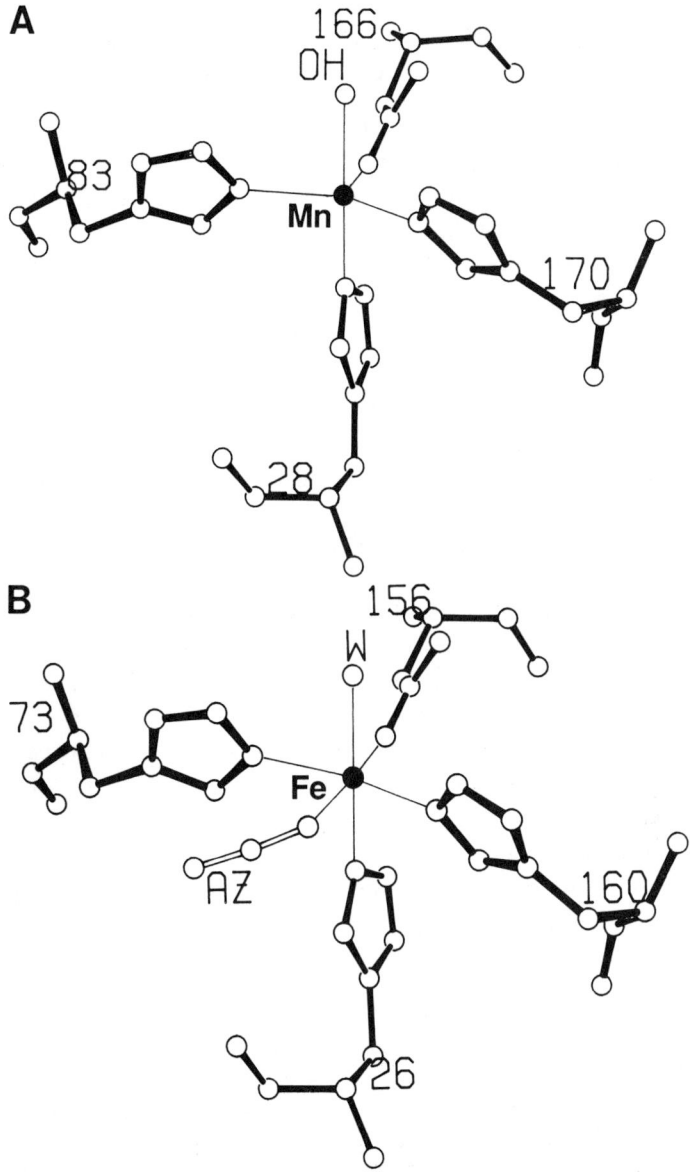

FIGURE 3 (A) Geometry of metal ligation in *T. thermophilus* MnSOD. The trigonal plane of the bipyramid is almost perpendicular to the page in this view, which is along the path by which substrate is thought to approach the metal center. (B) Metal ligation in the azide complex of *E. coli* FeSOD. The imidazole rings of His-83 and His-170 in A are oriented to allow coordination of sixth ligand; binding of azide (B) is accompanied by an increase of 20° in the NE2-Fe-NE2 bond angle involving His-73 and His-170.

THE STRUCTURES OF REDUCED SODS SUGGEST THAT THE SOLVENT LIGAND PLAYS A KEY ROLE IN CATALYSIS

Bull and Fee (1985) have demonstrated that reduction of Fe(III)SOD is accompanied by the uptake of one proton. By analogy with results from experiments on CuZnSOD (in which Co was substituted for Zn [Bertini et al. 1985]), we expected that the redox-linked proton acceptor would be one of the protein ligands and that the protonation would prompt dissociation of this ligand from the coordination sphere of the metal. To study the effects of reduction, we analyzed crystals that had been reduced in situ with dithionite. Upon addition of reductant, crystals of *T. thermophilus* MnSOD lose their color within minutes, do not disorder, and remain colorless for at least several days over the course of a diffraction experiment. Contrary to our expectations, the difference map comparing the oxidized and reduced enzymes was nearly featureless and gave no indication for displacement of any of the protein ligands. One simple interpretation of our results is that the solvent ligand, presumed to be hydroxide in the oxidized enzymes, is the group which is protonated on reduction of the metal center. Indeed, refinement of the Mn(II)SOD structure at 2.4 Å resolution and comparisons with the structure of the oxidized enzyme suggest a slight movement of the solvent ligand. Studies of the crystal structure of reduced FeSOD are in progress, with results that parallel those for MnSOD (Table 1); none of the ligands is lost upon reduction of crystals of FeSOD.

BINDING OF ANION LIGANDS INCREASES THE COORDINATION NUMBER FROM FIVE TO SIX

Another approach to understanding the structural chemistry of the dismutase reaction has exploited the binding of anionic ligands to the active site. The uncomplexed structures, in which the imidazole planes of the two equatorial histidine ligands are nearly parallel, may represent a transition-state-like geometry; although the N-Me(III)-N bond angle is distorted from ideality by only about 10° (from 120° to 130°), the angle between lines perpendicular to the imidazole rings is 165°.

This means that the coordination complex has an open face, which could facilitate the binding of substrates and inhibitors. Accordingly, we investigated the binding of azide to *E. coli* FeSOD by soaking the competitive inhibitor into the crystals. Binding occurs at the open face of the ligand cluster without displacing the solvent ligand, and increases the coordination number of the Fe(III) complex from five to six. Inhibitor binding is attended by movements of the equatorial imidazole ligands as they adopt a distorted octahedral geometry with the N-Me(III)-N bond angle increasing to 150° (Figs. 3B, 4).

ANION LIGANDS BIND IN A SOLVENT-INACCESSIBLE CAVITY

In the native enzymes, the imidazole ligands that form the open face of the coordination complex pack imperfectly against two other conserved histidine residues at positions 30 and 31 in *E. coli* FeSOD. The imperfection in the packing creates a small cavity (Fig. 4A) which difference Fourier ($|F_o|-|F_c|$) electron density maps show to be completely void of solvent. The structure of the Fe(III)SOD complex with azide shows the ligand bound in this cavity with coordination to the metal at a distance of about 2.3 Å (Fig. 4B). The walls of this binding pocket render its contents inaccessible to solvent with conserved residues His-26, Trp-77, and Tyr-34 further enclosing the cavity. Maintenance of a vacant, solvent-inaccessible cavity in the resting enzymes may confer catalytic advantage by avoiding competitive binding by solvent molecules at the open face of the metal complex. The binding pocket is located at the bottom of a deep funnel, some 15 Å from the surface of the enzyme. Conserved features of the quaternary structures of Fe- and Mn-dependent SODs create this channel.

THE QUATERNARY STRUCTURES OF FESODS AND MNSODS CONSERVE A DIMER INTERFACE

Fe- and Mn-dependent SODs are dimers and tetramers of identical subunits. Although human MnSOD and the enzyme from *T. thermophilus* form tetrameric structures in different ways (U. Wagner and J. Sussman, pers. comm.), they have a conserved dimer interface which is also present in the remaining three homodimeric structures (Table 1). The common interface separates the active-site metals by 18 Å and further stabilizes the environments of the metal ligand clusters by interdigitating surrounding side chains from one subunit into the active site of the other; this interface has been described previously (Stallings et al. 1985; Ludwig et al. 1991). The interface also forms the channel (Fig. 5) down which superoxide radical anions must migrate to the active site; the channel is filled with solvent molecules that interact with each of the two subunits involved in the interface. Residues forming the walls of the funnel are often conserved, but at a number of positions, interesting substitutions are observed (Stallings et al. 1991). Looking down this funnel along the putative path that substrates follow, the last features that come into view are the conserved tyrosine and histidine rings which block the entrance to the substrate-binding cavity (Y34 and H30 in *E. coli* FeSOD or Y36 and H32 in *T. thermophilus* MnSOD). Calculations of Sines et al. (1990) support the notion that these residues must move in a coordinated fashion to admit substrates and inhibitors to the active site. As illustrated in Figure 5, the mouth of the funnel has two lips, one from each of the subunits involved in forming the conserved interface. In all the reported structures, the lips are peppered with lysine and arginine side chains whose positive charges may serve to guide anions into the mouth of the funnel.

FEATURES OF THE CATALYTIC MECHANISM INFERRED FROM THE STRUCTURES

The complex with the competitive inhibitor, N_3^-, is proposed to mimic the inner sphere complexes of $\cdot O_2^-$ with FeSODs or

FIGURE 4 (A) The binding cavity in uncomplexed Fe(III)SOD from *E. coli*. Tyr-34 is included in this drawing to show how it caps the azide-binding site. To convey an impression of the packing, the residues of the protein are outlined with dot surfaces (green). Atoms from His-26 and Trp-122 (not shown) form the bottom surface, which completely encloses the site. (B) The orientation of azide in the Fe(III)SOD complex. The electron density corresponding to bound azide at 2.0 Å resolution is outlined in cyan. Tyr-34 (omitted for clarity) moves only slightly in the structure of the complex.

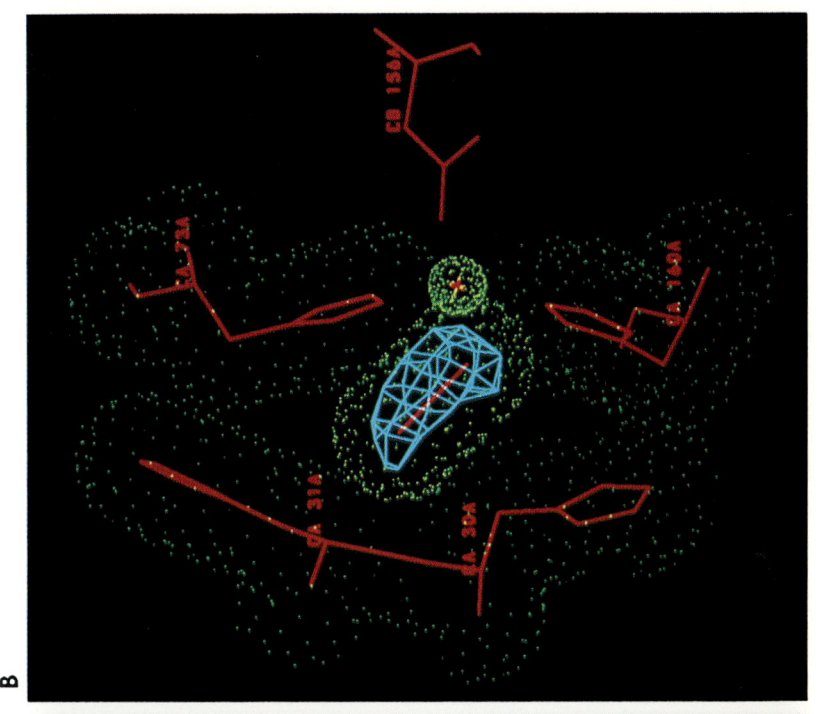

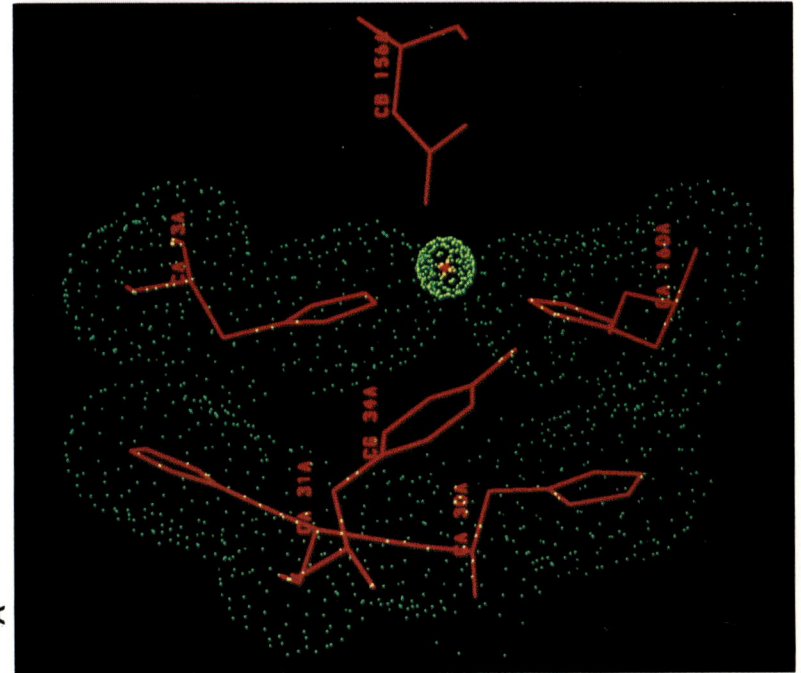

FIGURE 4 *(See facing page for legend.)*

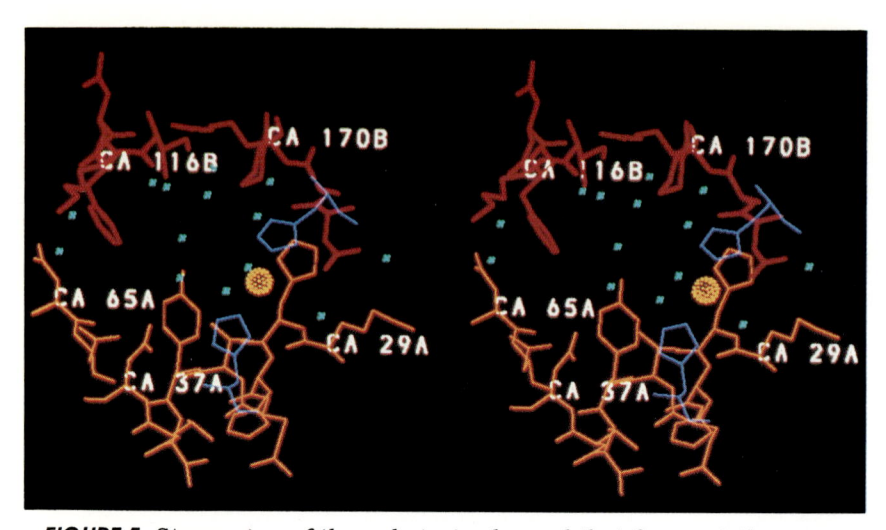

FIGURE 5 Stereo view of the substrate channel that forms at the interface of the A and B chains of *E. coli* FeSOD. The view is along the path from external solvent to the metal ion. The orange-gold residues (*below*) are K29-H30-H31-Q32-T33-Y34-V35-T36-N37 and N65-N66 from the A chain. The red residues (*above*) comprise I115-K116-N117-F118 and N168-A169-R170-P171 from the B chain. The Fe(III) ion of the A chain is stippled yellow and two of its ligands (H73A and H160A) are shown in blue. Crosses in cyan represent crystallographically ordered solvent molecules that fill the funnel. Y34 (near the center of the figure) appears to impede access to the active-site cavity. The two lips of the funnel contribute two positively charged residues (K29A and K116B), whose side chains extend into the medium; the guanidium moiety of R170B is more buried and is stabilized by a number of intramolecular hydrogen bonds to the protein (Borders et al. 1989; Chan et al. 1990; Ludwig et al. 1991).

SCHEME I Reaction scheme for FeSOD. The pentacoordinate Fe(III) (middle left) forms an inner-sphere complex with superoxide (upper right) resembling the six-coordinate complex with azide. Reduction to Fe(II) is coupled to proton uptake, with the OH$^-$ ligand serving as proton acceptor, and oxygen is released. Steps in which H$^+$ and e$^-$ are added have been combined for convenience. In the second half-reaction, the reduced enzyme (middle right) again forms a six-coordinate inner-sphere complex with substrate. The peroxo anion must be protonated to complete the reaction cycle; proton transfer from a general acid is rate-determining, as measured by solvent isotope effects on k_{cat} (Fee et al. 1986). The species that account for the high-pH behavior are shown in the upper left and lower right diagrams; the protein base, BH, whose titration is affected by the charge on the metal ion, is suggested to be Tyr-34 in FeSOD.

MnSODs. Thus, we envision reaction scheme I for FeSOD in which the coordination number of the metal increases to six on transient ligation of substrate to both the Fe(III) and the Fe(II) enzyme. The coordination geometry of the resting enzyme is slightly distorted in a direction that facilitates expansion of

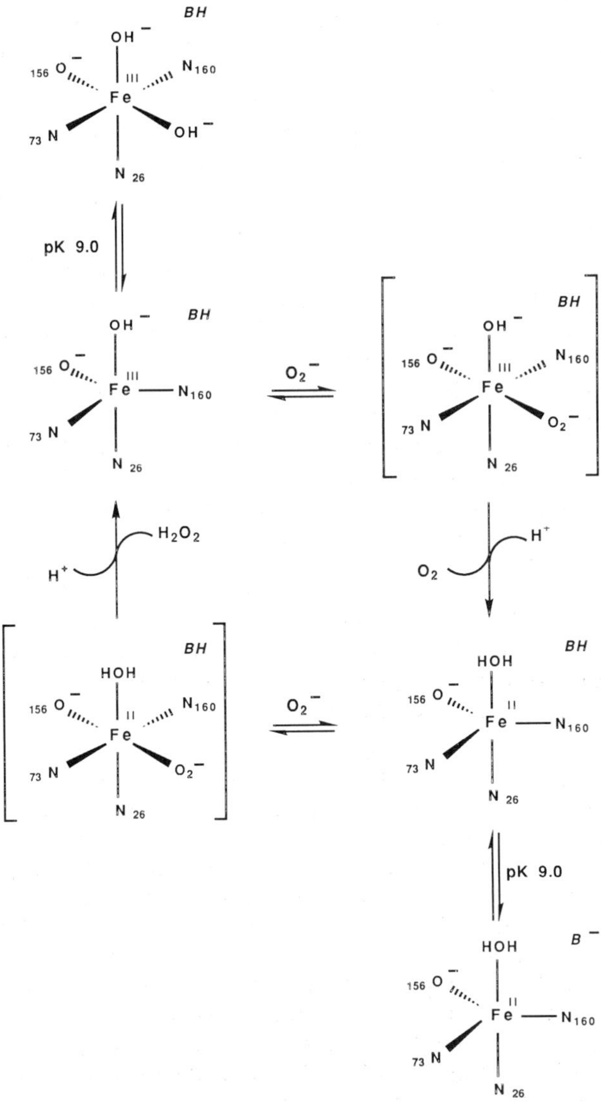

SCHEME I (See facing page for legend.)

the coordination sphere, but at the same time, the rings of His-73 and His-160 hinder formation of ideal six-coordinate geometry in the postulated intermediates. Substrate binding could be favored by the presence of a preexisting cavity between His-73 and the backbone of His-31, as shown in Figure 4. Displacement of the peroxo anion in the second half-reaction may be facilitated by the negatively charged carboxylate oxygen occupying a position *trans* to the product.

Retention of the solvent ligand in the substrate complexes is a key feature in our current mechanism (Stallings et al. 1991); from studies of the reduced enzyme, we postulate that this ligand accepts the proton that is bound on reduction of the Fe(III) to the Fe(II) species (see above). To account for the redox titration behavior and for the pH dependence of the kinetic parameters (Bull and Fee, 1985), the pK of ligated H_2O must be less than 6 in oxidized FeSOD and greater than 11 in reduced FeSOD (Niederhoffer et al. 1987). Scheme I has been extended to include the observed ionizations of Fe(III) and Fe(II)SOD near pH 9.0; the former is attributed to ligation of a second OH^- and the latter to ionization of Tyr-34. The network of interactions between the solvent ligand, Tyr-34, and Gln-69 provides a means by which the charge on the metal atom may affect the ionization of this tyrosine residue (Fig. 6).

SODs are interesting models for the thermodynamic study of redox-linked protonation reactions. Scheme I postulates that large changes in the pK values of ligated solvent and of Tyr-34 are induced by reduction of the metal ion. There may also be a kinetic linkage between reduction and protonation; however, it is known that the proton donor is not H_3O^+, but a general acid, probably water (Bull and Fee 1985). Structural and kinetic analyses of tyrosine mutants will test the proposed interaction between Tyr-34 and Fe; diffraction measurements at high pH may provide further tests of our model for e^-/H^+ linkage.

THE SPECIFICITY OF METAL BINDING

The similarity of the active sites of Fe- and Mn-dependent SODs extends well beyond their first coordination spheres (Carlioz et al. 1988). With few exceptions, there exists a 1:1

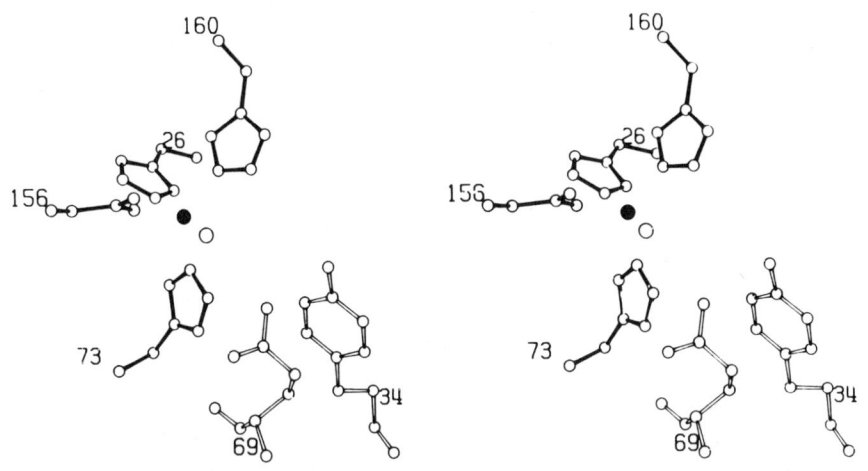

FIGURE 6 Stereo view of the metal-ligand environment of Fe(III)SOD from *E. coli*, showing the network of interactions between the metal (filled circle), the solvent ligand (unfilled circle), Gln-69, and Tyr-34. The protein ligands to Fe are shown with filled bonds, omitting backbone amide atoms. Hydrogen bonds connect the Tyr-34 hydroxyl group and the amide N of Gln-69, which in turn forms a hydrogen bond to the solvent ligand. The side chain O of Gln-69 interacts with the indole NH of Trp-122 (not shown).

correspondence in the chemical and three-dimensional structures of the local environments of the active-site clusters. The near identity of the active sites highlights a dilemma in the structural biology of MnSODs and FeSODs: How is the observed specificity of metal ion binding (Fee et al. 1976; Ose and Fridovich 1979; Whittaker and Whittaker 1991) encoded in the amino acid sequences of these enzymes? One distinction between the metal-ligand environments of Mn and Fe enzymes is an interchange of Gln-69 in FeSOD for Gln-151 in MnSOD; another is an exchange of tyrosine for phenylalanine at position 76 of *E. coli* SOD, but the three-dimensional structures fail to explain how these substitutions provide a chemical basis for specificity. Still another possibility is that initial binding (selection) of the metals occurs at alternate sites, possibly in partly folded structures. Nonspecific, more promiscuous dismutases that bind both Fe and Mn ions have been identified (Gregory 1985), and comparisons with the more selective enzymes are expected to be helpful in addressing this problem. However, the sequence of one of these enzymes, from *Bacteroides gingivalis* (Nakayama 1990), includes elements characteristic of both Mn and Fe enzymes but fails to provide an obvious explanation for metal selectivity.

FINAL COMMENTS

The structures of FeSOD from *P. ovalis* (Ringe et al. 1983; Stoddard et al. 1990b) and MnSOD from *B. stearothermophilus* (Parker and Blake 1988) have been determined (Table 1). Although the protein residues that serve as metal ligands are identical in these structures and in the structures we have described, a solvent ligand, which we believe plays a role in the chemistry and catalytic cycles of these enzymes, appears to be absent. The ligation geometry in MnSOD from *B. stearothermophilus* is very similar to that in *T. thermophilus*, but a distinct peak indicating a solvent ligand is not observed in maps at 2.4 Å resolution. The reported coordination geometry for *P. ovalis* FeSOD is distorted tetrahedrally with a very long bond (2.8 Å) to "axial" histidine. In the structure proposed for the azide complex of this FeSOD, determined at a resolution of 2.9 Å (Stoddard et al. 1990a), azide occupies the position we ascribe to solvent. We have confirmed the ligation geometry in our structures by an extensive series of crystallographic refinements, with data extending to 1.8 Å for the MnSOD and to 1.9 Å for the FeSOD, and we have also investigated crystallographically the stoichiometry (occupancy) of the binding of both the metal and the solvent ligand to the protein. In *E. coli* FeSOD, as well as in *T. thermophilus* MnSOD, the occupancies are very close to unity. Metal binding is suggested to be less than stoichiometric in the crystal structures of the proteins from *B. stearothermophilus* (Parker and Blake 1988) and *P. ovalis* (Ringe et al. 1983). As a result, electron density maps of these enzymes would present mixed images of the active sites with contributions from both the *apo* and *holo* forms of the enzymes.

ACKNOWLEDGMENTS

This research was supported by grants from the National Institutes of Health (GM-16429) to M.L.L. and (GM-35189) to

J.A.F. We thank Anita Metzger and Katherine Pattridge for sustained contributions to the structure elucidations. We are grateful to Dennis Riley of Monsanto Company for pointing out the importance of the *trans* effect in the catalytic mechanism.

REFERENCES

Barra, D., M.E. Schinina, W.H. Bannister, J.V. Bannister, and F. Bossa. 1987. The primary structure of iron-superoxide dismutase from *Photobacterium leiognathi*. *J. Biol. Chem.* **262:** 1001.

Bertini, I., C. Luchinat, and R. Monnanni. 1985. Evidence of the breaking of the copper-imidazolate bridge in copper/cobalt substituted superoxide dismutase upon reduction of the copper(II) centers. *J. Amer. Chem. Soc.* **107:** 2178.

Borders, C.L., P.J. Horton, and W.F. Beyer, Jr. 1989. Chemical modification of iron and manganese-containing superoxide dismutases from *Escherichia coli*. *Arch. Biochem. Biophys.* **268:** 74.

Bull, C. and J.A. Fee. 1985. Steady-state kinetic studies of superoxide dismutases: Properties of the iron-containing protein from *Escherichia coli*. *J. Amer. Chem. Soc.* **107:** 3295.

Bull, C., E.C. Neiderhoffer, T. Yoshida, and J.A. Fee. 1991. Kinetic studies of superoxide dismutases: Properties of the manganese-containing protein from *Thermus thermophilus*. *J. Amer. Chem. Soc.* **113:** 4069.

Carlioz, A., M.L. Ludwig, W.C. Stallings, J.A. Fee, H.M. Steinman, and D. Touati. 1988. Iron superoxide dismutase: Nucleotide sequence of the gene from *Escherichia coli* K12 and correlation with crystal structures. *J. Biol. Chem.* **263:** 1555.

Chan, V.W.F., M.J. Bjerrum, and C.L. Borders, Jr. 1990. Evidence that the chemical modification of a positively charged residue at position 189 causes the loss of catalytic activity of iron-containing superoxide dismutases. *Arch. Biochem. Biophys.* **279:** 195.

Fee, J.A., E.R. Shapiro, and T.H. Moss. 1976. Direct evidence for manganese(III) binding to the mangano-superoxide dismutase of *Escherichia coli* B. *J. Biol. Chem.* **251:** 6157.

Fee, J.A., T. Yoshida, C. Bull, P. O'Neill, and E.N. Fielden. 1986. On the mechanism of the iron-containing superoxide dismutase from *E. coli*. In *Superoxide and superoxide dismutases in chemistry, biology and medicine* (ed. G. Rotillo), p. 205. Elsevier, Amsterdam.

Gregory, E.M. 1985. Characterization of the O_2^- induced manganese-containing superoxide dismutase from *Bacteroides fragilis*. *Arch. Biochem. Biophys.* **238:** 83.

Harris, J.I., A.D. Auffret, F.B. Northup, and J.E. Walker. 1980. Structural comparisons of superoxide dismutases. *Eur. J. Biochem.* **106:** 297.

Lavelle, F., M.E. McAdam, E.M. Fielden, and P.B. Roberts. 1977. A pulse-radiolysis study of the catalytic mechanism of the iron-containing superoxide dismutase from *Photobacterium leiognathi*. *Biochem. J.* **161:** 3.

Ludwig, M.L., K.A. Pattridge, and W.C. Stallings. 1986. Manganese superoxide dismutases. In *Manganese in metabolism and enzyme function* (ed. V.L. Schramm and F.C. Wedler), p. 405, Academic Press, Orlando, Florida.

Ludwig, M.L., A.L. Metzger, K.A. Pattridge, and W.C. Stallings. 1991. Manganese superoxide dismutase from *Thermus thermophilus*. A structural model refined at 1.8 Å resolution. *J. Mol. Biol.* **219:** 335.

McAdam, M.E., R.A. Fox, F. Lavelle, and E.M. Fielden. 1977. A pulse-radiolysis study of the manganese-containing superoxide dismutase from *Bacillus stearothermophilus*. A kinetic model of the enzyme action. *Biochem. J.* **165:** 71.

Nakayama, K. 1990. The superoxide dismutase-encoding gene of the obligately anaerobic bacterium *Bacteroides gingivalis*. *Gene* **96:** 149.

Niederhoffer, E.C., J.A. Fee, V. Papaefthymiou, and E. Munck. 1987. Magnetic resonance studies involving iron superoxide dismutase from *Escherichia coli*. *Isotope and nuclear chemistry division*. Annual Report, Los Alamos National Laboratory, p. 79.

Ose, D.E. and I. Fridovich. 1979. Manganese-containing superoxide dismutase from *Escherichia coli*: Reversible resolution and metal replacements. *Arch. Biochem. Biophys.* **194:** 360.

Parker, M.W. and C.F. Blake. 1988. Crystal structure of manganese superoxide dismutase from *Bacillus stearothermophilus* at 2.4 Å resolution. *J. Mol. Biol.* **199:** 649.

Richardson, J.S. and D.C. Richardson. 1989. Principles and patterns of protein conformation. In *Prediction of protein structure and principles of protein conformation* (ed. G.D. Fasman), p. 1. Plenum Press, New York.

Ringe, D., G.A. Petsko, F. Yamakura, K. Suzucki, and D. Ohmori. 1983. Structure of iron superoxide dismutase from *Pseudomonas ovalis* at 2.8 Å resolution. *Proc. Natl. Acad. Sci.* **80:** 3879.

Sines, J., S. Allison, A. Wierzbicki, and J.A. McCammon. 1990. Brownian dynamics simulation of the superoxide-superoxide dismutase reaction: Iron and manganese enzymes. *J. Phys. Chem.* **94:** 959.

Stallings, W.C., K.A. Pattridge, R.K. Strong, and M.L. Ludwig. 1984. Manganese and iron superoxide dismutases are structural homologs. *J. Biol. Chem.* **259:** 10695.

———. 1985. The structure of manganese superoxide dismutase from *Thermus thermophilus* at 2.4 Å resolution. *J. Biol. Chem.* **260:** 16424.

Stallings, W.C., A.L. Metzger, K.A. Pattridge, J.A. Fee, and M.L. Ludwig. 1991. Structure-function relationships in iron and manganese superoxide dismutases. *Free Radicals Res. Commun.* **12-13:** 259.

Stallings, W.C., T.B. Powers, K.A. Pattridge, J.A. Fee, and M.L. Ludwig. 1983. Iron superoxide dismutase from *Escherichia coli* at 3.1 Å resolution: A structure unlike that of the copper/zinc protein at both the monomer and dimer levels. *Proc. Natl. Acad. Sci.* **80:** 3884.

Stoddard, B.L., D. Ringe, and G.A. Petsko. 1990a. The structure of iron superoxide dismutase from *Pseudomonas ovalis* complexed with the inhibitor azide. *Protein Eng.* **4:** 113.

Stoddard, B.L., P.L. Howell, D. Ringe, and G.A. Petsko. 1990b. The 2.1 Å structure of iron superoxide dismutase from *Pseudomonas ovalis*. *Biochemistry* **29:** 8885.

Tainer, J.A., E.D. Getzoff, K.A. Beem, J.S. Richardson, and D.C. Richardson. 1982. Determination and analysis of the 2 Å structure of copper, zinc superoxide dismutase. *J. Mol. Biol.* **160:** 181.

Texter, F.L., S.E. Radfords, E.D. Laue, R.N. Perham, S.J. Miles, and J.R. Guest. 1988. Site directed mutagenesis and ^{1}H NMR spectroscopy of an interdomain segment in the pyruvate dehydrogenase multi-enzyme complex of *Escherichia coli*. *Biochemistry* **27:** 289.

Wagner, U., M.M. Werber, Y. Beck, J.R. Hartman, F. Frolow, and J. Sussman. 1989. Characterization of crystals of genetically engineered human manganese superoxide dismutase. *J. Mol. Biol.* **206:** 787.

Whittaker, J.A. and M.M. Whittaker. 1991. Active site spectral studies on manganese superoxide dismutase. *J. Am. Chem. Soc.* **113:** 5528.

Superoxide Radical in Escherichia coli

S.I. Liochev and I. Fridovich
Department of Biochemistry
Duke University Medical Center
Durham, North Carolina 27710

E. coli can grow aerobically or anaerobically. The former condition allows greater efficiency in the extraction of useful energy from foodstuffs, but it necessitates self-protection against dangerous entities which can be, and are, produced from O_2. These are $\cdot O_2^-$ and H_2O_2, and they are eliminated by superoxide dismutases (SODs) and by hydroperoxidases (HPs), respectively. E. coli contains two of each of these enzymes. There are thus two SODs, one containing iron (FeSOD) (Yost and Fridovich 1973) and the other manganese (MnSOD) (Keele et al. 1970). There are also two hydroperoxidases, one of which (HPI) is active both as a catalase and as a peroxidase (Claiborne and Fridovich 1979), whereas the other (HPII) acts only as a catalase (Claiborne et al. 1979).

This duality and apparent redundancy has merit in that it allows regulation while maintaining a standby defense. This is particularly important in a facultative organism such as E. coli, which might otherwise perish, upon sudden exposure to O_2, before it had time to induce these defensive enzymes. In the case of the SODs, it is the FeSOD that provides the standby defense and the MnSOD that provides for regulation. FeSOD is thus produced under both aerobic and anaerobic conditions; MnSOD is not made anaerobically but can be induced to high levels under oxidative stress. Among the hydroperoxidases, it is HPI that appears constitutive and HPII that is inducible (Hassan and Fridovich 1978).

THE HABER-WEISS REACTION

The SODs and the hydroperoxidases constitute a defensive team that prevents $\cdot O_2^-$ and H_2O_2 from producing the terribly reactive hydroxyl radical ($\cdot OH$). Thus, $\cdot O_2^-$ can reduce Fe(III) to Fe(II) or Cu(II) to Cu(I), and these in turn can reduce H_2O_2 to $OH^- + \cdot OH$. In this process, which has been named the Haber-Weiss reaction (Beauchamp and Fridovich 1970), the metal acts catalytically and, if bound to DNA or membranes, causes selective damage to these critical targets (Czapski 1984; Goldstein and Czapski 1986). SODs inhibit this process by lowering the concentration of $\cdot O_2^-$, and hydroperoxidases do the same vis à vis H_2O_2.

There are many reasons for suspecting that the metal-catalyzed Haber-Weiss reaction can occur in vivo. Among these are the following: (1) Illuminated chloroplast lamellae converted methional to ethylene, and this conversion, which can be caused by $\cdot OH$, was inhibited by SOD, catalase, or o-phenanthroline (Elstner and Konz 1974). The inhibition by SOD exposes the role of $\cdot O_2^-$, whereas the inhibition by catalase does the same for H_2O_2, and the inhibition by o-phenanthroline indicates the need for iron. (2) Alloxan diabetes was blocked by scavengers of $\cdot OH$, such as alcohols or thiourea (Heikkila et al. 1976); alloxan can mediate increased production of $\cdot O_2^-$ and of H_2O_2 by cycles of reduction to dialuric acid followed by autoxidation. (3) The autoxidation of dialuric acid was seen to kill *E. coli*; SOD or catalase protected, whereas Cu(II) and H_2O_2 augmented, this lethality (van Hemmen and Meuling 1977). (4) The hydroxylation of benzene by microsomes plus NADPH, a known source of $\cdot O_2^- + H_2O_2$, was inhibited by SOD, catalase, or by scavengers of $\cdot OH$ (Johansson and Ingelman-Sundberg 1983). (5) The SOD-null strain of *E. coli* was hypersensitive to H_2O_2 although it contained normal levels of catalase (Carlioz and Touati 1986). (6) Iron appeared essential for expression of the toxicity of paraquat (Korbashi et al. 1986). (7) Rat pulmonary endothelial cells could slowly take up SOD from the suspending medium and when so loaded with SOD were rendered resistant to H_2O_2. Control loading of the cells with inactivated SOD did not protect (Markey et al. 1990). (8) A hamster fibroblast cell line, selected for resistance to H_2O_2, exhibited the expected increase in catalase plus a twofold elevation of SOD (Spitz et al. 1990). (9) SOD was delivered into hepatocytes, and this imparted resistance to tertiary butyl hydroperoxide (Nakae et al. 1990). (10) 8-Hydroxyguanine, one of the products of •OH attack on DNA, is present in nuclear and mitochondrial DNA, and its level in the mtDNA was increased fivefold following treatment with alloxan (Richter et al. 1988). Finally, there are several enzymes capable of participating in the repair of hydroxylated DNA (Breimer 1991), as would be expected were •OH a side product of aerobic metabolism.

ENZYME TARGETS FOR •O_2^-

Whereas the mutagenic and lethal effects of •O_2^- are probably due to •OH attack on DNA and cell membranes, the growth-inhibiting effects are due to direct actions of •O_2^- on susceptible enzymes. Several hydroxy acid dehydratases, which contain [4Fe-4S] clusters at their active sites, belong in this category. The first indication of the •O_2^--sensitivity of these enzymes arose from the work of Brown and Seither (1983), who were exploring the ability of branched chain amino acids to permit growth of E. coli under 4.2 atm of O_2. This hyperoxia-imposed auxotrophy was subsequently explained on the basis of the inactivation, by •O_2^-, of the α,β-dihydroxy acid dehydratase, which catalyzes the penultimate step in the biosynthesis of these amino acids (Kuo et al. 1987; Flint and Emptage 1990). There are also probably •O_2^--sensitive steps on the pathways of biosynthesis of aromatic and of sulfur-containing amino acids, since Carlioz and Touati (1986) noted that the SOD-null strain of E. coli exhibited dioxygen-dependent auxotrophies for these classes of amino acids.

The 6-phosphogluconate dehydratase and aconitase are also very sensitive to direct attack by •O_2^- (Gardner and Fridovich 1991a,b). The former of these enzymes is inactivated by •O_2^- with a rate constant of approximately 1×10^8 M^{-1} sec^{-1}, whereas for the E. coli aconitase, the corresponding rate constant is 1×10^9 M^{-1} sec^{-1}. The competitive inhibitor fluorocitrate protected aconitase against •O_2^-, decreasing this rate constant to 1×10^7 M^{-1} sec^{-1} and indicating that attack by $\cdot O_2^-$ was at the active site.

The possibility that the $\cdot O_2^-$ sensitivity of aconitase is an element of good design, rather than of unavoidable happenstance, has been considered (Gardner and Fridovich 1991b). Thus, aconitase could act as a "circuit breaker," which throttles back the supply of reducing equivalents and thus limits the production of $\cdot O_2^-$ until more satisfactory defenses can be put into place by induction of the *soxR* and *oxyR* regulons. Reduction of O_2 to $\cdot O_2^-$ requires reducing equivalents, and in the cells, these come from NAD(P)H. In the face of a sudden surge of $\cdot O_2^-$, inactivation of aconitase would limit the supply of NADH, and via the transhydrogenase, of NADPH also. The $\cdot O_2^-$-sensitive aconitase is viewed as a circuit breaker, rather than as a "fuse," because oxidatively inactivated aconitase can be reactivated by anaerobic exposure to Fe(II) plus thiols (Villafranca and Mildvan 1971), and we have noted reactivation of aconitase in anaerobically incubated extracts of *E. coli* (Gardner and Fridovich 1991a).

There are indications that inactivation of aconitase by $\cdot O_2^-$ can occur in vivo:

1. Enrichment of the growth medium with Mn(II) decreased the excretion of citrate from *Aspergillus niger*, while increasing the activity of aconitase (Kubicek and Rohr 1985). We may suppose that Mn(II) in the medium leads to elevation of MnSOD in *A. niger*, as it is known to do in *E. coli* (Moody and Hassan 1984; Pugh et al. 1984; Pugh and Fridovich 1985), and that elevated MnSOD elevated aconitase by protecting it against $\cdot O_2^-$.
2. Reperfusion injury to myocardium, which seems to be due to increased $\cdot O_2^-$ production upon reperfusion (McCord 1986, 1987), is accompanied by increased aerobic production of lactate (Schwaiger 1989). The aerobic fermentation could reflect inactivation of aconitase by $\cdot O_2^-$ produced during reperfusion.
3. WI-38 cells, exposed to an increase in pO_2 from 134 mm to 291 mm, secreted 4 to 6 times *more* lactate (Balin et al. 1976). This seemingly paradoxical increase in fermentation, in response to increased oxygenation, can be explained on

the basis of more $\cdot O_2^-$ and thus greater inactivation of aconitase at the higher pO_2.

HOW MUCH $\cdot O_2^-$ IN E. COLI?

This question was approached by the time-honored biochemical technique of "grind and find." The SOD-null strain of *E. coli* (Carlioz and Touati 1986) was used, since SOD would have interfered with measurement of $\cdot O_2^-$ in terms of SOD-inhibitable reduction of cytochrome *c*. Cytosolic and membrane fractions were incubated with NAD(P)H and with other cellular reductants, under conditions of pH, pO_2, temperature, and concentration appropriate to the intracellular milieu, and O_2 uptake and $\cdot O_2^-$ production were measured (Imlay and Fridovich 1991).

The membrane fraction was the predominant source of $\cdot O_2^-$, and membranes from a strain lacking the respiratory NADH dehydrogenase failed to generate $\cdot O_2^-$. Autoxidation of one or more components of the respiratory electron transport chain was therefore the source of $\cdot O_2^-$. Causing the components of this respiratory chain to accumulate in their reduced forms, by blocking electron outflow from the terminal oxidase with cyanide, amplified $\cdot O_2^-$ production.

The electron transport chain of *E. coli* was actually very good at conducting electrons to O_2 with minimal leakage. There were thus only three $\cdot O_2^-$ produced per 10^4 electrons conducted through the chain. Yet, given that each *E. coli* cell was consuming 6.2×10^6 O_2 per second, we calculate that each cell was producing 8100 $\cdot O_2^-$ per second. Intracellular volumes were also measured (Imlay and Fridovich 1991) and were found to be 3×10^{-15} liters per cell for LB-grown *E. coli* and 6.8×10^{-16} liters per cell for minimal medium-grown cells. Were $\cdot O_2^-$ infinitely stable, it would have accumulated in the LB-grown cells at a rate of 4.2 μM per second. When elimination of $\cdot O_2^-$ by the spontaneous dismutation was factored in, we calculated a steady-state level of 6.7 μM in these cells. The action of SOD, at the levels found in aerobic wild-type *E. coli*, would limit the steady-state concentration of $\cdot O_2^-$ to 2×10^{-10} M. Addition of viologens or quinones, or elevation of pO_2, would increase the production of $\cdot O_2^-$ and raise these steady-state levels; these compounds do exhibit the expected dioxygen-dependent toxicity.

Paraquat and $\cdot O_2^-$

Early studies of the herbicidal action of paraquat, and of related redox active compounds, indicated that diversion of electron flow, by redox cycling, might be the basis of that action. The herbicidal effect was thus seen to be dioxygen-dependent (Mees 1960) and a direct function of the standard redox potential of the compounds tested (Boon 1967). That some product of the redox cycling of paraquat might impose an oxidative stress was indicated by the lipid peroxidation which followed its application to leaves (Dodge et al. 1970).

The reduction of O_2 to $\cdot O_2^-$ by chemically or photochemically reduced diquat was shown with the aid of SOD (Stancliffe and Pirie 1971). Pulse radiolysis was used to show that the paraquat monocation radical (PQ$^+$) reduced O_2 to $\cdot O_2^-$ with a rate constant of 8×10^8 M^{-1} sec^{-1} and reduced $\cdot O_2^-$ to H_2O_2 with a comparably great rate constant of 6.5×10^8 M^{-1} sec^{-1} (Farrington et al. 1973). Since the concentration of O_2 in any aerobic and biologically relevant solution is certain to exceed the concentration of $\cdot O_2^-$ by several orders of magnitude, the predominant aerobic action of PQ$^+$ must be to generate $\cdot O_2^-$. That it indeed does so was shown in illuminated chloroplasts (Epel and Neuman 1973) and in mouse lung fibroblasts (Bus et al. 1974). In the chloroplast system, the identification of the product of paraquat action was solidified through spin trapping (Harbour and Bolton 1975).

PARAQUAT (PQ^{++}) AND E. COLI

PQ^{++} exerts a bacteriostatic effect on *E. coli* (Davison and Papirmeister 1971), which is dioxygen-dependent (Fisher and Williams 1976), as would be expected if cycles of reduction and autoxidation were essential for this effect. The reduced monocation radical (PQ$^+$) crosses the *E. coli* cell envelope more readily than does PQ^{++} (Jones et al. 1976). Indications of redox cycling by PQ^{++}, with production of $\cdot O_2^-$, include its increase in the cyanide-resistant respiration of suspensions of E. coli and its dramatic induction of MnSOD in aerobic cells (Hassan and Fridovich 1977). PQ^{++} is able to mediate electron flow not only to dioxygen, but also to anaerobic electron sinks such as nitrate (Jones and Garland 1977).

E. coli B was used to demonstrate that induction of MnSOD correlated with acquisition of tolerance to PQ^{++} (Hassan and Fridovich 1978b) and that the lethal action of PQ^{++} depended both on a carbon source, such as glucose, and on dioxygen. This bolstered the view that the deleterious effects of PQ^{++} are largely due to increased generation of $\cdot O_2^-$ and H_2O_2. That our understanding of the interactions of E. coli with PQ^{++} was incomplete was indicated by the greater growth inhibition seen in minimal, as compared with rich, medium. Moreover, enumeration of survivors from PQ^{++}-exposed cultures seemed greater on rich (TSY) than on minimal (VB) agar, and we had subsequently noticed that K12 strains were more resistant to the lethality of PQ^{++} than were B strains. More than a decade was to pass before these puzzlements could be explained.

In the interim, the association between PQ^{++} toxicity and $\cdot O_2^- + H_2O_2$ was broadened and strengthened. Thus, the application of PQ^{++} as an herbicide led to the appearance of resistant biotypes, and one of these, a rye grass, was found to have elevated levels of SOD and of catalase (Harper and Harvey 1978). A copper chelate with SOD-like activity was reported to diminish the phytotoxicity of PQ^{++} (Youngman et al. 1979), and exposure to a sublethal concentration of PQ^{++} induced biosynthesis of SOD in Chlorella (Rabinowitch et al. 1983). Growth of Chlorella in the presence of sulfite also induced SOD, and this correlated with enhanced resistance to paraquat (Rabinowitch and Fridovich 1985). A complex prepared from MnO_2 + desferrioxamine, which catalyzed the dismutation of $\cdot O_2^-$, protected Dunaliella salina against PQ^{++} (Rabinowitch et al. 1987). In rats, elevation of lung SOD, by pretreatment with endotoxin, correlated with resistance to PQ^{++} (Frank 1981). The mutagenicity of PQ^{++}, in a Salmonella tester strain, was shown to be dioxygen-dependent and diminished by elevated intracellular SOD (Hassan and Moody 1982). PQ^{++} was also mutagenic and lethal to yeast, and SOD-null mutants were ultrasensitive (Blaszczynski et al. 1985) whereas respiration-defective strains were resistant (Bilinski et al. 1985). In cultured CHO cells, elevation of SOD by scrape-loading imparted resistance to PQ^{++} (Bagley et al. 1986).

The action of PQ^{++} on *E. coli* must involve entry into the cell, reduction by an NADPH-dependent diaphorase (Hassan and Fridovich 1979), and spontaneous autoxidation. Toxicity could then follow both from the depletion of NAD(P)H and from the actions of •O$_2^-$, H$_2$O$_2$, and •OH. Although this view is probably correct, there are more facts to be accommodated. Thus, the bacteriostatic effect of PQ^{++} can be separated from its lethal effect. In a low-salt minimal medium, less than 1 μM PQ^{++} strongly inhibits the growth of *E. coli* (Kitzler and Fridovich 1986). Addition of yeast extract to the minimal medium eliminated this inhibition of growth, indicating that it was due to the imposition of nutritional auxotrophies by PQ^{++}. Significant lethality, on the other hand, was not observed until the concentration of PQ^{++} exceeded 100 μM, and yeast extract was then without effect. These studies (Kitzler and Fridovich 1986) were done with an *E. coli* B, and enumeration of survivors was done by plating on TSY agar. The significance of these details did not become clear until later on.

E. coli B was found to retain PQ^{++}, whereas *E. coli* K12 did not (Kitzler et al. 1990). A relatively brief (30 min) exposure of these two strains to PQ^{++}, followed by dilution and plating on TSY agar, made it appear that *E. coli* K12 was resistant to PQ^{++} whereas the B strain was not. This difference was more apparent than real. Loss of viability actually required several hours of incubation with PQ^{++}. The K strain was thus rescued during dilution and plating by washout of the PQ^{++}. The B strain, in contrast, retained the PQ^{++} and actually lost viability on the agar plates. Uptake and retention of PQ^{++} by *E. coli* B was very pH-dependent, being greater at pH 9.0 than at pH 5.0 (Minakami et al. 1990). It was also noted that the lethal effect of PQ^{++} could be seen in the rich TSY medium, but not in the minimal VB medium. There was a relationship between growth in the presence of PQ^{++} and cell death (Minakami and Fridovich 1990a,b). *E. coli*, in PQ^{++}-containing rich medium, continued to grow without dividing while simultaneously losing viability. A variety of antibiotics, which inhibited this growth, also protected against the loss of viability.

We can now contrast the bacteriostatic and the lethal actions of PQ^{++}. Growth inhibition by PQ^{++} is largely due to the imposition of auxotrophies and this, in turn, is largely due to inactivation by •O$_2^-$ of enzymes on biosynthetic and energy producing pathways. The dihydroxy acid dehydratase involved in biosynthesis of branched chain amino acids and the aconitase of the Krebs cycle are examples of •O$_2^-$-sensitive enzymes. The redox cycling of PQ^{++} in *E. coli* also depletes NADPH, and this contributes to inhibition of growth by diminishing reductive biosynthesis. The lethality of PQ^{++}, on the other hand, is probably due to damage to DNA and to the cell membrane by the metal-catalyzed Haber-Weiss reaction. This damage can be repaired in quiescent cells but becomes irreversible when the cell attempts to replicate a damaged genome and to grow.

ADAPTATION TO OXIDATIVE STRESS

Two groups of coordinately regulated genes in *E. coli* are activated by oxidative stress. One of these, induced by exposure to H$_2$O$_2$, has been named the *oxyR* regulon (Christman et al. 1989) and the other, induced by agents known to increase •O$_2^-$ production, has been called the *soxR* regulon (Greenberg et al. 1990; Tsaneva and Weiss 1990). In both, regulation is positive and is exerted on transcription. Alkylhydroperoxide reductase, glutathione reductase, and one of the hydroperoxidases are among the products of *oxyR*; MnSOD, glucose-6-phosphate dehydrogenase (G-6-PD), and endonuclease IV are among the proteins encoded by *soxR* (Greenberg et al. 1990; Storz et al. 1990b). OxyR, the positive regulator of the *oxyR* regulon, is a 305-residue polypeptide that exhibits sequence homologies with other positive regulators such as LysR and NodB (Christman et al. 1989). OxyR acts as an autogenous repressor. It can exist in reduced or oxidized states, both of which bind to the relevant operons, but only the oxidized form activates transcription (Storz et al. 1990a). Redox-sensitive changes in its footprint suggest that OxyR undergoes a conformational change when oxidized or reduced. OxyR thus exhibits the properties needed in a redox-controlled regulator of gene action; however, both the nature of its redox center and the cellular redox couples with which it interacts remain to be identified.

The *soxR* regulon seems to be controlled by two divergently transcribed genes, which have been named *soxR* and *soxS* (Wu and Weiss 1991). *soxR* encodes a 17-kD polypeptide and *soxS* a 13-kD polypeptide. Both are required for induction of the regulon by PQ^{++}, and both contain sequences characteristic of the helix-turn-helix domains of one class of DNA-binding proteins. *soxR* contains four cysteine residues, and these may be part of the redox sensor.

The *soxR-soxS* combination may indeed be responsive to •O$_2^-$, but this cannot be its only mode of sensing the redox status of the cell. Thus, a variety of oxidants have been seen to induce MnSOD in *E. coli* under anaerobic conditions, when •O$_2^-$ cannot be formed (Smith and Neidhardt 1983; Miyake 1986; Hassan and Moody 1987; Privalle and Fridovich 1988, 1990; Schiavone and Hassan 1988). The same statement can be made with respect to G-6-PD, another product of the *soxR* regulon (Privalle and Fridovich 1990). It should also be recalled that the in vitro transcription of *sodA*, which codes for MnSOD, was suppressed by specific thiols and by NADPH, but not by NADH (Gardner and Fridovich 1987).

EXPLAINING A PARADOX

If PQ^{++} induces the biosynthesis of MnSOD and if this has been shown to be an adaptive response, then how can overproduction of SOD, by genetic manipulation, do other than protect against PQ^{++}? Yet Bloch and Ausubel (1986), in achieving gross overproduction of MnSOD by virtue of a multicopy plasmid bearing *sodA*, noted that this made *E. coli* more sensitive to growth inhibition by PQ^{++}. Scott et al. (1987) later showed a similar effect of overproduction of FeSOD. Being unaware of the *soxR* and *oxyR* regulons, these workers were unable to adequately explain these observations. We can now appreciate that an effective defense against PQ^{++} requires the balanced induction of these families of enzymes, not simply of SOD. Moreover, we can imagine that marked overproduction of

SOD might prevent induction of the other members of *soxR*. G-6-PD was selected as an easily assayable product of *soxR*, and overproduction of MnSOD was indeed shown to diminish induction of G-6-PD in response to PQ^{++} (Liochev and Fridovich 1991).

Were *soxR* specifically responsive to $\cdot O_2^-$, then overproduction of SOD could prevent its activation by keeping [$\cdot O_2^-$] below the level needed for this activation. Alternatively, we can anticipate that gross overproduction of SOD, or of any other protein, should suppress induction of *soxR* because of simple competition for the cellular resources needed for such induction. Consider that the MnSOD overproducer, when exposed to 50 µM PQ^{++} in a medium enriched with Mn(II), accumulated *active* MnSOD until it constituted 33% of the water-soluble protein. In the absence of enrichment with Mn(II), the level of *active* MnSOD was only one-fifth as great, but we have previously seen that MnSOD, when made in excess of available manganese, becomes incorrectly substituted with iron and accumulates in this inactive form (Privalle et al. 1989; Beyer and Fridovich 1991). It is thus clear that the same amount of MnSOD protein was made whether or not the medium was enriched with Mn(II). Yet, enrichment of the medium with Mn(II) did not influence the induction of G-6-PD by PQ^{++} in the MnSOD-overproducing strain (Liochev and Fridovich 1991). It follows that suppression of [$\cdot O_2^-$] was not the means by which overproduction of MnSOD prevented induction of G-6-PD and, by extension, of *soxR*. This result suggests that overproduction of MnSOD, or of FeSOD, decreased induction of *soxR* by competition for cellular resources. Were this the case, stationary phase inocula would suffer more from the overproduction of SOD than log phase inocula. This is supposed because stationary phase cells need to replace ribosomes and the other components of the protein biosynthetic apparatus, whereas log phase cells have this already in place. The stationary phase cells would thus be more strongly affected by a dearth of cellular resources. We have observed that induction of G-6-PD by PQ^{++} is suppressed by overproduction of MnSOD to a much greater extent with stationary phase than with log phase inocula (S.I. Liochev and I. Fridovich, unpubl.).

The discovery of SOD more than two decades ago (McCord

and Fridovich 1968, 1969) ignited widespread interest in oxidative stress and in the defenses thereto. This interest, and the work of a host of dedicated and talented investigators, has led in turn to a flood of new knowledge. This field of endeavor has moved from infancy to early adulthood. Among the things we have learned is that SOD is not the only kid on the block and that the only simple aspects of living cells are those which have not been carefully examined.

ACKNOWLEDGMENTS

This work was supported by grants from the National Science Foundation, the Council for Tobacco Research-U.S.A., Inc., and the National Institutes of Health.

REFERENCES

Bagley, A.C., J. Krall, and R.E. Lynch. 1986. Superoxide mediates the toxicity of paraquat for Chinese hamster ovary cells. *Proc. Natl. Acad. Sci.* **83:** 3189.

Balin, A.K., D.B.P. Goodman, H. Rasmussen, and V. J. Cristofalo. 1976. The effect of oxygen tension on the growth and metabolism of WI-38 cells. *J. Cell. Physiol.* **89:** 235.

Beauchamp, C. and I. Fridovich. 1970. A mechanism for the production of ethylene from methional: The generation of hydroxyl radical by xanthine oxidase. *J. Biol. Chem.* **245:** 4641.

Beyer, W.F., Jr. and I. Fridovich. 1991. *In vivo* competition between iron and manganese for occupancy of the active site of the manganese superoxide dismutase of *Escherichia coli*. *J. Biol. Chem.* **266:** 303.

Bilinski, T., J. Litwinska, and M. Blaszczynski. 1985. Selective killing of respiratory sufficient yeast cells by paraquat. *Acta Microbiol. Pol.* **34:** 15.

Blaszczynski, M., J. Litwinska, D. Zaborowska, and T. Bilinski. 1985. The role of the respiratory chain in paraquat toxicity in yeast. *Acta Microbiol. Pol.* **34:** 243.

Bloch, C.A. and F.M. Ausubel. 1986. Paraquat-mediated selection for mutation in the manganese-superoxide dismutase gene sodA. *J. Bacteriol.* **168:** 795.

Boon, W.R. 1967. The quaternary salts of bipyridylium: A new agricultural tool. *Endeavour* **26:** 27.

Breimer, L.H. 1991. Repair of DNA damage induced by reactive oxygen species. *Free Rad. Res. Commun.* **14:** 159.

Brown, O.R. and R.L. Seither. 1983. Oxygen and redox-active drugs: Shared toxicity sites. *Fundam. Appl. Toxicol.* **3:** 209.

Bus, J.S., S.D. Aust, and J.E. Gibson. 1974. Superoxide and singlet oxygen catalyzed lipid peroxidation as a possible mechanism for paraquat (methyl viologen) toxicity. *Biochem. Biophys. Res. Commun.* **58:** 749.

Carlioz, A. and D. Touati. 1986. Isolation of superoxide dismutase mutants in *Escherichia coli*: Is superoxide dismutase strictly necessary for aerobic life? *EMBO J.* **5:** 623.

Christman, M.F., G. Storz, and B.N. Ames. 1989. OxyR, a positive regulator of hydrogen peroxide-inducible genes in *Escherichia coli* and *Salmonella typhimurium*, is homologous to a family of bacterial regulatory proteins. *Proc. Natl. Acad. Sci.* **86:** 3484.

Claiborne, A. and I. Fridovich. 1979. Purification of the o-dianisidine peroxidase from *Escherichia coli* B. Physicochemical characterization and analysis of its dual catalatic and peroxidatic activities. *J. Biol. Chem.* **254:** 4245.

Claiborne, A., D.P. Malinowski, and I. Fridovich. 1979. Purification and characterization of hydroperoxidase II of *Escherichia coli* B. *J. Biol. Chem.* **254:** 11664.

Czapski, G. 1984. On the use of hydroxyl radical scavengers in biological systems. *Isr. J. Chem.* **24:** 29.

Davison, L.C. and B. Papirmeister. 1971. Bacteriostasis of *Escherichia coli* by the herbicide paraquat. *Proc. Soc. Exp. Biol. Med.* **136:** 359.

Dodge, A.D., N. Harris, and B.C. Baldwin. 1970. The mode of action of paraquat and diquat. *Biochem. J.* **118:** 43P.

Elstner, E.F. and J.R. Konz. 1974. Light-dependent ethylene production by isolated chloroplasts. *FEBS Lett.* **45:** 18.

Epel, B.L. and J. Neuman. 1973. The mechanism of the oxidation of ascorbate and Mn(II) by chloroplasts. The role of the radical superoxide. *Biochim. Biophys. Acta* **325:** 520.

Farrington, J.A., M. Ebert, E.J. Land, and K. Fletcher. 1973. Bipyridylium quaternary salts and related compounds. V. Pulse radiolysis studies of the reaction of paraquat radical with oxygen. Implications for the mode of action of bipyridyl herbicides. *Biochim. Biophys. Acta* **314:** 372.

Fisher, H.K. and G. Williams. 1976. Paraquat is not bacteriostatic under anaerobic conditions. *Life Sci.* **19:** 421.

Flint, D.H. and H. Emptage. 1990. Dihydroxy acid dehydratase: Isolation, characterization as an iron-sulfur protein and sensitivity to inactivation by oxygen radicals. In *Biosynthesis of branched chain amino acids* (ed. D. Chipman et al.), p. 285. Balaban, Philadelphia.

Frank, L. 1981. Prolonged survival after paraquat. Role of lung antioxidant enzyme systems. *Biochem. Pharmacol.* **30:** 2318.

Gardner, P.R. and I. Fridovich. 1987. Controls on the biosynthesis of the manganese-containing superoxide dismutase of *Escherichia* coli: Effects of thiols. *J. Biol. Chem.* **262:** 17591.

Gardner, P.R. and I. Fridovich. 1991a. Superoxide sensitivity of the *Escherichia coli* 6-phosphogluconate dehydratase. *J. Biol. Chem.* **266:** 1478.

———. 1991b. Superoxide sensitivity of the *Escherichia coli* aconitase. *J. Biol. Chem.* **266:** 19328-19333.

Goldstein, S. and G. Czapski. 1986. The role and mechanism of metal ions and their complexes in enhancing damage in biological systems or in protecting these systems from the toxicity of superoxide radical. *J. Free Radicals Biol. Med.* **2:** 3.

Greenberg, J.T., P. Monach, J.H. Chou, P.D. Josephy, and B. Demple. 1990. Positive control of a global antioxidant defense regulon activated by superoxide-generating agents in *Escherichia coli*. *Proc. Natl. Acad. Sci.* **87:** 6181.

Harbour, J.R. and J.R. Bolton. 1975. Superoxide formation in spinach *chloroplasts*: Electron spin resonance detection by spin trapping. *Biochem. Biophys. Res. Commun.* **64:** 803.

Harper, D.B. and B.M.R. Harvey. 1978. Mechanism of paraquat resistance in perennial rye grass. II. Role of superoxide dismutase, catalase and peroxidase. *Plant Cell Environ.* **1:** 211.

Hassan, H.M. and I. Fridovich. 1977. Regulation of the synthesis of superoxide dismutase in *Escherichia coli*. *J. Biol. Chem.* **252:** 7667.

———. 1978a. Regulation of the synthesis of catalase and peroxidase in *Escherichia coli*. *J. Biol. Chem.* **253:** 6445.

———. 1978b. Superoxide radical and the oxygen enhancement of the toxicity of paraquat in *Escherichia coli*. *J. Biol. Chem.* **253:** 8143.

———. 1979. Mechanism of production of extracellular superoxide radical. *J. Biol. Chem.* **254:** 10846.

Hassan, H.M. and C.S. Moody. 1982. Superoxide dismutase protects against paraquat-mediated dioxygen toxicity and mutagenicity: Studies in *Salmonella typhimurium*. *Can. J. Physiol. Pharmacol.* **60:** 1367.

———. 1987. Regulation of manganese-containing superoxide dismutase in *Escherichia coli*. Anaerobic induction by nitrate. *J. Biol. Chem.* **262:** 17173.

Heikkila, R.E., B. Winston, G. Cohen, and H. Barden. 1976. Alloxan-induced diabetes: Evidence for hydroxyl radical as a cytotoxic intermediate. *Biochem. Pharmacol.* **25:** 1085.

Imlay, J.A. and I. Fridovich. 1991. Assay of metabolic superoxide production in *Escherichia coli*. *J. Biol. Chem.* **266:** 6957.

Johansson, I. and M. Ingelman-Sundberg. 1983. Hydroxyl radical-mediated, cytochrome P-450 dependent metabolic activation of benzene in microsomes and reconstituted enzyme systems from rabbit liver. *J. Biol. Chem.* **258:** 7311.

Jones, R.W. and P.B. Garland. 1977. Sites and specificity of the reaction of bipyridylium compounds with the anaerobic respiratory enzymes of *Escherichia coli. Biochem. J.* **164:** 199.

Jones, R.W., T.A. Gray, and P.B. Garland. 1976. A study of the permeability of the cytoplasmic membrane of *Escherichia coli* to reduced and oxidized benzyl viologen and methyl viologen cations. Complications in the use of viologens as redox mediators for membrane-bound enzymes. *Biochem. Soc. Trans.* **4:** 671.

Keele, B.B., Jr., J.M. McCord, and I. Fridovich. 1970. Superoxide dismutase from *Escherichia coli* B. *J. Biol. Chem.* **245:** 6176.

Kitzler, J. and I. Fridovich. 1986. The effects of paraquat on *Escherichia coli*: Distinction between bacteriostasis and lethality. *J. Free Radicals Biol. Med.* **2:** 245.

Kitzler, J.W., H. Minakami, and I. Fridovich. 1990. Effects of paraquat on *Escherichia coli*: Differences between B and K-12 strains. *J. Bacteriol.* **172:** 686.

Korbashi, P., J. Katzhendler, and M. Chevion. 1986. Iron mediates paraquat toxicity in *Escherichia coli. J. Biol. Chem.* **261:** 12472.

Kubicek, C.P. and M. Rohr. 1985. Aconitase and citric acid fermentation by *Aspergillus niger. Appl. Environ. Microbiol.* **50:** 1336.

Kuo, C.F., T. Mashino, and I. Fridovich. 1987. α,β-Dihydroxyisovalerate dehydratase: A superoxide-sensitive enzyme. *J. Biol. Chem.* **262:** 4724.

Liochev, S.I. and I. Fridovich. 1991. Effects of overproduction of superoxide dismutase on the toxicity of paraquat toward *Escherichia coli. J. Biol. Chem.* **266:** 8747.

Markey, B.A., S.H. Phan, J. Varani, U.S. Ryan, and P.A. Ward. 1990. Inhibition of cytotoxicity by intracellular superoxide dismutase supplementation. *J. Free Radicals Biol. Med.* **9:** 307.

McCord, J.M. 1986. Superoxide dismutase: Rationale for use in reperfusion injury and inflammation. *J. Free Radicals Biol. Med.* **2:** 307.

_____. 1987. Oxygen-derived free radicals: A link between reperfusion injury and inflammation. *Fed. Proc.* **46:** 2402.

McCord, J.M. and I. Fridovich. 1968. The reduction of cytochrome c by milk xanthine oxidase. *J. Biol. Chem.* **243:** 5753.

_____. 1969. Superoxide dismutase: An enzymic function for erythrocuprein. *J. Biol. Chem.* **244:** 6049.

Mees, G.C. 1960. Experiments on the herbicidal action of 1,1'-ethylene-2,2'-dipyridylium dibromide. *Ann. Appl. Biol.* **48:** 601.

Minakami, H. and I. Fridovich. 1990a. Effects of paraquat on cultures of *Escherichia coli*: Turbidity versus enumeration. *J. Free Radicals Biol. Med.* **8:** 387.

_____. 1990b. Relationship between growth of *Escherichia coli* and susceptibility to the lethal effect of paraquat. *FASEB J.* **4:** 3239.

Minakami, H., J.W. Kitzler, and I. Fridovich. 1990. Effects of pH, glucose, and chelating agents on the lethality of paraquat to *Escherichia coli. J. Bacteriol.* **172:** 691.

Miyake, K. 1986. Effect of nitrate on the level of superoxide dismutase in anaerobically grown Escherichia coli. J. Gen. Appl. Microbiol. **32:** 527.

Moody, C.S. and H.M. Hassan. 1984. Anaerobic biosynthesis of the manganese-containing superoxide dismutase in Escherichia coli. J. Biol. Chem. **259:** 12821.

Nakae, D., H. Yoshiji, T. Amanuma, T. Kinugasa, J.L. Farber, and Y. Konishi. 1990. Endocytosis-independent uptake of liposome-encapsulated superoxide dismutase prevents the killing of cultured hepatocytes by t-butylhydroperoxide. Arch. Biochem. Biophys. **279:** 315.

Privalle, C.T. and I. Fridovich. 1988. Inductions of superoxide dismutases in Escherichia coli under anaerobic conditions. Accumulation of an inactive form of the manganese enzyme. J. Biol. Chem. **263:** 4274.

————. 1990. Biosynthesis of the manganese-containing superoxide dismutase in Escherichia coli: Effects of diazene dicarboxylic acid (N,N'-dimethylamide). J. Biol. Chem. **265:** 21966.

Privalle, C.T., W.F. Beyer, Jr., and I. Fridovich. 1989. Anaerobic induction of proMn-superoxide dismutase in Escherichia coli. J. Biol. Chem. **264:** 2758.

Pugh, S.Y. and I. Fridovich. 1985. Induction of superoxide dismutase in Escherichia coli by metal chelators. J. Bacteriol. **162:** 196.

Pugh, S.Y., J.L. DiGuiseppi, and I. Fridovich. 1984. Induction of superoxide dismutases in Escherichia coli by manganese and iron. J. Bacteriol. **160:** 137.

Rabinowitch, H.D. and I. Fridovich. 1985. Growth of Chlorella sorokiniana in the presence of sulfite elevates cell content of superoxide dismutase and imparts resistance towards paraquat. Planta **164:** 524.

Rabinowitch, H.D., C.T. Privalle, and I. Fridovich. 1987. Effects of paraquat on the green alga Dunaliella salina: Protection by the mimic of superoxide dismutase, Desferal-Mn(IV). J. Free Radicals Biol. Med. **3:** 125.

Rabinowitch, H.D., D.A. Clare, J.D. Crapo, and I. Fridovich. 1983. Positive correlation between superoxide dismutase and resistance to paraquat toxicity in the green alga Chlorella sorokiniana. Arch. Biochem. Biophys. **225:** 640.

Richter, C., J.-W. Park, and B.N. Ames. 1988. Normal oxidative damage to mitochondrial and nuclear DNA is extensive. Proc. Natl. Acad. Sci. **85:** 6465.

Schiavone, J.R. and H.M. Hassan. 1988. The role of redox in the regulation of manganese-containing superoxide dismutase biosynthesis in Escherichia coli. J. Biol. Chem. **263:** 4269.

Schwaiger, M. 1989. Sustained nonoxidative glucose utilization and depletion of glycogen in reperfused canine myocardium. J. Am. Coll. Cardiol. **13:** 745.

Scott, M.D., S.R. Meshnick, and J.W. Eaton. 1987. Superoxide dismutase-rich bacteria. Paradoxical increase in oxidant toxicity. *J. Biol. Chem.* **262:** 3640.

Smith, M.W. and F.C. Neidhardt. 1983. Proteins inducible by aerobiosis in *Escherichia coli. J. Bacteriol.* **154:** 344.

Spitz, D.R., J.H. Elwell, Y. Sun, L.W. Oberley, T.D. Oberley, S.J. Sullivan, and R.J. Roberts. 1990. Oxygen toxicity in control and H_2O_2-resistant Chinese hamster fibroblast cell lines. *Arch. Biochem. Biophys.* **279:** 315.

Stancliffe, T.C. and A. Pirie. 1971. The production of superoxide radicals in reactions of the herbicide diquat. *FEBS Lett.* **17:** 297.

Storz, G., L.A. Tartaglia, and B.N. Ames. 1990a. Transcriptional regulator of oxidative stress-inducible genes: Direct activation by oxidation. *Science* **248:** 189.

Storz, G., L.A. Tartaglia, S.B. Farr, and B.N. Ames. 1990b. Bacterial defenses against oxidative stress. *Trends Genet.* **6:** 363.

Tsaneva, I.R. and B. Weiss. 1990. *SoxR*, a locus governing a superoxide response regulon in *Escherichia coli* K12. *J. Bacteriol.* **172:** 4197.

van Hemmen, J.J. and W.J.A. Meuling. 1977. Inactivation of *Escherichia coli* by superoxide radicals and their dismutation products. *Arch. Biochem. Biophys.* **182:** 743.

Villafranca, J.J. and A.S. Mildvan. 1971. The mechanism of aconitase action. I. Preparation, physical properties of the enzyme, and activation by iron(II). *J. Biol. Chem.* **246:** 772.

Wu, J. and B. Weiss. 1991. Two divergently transcribed genes, *soxR* and *soxS*, control a superoxide response regulon of *Escherichia coli. J. Bacteriol.* **173:** 2864.

Yost, F.J., Jr. and I. Fridovich. 1973. An iron-containing superoxide dismutase from *Escherichia coli. J. Biol. Chem.* **248:** 4905.

Youngman, R.J., A.O. Dodge, E. Lengfelder, and E.F. Elstner. 1979. Inhibition of paraquat phytotoxicity by a novel copper chelate with superoxide dismutase activity. *Experientia* **35:** 1295.

Regulation and Protective Role of the Microbial Superoxide Dismutases

D. Touati
Institut Jacques Monod, CNRS, Université Paris VII
75251 Paris cedex 05, France

To face the permanent aggression that constitutes the production of active oxygen species during aerobic metabolism, microorganisms mobilize considerable resources for eliminating the threatening oxygen species, repairing the oxidative damage, and bypassing or compensating for the damaged functions. This is the oxidative stress response, which incorporates multiple global responses to various oxidative stimuli. Superoxide dismutases (SODs) play a key role in this response (Fridovich 1986; Beyer et al. 1991).

Superoxide, the first activated oxygen species to be formed, is broken down into hydrogen peroxide and oxygen. The hydrogen peroxide is removed by hydroperoxidases, but it can give rise through a Fenton reaction (the so-called Haber-Weiss reaction) to the extremely reactive and deleterious hydroxyl radical in the presence of reduced iron (Halliwell and Gutteridge 1984), which can be provided by the superoxide-mediated reduction of ferric iron (Gutteridge 1985). To prevent or minimize hydroxyl radical formation, a subtle equilibrium must be maintained between superoxide flux, hydrogen peroxide formation, and the intracellular availability of reduced iron (Fig. 1).

SOD is the key to maintaining this balance by eliminating superoxide and consequently inhibiting the superoxide-mediated reduction of iron, thus limiting the amount of hydrogen peroxide that is broken down via the Fenton reaction. Clearly, any lack or excess of SOD will disrupt this equilibrium, resulting in hydroxyl radical formation and injury to the cells. Too little SOD will result in superoxide accumula-

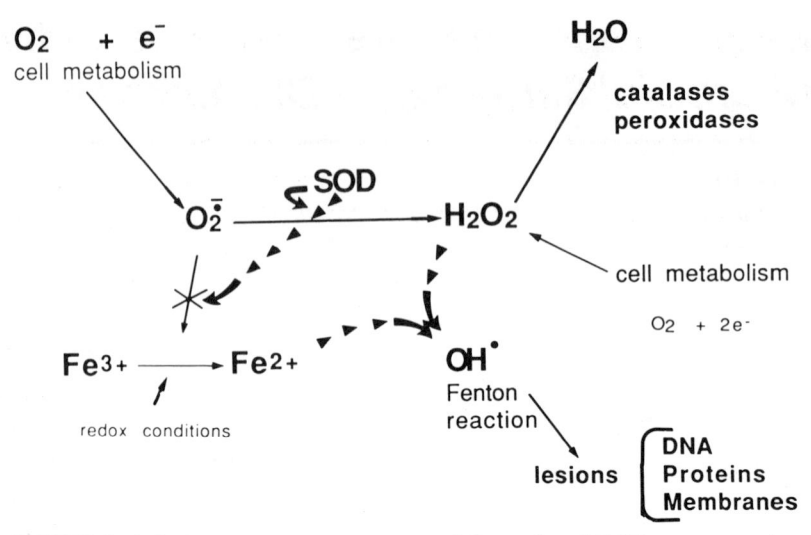

FIGURE 1 Schematic representation of the role of SOD in preventing hydroxyl radical formation. (Reprinted, with permission, from Touati et al. 1992 [copyright Pergamon Press].)

tion, causing direct damage, but also favoring iron reduction and its subsequent reaction with the hydrogen peroxide generated by spontaneous dismutation and other metabolic pathways; high levels of SOD in excess may generate more hydrogen peroxide than can be removed.

Recent studies, mainly in *Escherichia coli*, support the view that SOD has a central role in directly protecting the cells against oxidative stress and indirectly maintaining a balance between the various toxic oxygen species. The physiology of SOD-lacking mutants renders their survival very unlikely in natural conditions, showing that SOD is essential for normal aerobic life, whereas the strict regulation of SOD in response to the environment strongly suggests that too much SOD can be dangerous under certain conditions.

SOD FORMS IN MICROORGANISMS: AN ADAPTIVE RESPONSE?

SODs have been found in all microorganisms, including archaebacteria (see Salin et al. 1988; May and Dennis 1990; Takao et al. 1990) and some obligate anaerobes (see Gregory et al. 1978; Abdollahi and Wimpenny 1990). The forms of SOD appear to be intimately related to the natural microorganism environment. Prokaryotic SODs are usually iron and/or manganase SOD, which are highly homologous and presumably derived from the same ancestor. FeSOD is the predominant anaerobic form, whereas MnSOD is usually found in the presence of oxygen.

The discovery of bacteriocuprein SOD (CuZnSOD) has provided an exception to the general distribution scheme of SOD. CuZnSOD has been found in *Photobacterium leiognathi* (Puget and Michelson 1974), *Caulobacter crescentus* (Steinman 1982), *Pseudomonas diminuta* and *Pseudomonas maltophilia* (Steinman 1985), *Paracoccus denitrificans* (Vignais et al. 1982), and *Brucella abortus* (Beck et al. 1990). The cloning and sequencing of two of those SOD genes, one from *P. leiognathi* (Steinman 1987) and the other from *C. crescentus* (Steinman and Ely 1990), showed that they are related to eukaryotic CuZnSODs but differ from them by the presence of a 22–23-residue leader sequence (also found in the amino acid sequence of the CuZnSOD from *B. abortus*; Beck et al. 1990), suggesting that the mature protein is extracytoplasmic. Investigation of the intracellular localization of *Caulobacter* CuZnSOD indeed demonstrated that the enzyme was periplasmic. This periplasmic location of SOD may reflect the need for some microorganisms to be protected from an extracellular oxidative stress, against which a cytoplasmic SOD may not be effective. Steinman and Ely (1990) proposed that, for *C. crescentus*, this stress may arise from its attachment to the surface of photosynthetic cyanobacteria that produce large fluxes of oxygen during the daylight hours. Beck et al. (1990) suggested that the CuZnSOD of *B. abortus* helps the bacterium evade the host's phagocytic defenses.

Cyanobacteria generally possess both iron and manganese SODs, each located in a different compartment. The majority of the SOD activity comes from the FeSOD in the cytosol under normal growth conditions, whereas the MnSOD is in the thylakoid membranes. Photooxidative damage occurs during the process of photoinhibition (inhibition of photosynthesis by excess visible light), and SOD activity level drops, presumably because of FeSOD inactivation by H_2O_2. The MnSOD content of an uncharacterized mutant of *Plectonema boryanum* that is resistant to photooxidation is particularly high, so that it contains more MnSOD than FeSOD (Steinitz et al. 1979). Cyanobacteria, such as *Anacystis nidulans*, which possess only FeSOD, are extremely sensitive to photooxidative stress. The introduction of a plasmid expressing the *E. coli* MnSOD gene into this cyanobacterium confers some protection against paraquat-mediated photooxidative stress (Gruber et al. 1990). The recent cloning and sequencing of the FeSOD gene of *A. nidulans* should allow studies on mutants that may shed light on the role of SOD in oxygen-evolving photoautotrophs (Laudenbach et al. 1989).

The presence of both MnSOD and FeSOD in the cytoplasm of several bacteria, such as *E. coli*, appears redundant (Britton and Fridovich 1977). The evolutionary persistence of those two isoenzymes in the cell, even in the same cellular compartment, despite their relatedness, is presumably because they are regulated differently (see below). In a preliminary study in *E. coli*, Hopkin and Steinman (1992) report that MnSOD and FeSOD may not be equivalent in vivo. Their conclusions are based on differences in the growth and tolerance to paraquat of strains bearing plasmids overproducing either MnSOD or FeSOD under the control of the oxidative stress-independent *tac* foreign promoter. However, further studies are necessary to determine whether this nonequivalence is effectively related to specific in vivo functions of the isoenzymes, or is just an indirect consequence of enzyme overproduction, such as metal consumption by the overproduced protein.

Despite their strong homologies, most of the MnSODs and FeSODs studied have strict metal specificity. This is particularly puzzling when both enzymes show the same structure by X-ray analysis, share the same metal ligands, and are capable of forming in vivo, as in *E. coli*, an active heterodimer hybrid protein (Dougherty et al. 1978; Touati 1989) but form totally inactive homodimers when one original metal is replaced by the other (Beyer and Fridovich 1991). The cloning and knowledge of the nucleotide sequence of both *E. coli* SOD genes now offers the possibility of using site-directed mutagenesis to investigate the structure-function relationships of these enzymes. This metal specificity has some biological significance, because it provides a way of modulating, at a posttranslational level, the amount of active protein (see below).

There are microorganisms in which this strict metal cofactor specificity does not exist. Bacteria such as *Bacteroides fragilis* (Gregory 1985), *Streptococcus mutans* (Martin et al. 1986), *Bacteroides thetaiotaomicron* (Pennington and Gregory 1986), *Propionibacterium shermanii* (Meier et al. 1982), and *Bacteroides gingivalis* (Amano et al. 1990) use the same protein moiety to form active MnSOD or FeSOD ("cambialistic" SOD), depending on the metal supplied in the growth medium. It was suggested that this capacity to use one or the other metal allows these microorganisms, which are generally anaerobic, to survive a transient period of aerobiosis (Privalle and Gregory 1979; Amano et al. 1990). In vivo, the anaero-SOD has the characteristics of an FeSOD, whereas the aero-SOD has characteristics of an MnSOD. Under conditions of oxygen stress, where iron is unavailable (as ferric iron precipitates), the cell's ability to substitute Mn as the active metal permits it to survive. The amino acid sequencing and gene cloning and sequencing of the *B. gingivalis* SOD have shown that the same apoprotein is used. This protein has an amino acid sequence that is intermediate between those of FeSOD and MnSOD in a limited region around the putative second metal ligand (Nakayama 1990).

SOD AND MICROBIAL PATHOGENICITY

Microorganism pathogenicity may be caused by a variety of factors, including toxin secretion, invasion of an unusual ecological niche, and the ability to escape phagocytosis or to survive ingestion by phagocytic cells. There is now some evidence that SOD may be associated with pathogenesis in several cases. Thus, cambialistic SOD in the presence of manganese permits *Streptococcus* mutants to survive in oral regions of elevated oxygen tension, such as the tooth surface, causing dental caries (Martin et al. 1986; Arai and Oguchi 1990).

The generation of active oxygen species is a major component of the bactericidal mechanism employed by phagocytes (Hassett and Cohen 1989). The mechanisms by which resistant bacteria combat the oxidative burst during phagocytosis is still poorly understood. It has been proposed that the SODs and/or catalases of pathogenic bacteria are important for resistance to oxidative killing. Several lines of evidence suggest that there is a correlation between the content or nature of the SOD in microorganisms and their virulence (Welch et al. 1979; Beaman and Beaman 1990).

A *SodB* mutant (FeSOD-negative) of *Shigella flexneri*, constructed by transduction of the *E. coli* mutated gene, was found to be much more sensitive to killing by phagocytes than the wild-type parent and lost its virulence (Franzon et al. 1990). The periplasmic CuZnSOD of *B. abortus* may permit this intracellular parasite to survive in pathocytes (Smith 1977). *Mycobacterium tuberculosis* and *Nocardia asteroides* are resistant to phagocytic attack (Lowrie 1983) and secrete SOD into the extracellular fluid during logarithmic phase growth (Beaman et al. 1983, 1985). Studies using monoclonal antibodies directed against nocardial SOD have demonstrated that SOD helps protect *N. asteroides* within its host (Beaman and Beaman 1990).

Molecular biology has provided a new range of tools for studying pathogenic bacteria. SOD genes from *M. tuberculosis* (Zhang et al. 1991) and *Mycobacterium leprae* (Thangaraj et al. 1990) have been cloned and sequenced, and the gene from *Listeria ivanovii* (Haas and Goebel 1990) has been cloned. These are the first steps in a new approach to the question of the role of SOD in pathogenesis.

BIOLOGICAL CONSEQUENCES OF A LACK OF SOD IN E. COLI

The only SOD-negative mutants that have been obtained to date among microorganisms are from *E. coli*. Mutants lacking either MnSOD (*sodA*) or FeSOD (*sodB*) are first obtained by transposon insertions into the cloned structural genes, followed by an exchange between the mutated SOD allele carried by a plasmid and the corresponding chromosomal wild-type allele (Touati 1983; Sakamoto and Touati 1984; Carlioz and Touati 1986). Deletion mutants have also been constructed (I. Compan and D. Touati, unpubl.).

Amino Acid Auxotrophies and Membrane Damage

Physiological studies on a double mutant completely lacking SOD activity revealed that *E. coli* can survive in the absence of SOD only when all the amino acids are supplied by the growth medium. Growth is slightly impaired in a rich medium, whereas there is no growth in a minimal medium. The molecular basis for this inability to grow in aerobic minimal medium is still unclear, but presumably it results from several defects. Growth is restored by providing the 20 amino acids. Although a lack of certain amino acids, such as any branched amino acid, completely inhibits growth, the lack of others is less effective (Carlioz and Touati 1986). This suggests that there are targets with different sensitivities to superoxide in the amino acid biosynthetic pathways. One target is the enzyme dihydroxy acid dehydratase, an enzyme in the branched-chain amino acid pathway that has been shown to be superoxide sensitive (Brown and Yein 1978; Kuo et al. 1987; Flint and Emptage 1990; Liochev and Fridovich, this volume). External suppressors that relieve the requirement for amino acids have been isolated (Fee et al. 1988; Imlay and Fridovich 1991a; D. Touati, unpubl.). The recent study of Imlay and Fridovich may shed light on the SOD-negative mutant auxotrophies. While characterizing and mapping one class of pseudorevertants at the *ssa* locus, located at 4 minutes on the *E. coli* chromosome, Imlay observed that SOD mutants are unable to assimilate diaminopimelic acid from the growth medium, whereas the *ssa* pseudorevertants can do so, suggesting that superoxide accumulation damages the membrane and that this damage is partially circumvented by *ssa* mutation. The authors propose the attractive hypothesis that the auxotrophies of SOD-negative mutants are due to inactivation of superoxide-sensitive enzymes in the amino acid pathway, resulting in a poor biosynthesis of amino acids, together with a membrane permeability defect that exacerbates the amino acid scarcity.

SOD-negative mutants may have a damaged envelope, so that the turgor imposed by the high internal solute concentration results in the escape of valuable products, such as growth-limiting metabolites. The growth of SOD-lacking mutants in minimal medium supplemented with all amino acids remains very poor (100 min doubling time), whereas the growth of pseudorevertant mutants is normal (35 min). The addition of a variety of osmolytes to the growth medium stimulates growth of SOD mutants (reducing the doubling time to 50 min) and partially eliminates the apparent amino acid auxotrophy (Imlay and Fridovich 1992). This strongly reinforces the idea of envelope lesions generated by superoxide in SOD mutants.

Decreases in proton motive force-dependent and -independent transports have also been observed under conditions of oxidative stress (Farr et al. 1988). Expression of hydroperoxidase I restores normal transport, showing that damage involves formation of hydrogen peroxide. It thus seems likely that the membrane lesions are not directly due to superoxide damage.

Sensitivity to Oxidizing Agents

SOD-negative mutants are very sensitive to oxygen and intracellular superoxide generators, such as paraquat, plumbagin, and methylene blue (Carlioz and Touati 1986). They are hypersensitive to hydrogen peroxide, presumably because the accumulation of superoxide permits the formation of hydroxyl radicals from hydrogen peroxide (Carlioz and Touati 1986; Imlay and Lin 1987).

Although certain effects, such as hydrogen peroxide sensitivity, are clearly due to the participation of superoxide in the Haber-Weiss reaction, there is evidence of direct deleterious effects of superoxide (Fridovich 1986; see also Liochev and Fridovich, this volume). Superoxide is moderately reactive and appears to attack selective targets. These targets remained unidentified until recently because the damage is usually reversible. The use of SOD-deficient mutants, in which the superoxide steady state is much higher (Imlay and Fridovich 1991b), has made identification of superoxide-sensitive enzymes much easier. All enzymes characterized so far contain [4Fe-4S] clusters, which appear to be the target of the oxidative damage (Flint and Emptage 1990). They are α-β-dihydroxy acid dehydratase, 6-phosphogluconate dehydratase, and aconitase (Kuo et al. 1987; Gardner and Fridovich 1991a,b). Under oxidative stress, inactivation of these enzymes results in a more or less drastic decrease in metabolism due to breaks in branched amino acid biosynthetic pathways or in the Entner-Doudoroff pathway, and closing down of the TCA cycle. This reversible drop in metabolism may permit the cell to escape greater damage if oxidative stress is transient, sparing DNA and cell membranes from a potentially lethal oxidative attack.

Other situations in which superoxide inactivates or activates enzymes have been reported (for review, see Touati 1988b). The role of SOD in permitting the activation of ribonucleotide reductase under normal, in vivo growth conditions is particularly interesting. The enzymatic activation of ribonucleotide reductase in *E. coli* includes an essential step of tyrosyl radical formation. Superoxide radicals are generated as by-products during activation in situ, and in the absence of SOD (SOD-negative mutants or in vitro experiments), they inactivate ribonucleotide reductase (Fontecave et al. 1987).

Mutagenesis

SOD-deficient mutants exhibit an increased spontaneous oxygen-dependent mutagenesis (Farr et al. 1986). Their mutagenesis in air is 40-fold higher than in wild-type cells and is further increased about 10-fold by exposure to oxygen or redox active compounds. The superoxide-induced mutagenesis is specific and does not result from induction of the repair-induced error-prone SOS response by the DNA lesions (*recA* and *umuC* independent) (Table 1). Hydrogen peroxide and near-UV light exacerbate the mutagenesis, presumably by generating hydrogen peroxide (Hoerter et al. 1989). The elevated level of mutagenesis is presumably not due to lesions directly produced by superoxide radicals. Indeed, superoxide has no effect on DNA in vitro, and the lesions are more likely to be produced by hydroxyl radicals, which are known to attack DNA and whose formation is favored by the excess of superoxide. Oxygen free radicals in vitro produce a wide range of base modifications (Cadet and Berger 1985; Hutchinson 1985; Aruoma et al. 1989). The nature of the oxidative DNA lesions resulting from the accumulation of superoxide in SOD-defi-

TABLE 1 TET ^{S} TO TET ^{R} FORWARD MUTATIONS

Strain relevant genotype	Anaerobiosis		Aerobiosis		Oxygenated	
Wild-type	7	+4	8	+4	7	+4
sodA sodB	8	+4	122	+32	1550	+70
sodA sodB xthA	--		28	+7	--	
sodA sodB umuC	--		139	+30	1480	+70

Strains to be tested were transformed with the plasmid pPY98, a derivative of pBR322 in which the *tet* gene is under the control of the *mnt*-regulated *ant* promoter of P22. Cells containing the wild-type plasmid are Amp^R, Tet^R. Mutations in the *mnt* gene or in its operator confer tetracycline resistance on the cell. Cultures in early stationary phase were plated onto LB containing 5 µg/ml tetracycline (and 25 µg/ml ampicillin when needed) for Tet^R screening. Numbers are Tet^R colonies for 10^8 cells.

cient strains remains to be determined; premutagenic lesions must be identified, and the spectrum of mutations has to be established and compared with those obtained in vitro by exposing DNA to oxygen free radicals (Loeb et al. 1991).

The mechanism of oxidative mutagenesis is still unknown. Both error-prone bypass of a DNA lesion by the DNA polymerase and error during repair of the damaged DNA can occur. The discovery that much of the observed mutagenesis depends on a functional exonuclease III (encoded by the *xthA* gene), an enzyme involved in the specific repair of oxidative DNA lesions (Kow and Wallace 1985), suggests that at least some of the mutagenic events occur during the repair process.

All deficiencies of SOD-negative *E. coli* mutants are suppressed during anaerobic growth or by the production of SOD from a plasmid carrying a *sod*$^+$ gene. The plasmid may produce an *E. coli* SOD, or any other SOD originating from a different species, including the evolutionarily unrelated human CuZnSOD (Natvig et al. 1987), the *Bacillus stearothermophilus* MnSOD (Bowler et al. 1990), the *A. nidulans* FeSOD (Laudenbach et al. 1989), or the plant *Nicotiana plumbaginifolia* MnSOD (Bowler et al. 1989). This functional complementation has been used successfully to clone SOD genes from other species, such as *B. gingivalis* SOD (Nakayama 1990), *L. ivanovii* SOD (Haas and Goebel 1990), and plant FeSOD (Van Camp et al. 1990). Last, SOD-lacking mutants furnish a unique genetic background for studying any biological process that may include a role for superoxide or SOD.

IS EXCESS SOD DANGEROUS?

There is evidence that a relatively small increase in SOD activity is dangerous in eukaryotes (see Groner et al., this volume). Deleterious effects of excess SOD have been observed in *E. coli* only when there is conjugate overexpression of SOD (SOD expressed from a multicopy plasmid) together with a larger flux of superoxide radicals (in the presence of an intracellular superoxide generator, such as paraquat), suggesting that the natural balance has been upset in these extreme conditions (Bloch and Ausubel 1986; Scott et al. 1987; Laudenbach et al. 1989; Liochev and Fridovich 1991). Indeed, functions involved in defense against oxidative stress are expressed in coordinated responses in *E. coli*. An increase in superoxide flux triggers the induction of numerous proteins (Greenberg and Demple 1989; Walkup and Kogoma 1989), several of which, including MnSOD, are under the positive control of SoxRS (Greenberg et al. 1990; Tsavena and Weiss 1990). Another set of proteins is induced in response to the increased hydrogen peroxide concentration due to superoxide dismutation. Several of them, including the hydroperoxidases that prevent hydrogen peroxide accumulation, belong to the positive regulon *oxyR* (Storz et al. 1990).

Excess of SOD and superoxide may result in accumulation of hydrogen peroxide and a higher level of reduced iron, leading to increased hydroxyl radical formation. Alternatively, or concomitantly, the multicopy plasmid expressing SOD can titrate an effector of SOD regulation, such as SoxR, disrupting the balance between genes under the control of this effector. Liochev and Fridovich (1991) reported that overproduction of SOD from a multicopy plasmid produces growth inhibition and interferes with induction of glucose-6-phosphate dehydrogenase, a member of the *soxRS* regulon. SOD regulation prevents such situations from happening under natural conditions.

SOD REGULATION IN RESPONSE TO THE ENVIRONMENT: MNSOD REGULATION IN E. COLI

A correlation has been established between oxygen tolerance and the SOD content of microorganisms in several situations.

SOD activity is modulated at the posttranslational or the transcriptional level, depending on the organism, the type of SOD, and the environment. Both modes of regulation often occur in the same organism. Thus, the substitution of manganese for iron when the latter becomes scarce is accompanied by an increase in SOD activity in cambialistic SOD. Increased oxidative stress results in increased SOD in numerous bacteria (for review, see Hassan 1989). Although variations in the FeSOD level have been observed in some bacteria, it is the MnSOD that is usually responsible for modulating the total level of SOD in bacteria. The molecular basis of such modulation is, as yet, well documented only in *E. coli*. The following sections therefore concentrate on MnSOD regulation in *E. coli*.

The FeSOD level in *E. coli* is relatively unaffected by a range of environmental perturbations and is generally thought to provide the cell with a first line of defense against superoxide. MnSOD is normally not expressed in anaerobiosis but is induced by aerobiosis and is overexpressed in conditions referred to as "oxidative stress," i.e., increased oxygenation, and challenge with redox cycling compounds (Gregory and Fridovich 1973; Hassan and Fridovich 1977, 1979). MnSOD thus appears to be a supplementary protectant, permitting the adjustment of total SOD to the oxidative stress. However, the extent of this response is limited; the total level of SOD varies only tenfold between extreme conditions of anaerobiosis (where there is presumably no need for SOD) and challenge with high concentrations of redox cycling compounds (100 μM paraquat), suggesting a rigorous control.

Fusions with Lactose Operon Genes: A Useful Tool for Studying MnSOD Regulation

During the past few years, studies on MnSOD induction, mainly by the teams of I. Fridovich and H. Hassan, have demonstrated that MnSOD is induced aerobically by oxygen, redox cycling compounds, and iron chelators, and anaerobically by iron chelators and compounds like nitrate, which act as alternative terminal electron acceptors and permit anaerobic respiration (Pugh and Fridovich 1985; Pugh et al. 1984; Miyake 1986; Moody and Hassan 1984; Hassan and Moody

1987; Privalle and Fridovich 1988). Several models have been proposed to explain the anaerobic induction of MnSOD: Moody and Hassan (1984) hypothesized that MnSOD is regulated by an iron-containing allosteric repressor protein which is active only if it contains Fe^{++} and inactive when it contains Fe^{+++} or no iron; Pugh and Fridovich (1985) favored a posttranslational regulation by metal loading.

Protein and operon fusions have been constructed with the genes of the lactose operon: *sodA-lacZ* protein fusions, *tac-sodA*+ operon fusion to investigate the molecular basis of these observations (Touati 1988a). The recent engineering of a *sodA* deletion strain, together with a *sodA-lacZ* operon fusion integrated at the attλ site on the bacterial chromosome (I. Compan and D. Touati, unpubl.), permits further analysis of the transcriptional events in the presence of an intact or deleted *sodA*+ allele (merodiploids). These fusions, together with various plasmids carrying part or all of the *sodA* locus, led to the conclusion that MnSOD is multiregulated in a more elaborate fashion than that proposed in earlier models, and supported both transcriptional and posttranslational control. The studies indicate that a positive control, via superoxide, and a negative control, via Fe, were exerted at the transcriptional level. An autogenous regulation necessitating only the amino-terminal part of the protein was also suggested by the inhibition of β-galactosidase (β-Gal) induction by paraquat from a chromosomal *sodA-lacZ* fusion in the presence of a plasmid expressing hybrid protein (*sodA-kan*). Last, there was posttranslational modulation of the activity that depended on the metal concentration in the growth medium. The effectors of these multiple controls were identified by isolating and characterizing regulation mutants.

MnSOD Is Part of the soxR Regulon

Evidence for a positive regulation of MnSOD has been recently provided. The biosynthesis of about 40 proteins is enhanced in response to increased superoxide flux (Walkup and Kogoma 1989; Greenberg and Demple 1989). Nine of these proteins belong to the same regulon, controlled by the locus *sox*, located at 92 min on the linkage map. The *sox*-dependent induced gene products are MnSOD, the DNA repair enzyme endonuclease IV (*nfo*), glucose-6-phosphate dehydrogenase (*zwf*), a NADPH paraquat-diaphorase, a modified ribosomal protein, the uncharacterized products of the *soi* genes (Kogoma et al. 1988), and an antisense inhibitor of the gene for a major porin, *ompF* (Greenberg et al. 1990; Tsavena and Weiss 1990). Interestingly, two enzymes, glucose-6-phosphate dehydrogenase and paraquat diaphorase of the *soxR-soxS* regulon, enhance superoxide production from paraquat, the first enzyme producing the NADPH used by the second. Therefore, Weiss and Wu (1992) proposed that the regulon may provide antimicrobial offense as well as a defense.

The *sox* locus exerts positive control over the regulon genes. It has been sequenced (Wu and Weiss 1991) and contains two adjacent divergently transcribed genes, *soxR* and *soxS*. The *soxS* promoter is within the 85-nucleotide intergenic region, whereas the *soxR* promoter is within *soxS*. The *soxS* mRNA increased after superoxide induction, and the *soxS* protein resembles members of the AraC family of positive regulators. The *soxR* gene product is not inducible; it contains four cysteines that may be involved in a redox sensing activity of *soxR*. The way in which Sox proteins activate transcription, and the individual roles of SoxR and SoxS, remain to be discovered. The present data suggest that *soxR* and *soxS* constitute a two-component system, in which SoxR is the sensor protein and SoxS the regulatory protein. Motifs that may give binding to DNA are found in the sequences of both genes.

MnSOD, as monitored by β-Gal expression in a *sodA-lacZ* fusion, is expressed at a low level in aerobiosis in a *sox*-deleted strain (Δ*sox*) and is no longer inducible by paraquat. It is expressed at a high level in the *sox* constitutive mutant (*sox^c*) and is not further (or very slightly) inducible by paraquat. *sox* mutations have no effect on the *sodA-lacZ* fusion expression in anaerobiosis, in contrast with other members of the *sox* regulon. Thus, the *nfo* gene is overexpressed in a *sox* constitutive mutant in anaerobiosis, whereas the *sodA* gene is repressed (Table 2) (Tsavena and Weiss 1990; Touati et al. 1992). This suggests that additional controls, active in anaerobiosis, preclude activation of *sodA* by *soxR-soxS*.

The sequences of other *soxR-soxS* regulated genes have

TABLE 2   EFFECT OF MUTATIONS IN FUR, ARCA, AND SOX ON MNSOD EXPRESSION IN VARIOUS CONDITIONS, AS MONITORED BY β-GAL ACTIVITY IN A SODA-LACZ FUSION

	β-Gal units			
	anaerobiosis		aerobiosis	
Strain relevant genotype	no addition	1-10 phe 100 μM	no addition	paraquat 50 μM
Wild-type	8	40	850	7300
fur	35	80	1900	8500
arcA	16	530	850	7250
fur arcA	540	1100	2650	8550
Δsox	7	18	650	680
sox^c	7	17	2650	6000
fur arcA sox^c	630	1300	4800	5000

Values in aerobiosis were taken from kinetics at OD 1.0; measurements in anaerobiosis were done on cultures grown for 3 hr from overnight anaerobic precultures diluted to OD 0.1. All strains were isogenic except for the indicated mutations. Wild type was QC772 (Carlioz and Touati 1986); mutant alleles were: fur = fur::Tn5 (Bagg and Neilands 1987b), arcA = sodZ (Tardat and Touati 1991), Δsox = Δsox8::cat and sox^c = soxR4 (Tsavena and Weiss 1990). 1-10 phe = 1-10 phenanthroline.

been established (zwf, Rowley and Wolf 1991; nfo, Saporito and Cunningham 1988). The zwf gene is regulated by growth rate, but no relationship between this regulation and the mechanism for induction by superoxide has yet been established. R. Cunningham reported a palindromic sequence in the −90, −110 region of nfo as a possible Sox-binding site (Cunningham and Saporito 1990).

Two Negative Controls of MnSOD Expression

Analysis of trans-acting regulation mutants, derepressed for β-Gal expression of a sodA-lacZ fusion in anaerobiosis, revealed that they were double mutants carrying mutations in each of the global systems, Fur (ferric uptake regulation) and Arc (aerobic respiration control) (Tardat and Touati 1991). Two double mutants harboring four distinct mutations were characterized (see below). Mutations affecting a single control did not relieve anaerobic repression (6-30 units of β-Gal as monitored by a sodA-lacZ fusion); mutations in the two controls had a synergistic effect (350-600 units of β-Gal), demonstrating two independent negative controls.

MnSOD is negatively controlled by Fur protein. With the appearance of oxygen on earth, iron became insoluble and cells had to develop systems for its assimilation. About 30 genes are involved in iron uptake metabolism in *E. coli* (Braun 1985; Neilands 1990). They all are under the control of the product of the *fur* gene (Hantke 1982, 1984). Under conditions of iron sufficiency, the Fur protein acts as a transcriptional repressor (Bagg and Neilands 1987a). Active Fur protein binds to a specific DNA sequence (GATAATGATAATCATTATC), termed the iron box (de Lorenzo et al. 1987). Fur can use several divalent metals as cofactors in vitro, but it seems likely that Fe^{++} is used in vivo. The binding of Fur to DNA appears to be an unusual association in which the repressor sandwiches and/or wraps around the DNA rather than interacting with it on only one side of the helix (de Lorenzo et al. 1988).

The *sodA* sequence contains two putative promoters (indicated -35, -10 and -35', -10' on Fig. 2), but only the -35, -10 is used in vivo (and in vitro) under normal growth conditions (Takeda and Avila 1986). The *sodA* promoter region contains two possible overlapping Fur consensus sequences (Fig. 2); they overlap the -35 (and -35') RNA polymerase binding sites, as usually do the iron boxes (Neilands 1990). Binding of Mn^{++}-holoFur to *sodA* DNA has been demonstrated by Niederhoffer et al. (1990) using purified protein and band-shift gel electrophoresis analysis. The exact site at which Fur is bound to *sodA* remains to be established.

Mutations affecting *sodA* expression have been identified: *fur(sodH)* is, by mapping, complementation test, and sequencing, in the *fur* gene, and *sodX* is an unknown gene, mapped in the *ompB* region and presumably involved in iron uptake. Mutations in *sodX* have the same effects on *sodA* as do *fur(sodH)* and previously isolated mutations in the *fur* gene (Hantke 1984; Bagg and Neilands 1987b). Conversely, mutations in the *fur(sodH)* and *sodX* genes derepress other *fur* regulated genes, as do *fur* mutations. *fur* and *sodX* affect *sodA* expression only slightly in anaerobiosis, unless they are associated with a mutation in one of the *arc* genes. A large derepression of *sodA* occurs when both *arc* and *fur* genes are mutated (Table 2). In aerobiosis, the *sodA* concentration is two- to threefold higher in the *fur* mutant than in *fur*$^+$ cells, showing

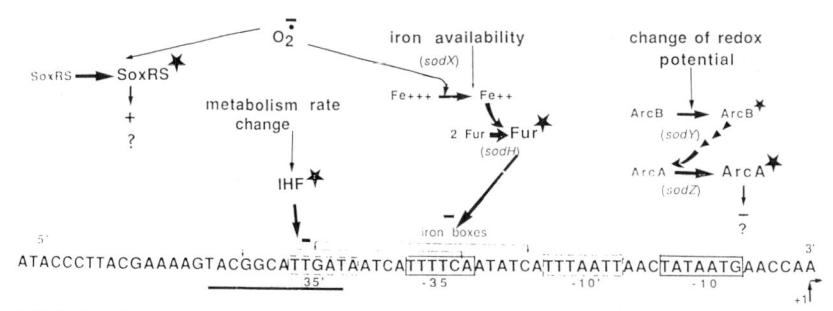

FIGURE 2 Hypothetical regulation of *sodA* transcription. The 5′ nucleotide sequence of *sodA* is shown. The two possible iron boxes are indicated. One IHF consensus sequence is underlined. Question marks indicate that effector DNA targets are not known. Hypothetical stimuli are indicated.

that *fur* is still operating aerobically. Aerobic repression by Fur in conditions of iron sufficiency does not interfere with *sox* activation, suggesting that the two effectors have different DNA targets (Tardat and Touati 1991).

Iron chelators render Fur inactive, so that adding iron chelators to the growth medium mimics the effects of *fur* mutations. However, iron chelators have additional effects in *fur* mutants (Tardat and Touati 1991) that may be posttranslational (see below).

ArcA represses MnSOD in anaerobiosis. ArcA and ArcB genes are involved in the anaerobic repression of several genes for aerobic respiration, such as succinate dehydrogenase, L-lactate dehydrogenase, and aconitase (Iuchi and Lin 1988; Iuchi et al. 1989a). They form a two-component system (Stock et al. 1989) in which the sensor protein, ArcB, perceives the unfavorable respiratory state and activates the ArcA regulatory protein (Iuchi et al. 1990a). ArcA usually acts as a repressor, although it may occasionally act as an activator (Iuchi et al. 1989b, 1990b). Double mutants, in which *sodA* was derepressed in anaerobiosis, contained mutations in the *fur* system and mutations in either *arcA* or *arcB*. The mutations have been characterized by mapping, by complementation tests, and by their effects on other *arc*-regulated genes (Tardat and Touati 1991). Mutations in *arc* had negligible effect on *sodA-lacZ* expression in a Fur⁺ context. When *fur* was inactivated by mutation or by adding chelating agents to the growth medium, mutations in *arc* derepressed *sodA*, showing that single repressions by Fur or Arc are fully effective and act independently of each other. Mutations in *arc* had no effect in aerobiosis, in agreement with the observation of Iuchi and Lin (1988) that ArcA has no action on respiratory functions in aerobiosis.

arc genes have been sequenced (Drury and Buxton 1985; Iuchi et al. 1990a). Nothing is known of the mechanism of ArcA repression or activation. Genetic studies indicate that it acts at the transcriptional level; both *arcA* and *arcB* genes contain possible DNA-binding motifs. The *arcA(sodZ)* mutation (Tardat and Touati 1991), which has a leaky phenotype compared to other *arc* mutants (*arcA1*, Iuchi and Lin 1988), has been shown by sequencing of the mutated gene to form a 45-amino-acid carboxy-terminally truncated protein, altering the presumed DNA-binding region (B. Tardat and D. Touati, unpubl.). A DNA consensus sequence for Arc binding cannot yet be identified because the sequences of other genes under *arc* control are not available.

The pleiotropic phenotype of regulatory mutants sometimes makes them difficult to use. For instance, interpretation of genetic studies using available *arcA* mutants is complicated by the fact that these mutants do not contain absolute mutations or contain deletions that are too large, covering several kilobases (many genes). Thus, the inability of *sox* constitutive mutants to activate *sodA* in anaerobiosis in a *fur*::Tn5 *arcA1* background (Table 2) may be due to a residual ArcA binding to DNA, and thus competing with Sox. We are attempting to construct new tools, such as Δ*arcA* internal deletion, to test this possibility.

The signal recognized by ArcB is still not identified. Whereas the *arcA* gene product is a soluble protein (Drury and Buxton 1985), the ArcB protein is probably anchored to the membrane, close to the inner surface where the signal is generated. ArcB contains cysteine residues that are good candidates for sensing signals from a redox reaction (Iuchi et al. 1990a). The requirement of terminal cytochromes for generating the aerobic signal for the *arc* regulatory system suggests that ArcB senses the presence of O_2 by the concentration of a reduced form of a terminal electron transport component or that of a non-autooxidizable compound linked to the process by a redox reaction (Iuchi et al. 1990b). Like other sensor proteins of two-component systems (Stock et al. 1989), ArcB is believed to perceive and transmit signal by phosphorylation reactions (Iuchi et al. 1990a).

Effects of DNA Supercoiling and Integration Host Factor on MnSOD Expression

DNA topology plays an important role in *sodA* expression. Changes in DNA supercoiling have been shown to have pleiotropic effects. They regulate the transcriptional expression of various promoters in response to different environmental signals triggered by growth in adverse conditions, including osmotic and anaerobic stresses. Two enzymes, DNA gyrase and DNA topoisomerase I, are primarily responsible for determining the level of DNA supercoiling. Mutations in another gene, *oxrC*, originally considered to be an aerobic regulatory locus, are now recognized as affecting the level of negative supercoiling (Ni Bhriain et al. 1989).

During their studies on transcription of the MnSOD gene, Takeda and Avila (1986) noticed that in vitro transcription was much less efficient (one-fifth to one-tenth) when supercoiled plasmid was used as template rather than linear DNA. L.W. Schrum and H.M. Hassan (in prep.) have shown that the expression of MnSOD, monitored by measuring β-Gal in the protein fusion *sodA-lacZ*, is stimulated by the gyrase inhibitors, nalidixic acid and coumermycin A. The stimulatory effect is seen when SOD is derepressed and is additive to the Sox-mediated stimulation by paraquat.

Takeda and Avila (1986) analyzed the MnSOD nucleotide sequence and identified several possible integration host factors (IHFs), binding sites within and around a palindrome in front of and overlapping the −35 RNA polymerase binding site (Fig. 2). An IHF is a DNA-binding protein that interacts with specific DNA sequences (Craig and Nash 1984). Biochemical experiments have shown that an IHF interacts primarily with the bases and deoxyribose sugars of the DNA backbone in the minor growth of the DNA helix (Yang and Nash 1989). It induces a bend, estimated to be greater than 140°, in the DNA target of the binding site, thus altering regulation. The IHF complex is formed by two subunits encoded by *himA* and *himD* genes. Mutations in either gene destroy IHF action.

L.W. Schrum and H.M. Hassan (in prep.) have shown that *himA* mutations increase aerobic *sodA* expression but have no effect under anaerobic conditions. The effect of gyrase inhibitors was found to be independent of the *himA* gene product. One IHF major consensus site on the *sodA* regulatory sequence overlaps with the iron boxes. IHF binding therefore probably interferes with Fur binding. Modifying a potential IHF site by directed mutagenesis (Naik and Hassan 1990) renders constitutive the expression of *sodA* on the multicopy modified plasmid (5 bp were changed). It is indeed independent of the presence of iron chelators in the growth medium. It is also independent of paraquat in the medium, suggesting that the DNA modification precludes *sox* activation either directly, by changing the Sox DNA-binding site, or indirectly, by changing the DNA topology. The decreased expression of SOD from the mutated plasmid in anaerobiosis also suggests that *arc* repression still operates on the modified promoter region.

Posttranslational Modulation of MnSOD Activity

During the investigation of the MnSOD activity expressed by a multicopy plasmid carrying a *sodA*[+] gene in front of which the *sodA* promoter has been replaced by the Lac promoter, it was noticed that the level of activity depended on the growth medium's being supplemented with manganese. Similarly, cells exposed to paraquat, which at low concentration parallels the superoxide challenge, showed a much higher SOD activity only if manganese (100 µM) was added to the medium (Touati 1988a). This strongly suggests that there was too little manganese in the medium to fully occupy the active sites when the apoprotein was synthesized in large amounts.

Privalle et al. (1989) studied this phenomenon in detail in anaerobiosis. They showed that inactive, reconstitutable MnSOD was induced by anaerobic respiration. Reconstitution experiments suggested that the lack of activity was because the active site was occupied by a metal other than manganese.

This metal is presumably iron, which has been found to compete with manganese for the active site under aerobic conditions (Hassan and Moody 1987; Beyer and Fridovich 1991).

Privalle and Fridovich (1990) also showed that anaerobically grown E. coli accumulate active MnSOD upon exposure to diamide. The oxidative action of diamide is proposed to occur at the transcriptional and posttranslational levels. Posttranslationally, it facilitates the insertion of manganese, rather than iron, into the nascent MnSOD polypeptide. This posttranslational modulation explains an earlier observation that anaerobic expression of active MnSOD from a plasmid carrying the superoxide-insensitive tac-$sodA^+$ fusion was enhanced by the iron chelator 1-10-phenanthroline (Touati 1988a).

Biological Significance of the Intricate MnSOD Multiregulation

MnSOD is mainly responsible for modulating total cell SOD in E. coli. The concentration of FeSOD usually remains constant, although there is a recent report that it can also vary and seems to be positively controlled by *fur* (Niederhoffer et al. 1990). A lack or excess of SOD results in cell damage, either because a defense against oxygen poisoning is directly impaired or because the balance between the components of the defense system is perturbed. Oxidative stresses vary with the environment, and their potential for damage depends on the environment (e.g., iron concentration). Damage may occur to DNA, membranes, or proteins. The availability of active SOD should rapidly adapt to the stress, but without overshoot, to provide maximum protection for the cell, in cooperation with other components of the defense system. Therefore, several signals can trigger SOD regulators to produce a finely tuned response. The roles of the various factors controlling SOD activity are still far from completely understood; there is, however, some coherence in the MnSOD expression in response to the various signals.

MnSOD is repressed in anaerobiosis, where there is usually no oxidative stress. Oxygen or oxidative conditions trigger an initial signal transmitted by ArcB to ArcA. As oxidative stress becomes more drastic, Arc repression is totally released and Sox activation begins and increases with the magnitude of the oxidative stress. The Arc shutoff of MnSOD in anaerobiosis, like the shutoff of numerous aerobic functions, is probably due to a strategy of economy. *ArcA*-controlled aerobic functions are derepressed when they become necessary.

The regulations by iron and Fur occur in both anaerobiosis and aerobiosis, and superimpose on the primary scheme in a more subtle way. In iron excess, an environment which favors hydroxyl radical formation, MnSOD activity is decreased by Fur transcriptional repression and inactivation of the holoenzyme by replacing the manganese by iron. Iron unavailability, as a result of starvation or oxidation, gives rise to an increase in MnSOD when FeSOD may become inactive. Since anaerobic Arc repression is sufficient to shut off MnSOD expression, it is possible that the major role of Fur is to sense iron rather than the redox state of the cells, although the Fe^{+++}/Fe^{++} balance depends on both.

Osmotic stress and aerobiosis-anaerobiosis shifts produce changes in DNA supercoiling. The stress signals leading to the change in topology are unknown. They may well originate from the membrane, and the topological regulation of MnSOD could then be a response correlated to (potential) membrane damage. An increase in metabolism requires more SOD (Imlay and Fridovich 1991b). The message detected by IHF might be a change in growth rate. Yet in vivo, the level of the IHF α subunit increases 15-fold with a 4-fold increase in growth rate (Pagel and Hatfield 1991).

There are two particularly striking features of this system: First, MnSOD regulators are all global regulators, and second, a lack of SOD has pleiotropic effects. The multiple routes by which MnSOD may be induced thus permit the induction of other functions together with SOD. The particular group of functions induced will depend on the threat to the cells and the signal detected, such as iron scarcity, oxygen, excess superoxide, membrane alteration, or a change in metabolism.

PERSPECTIVES

The work reviewed above describes the early phases in the molecular characterization of the MnSOD gene from *E. coli* and the regulation of its expression. As our understanding of MnSOD regulation progresses, the topic becomes more and more fascinating. The molecular role of each of the regulatory factors remains to be determined. Few genes, if any, are regulated by so many different global regulators. No other genes have to date been shown to be regulated by even two of the Fur, Arc, or Sox proteins. Thus, study of MnSOD regulation is not only of enormous importance for comprehending the biological role of SOD, it is also an exciting model for studies on protein-DNA and protein-protein interactions.

Relatively little is known of the molecular biology of bacteria other than *E. coli*. Further insight into the nature and regulation of various bacterial SODs should provide invaluable information on the relationship between oxidative stress and such fundamental bacterial properties as pathogenicity and photosynthesis.

ACKNOWLEDGMENTS

I thank I. Fridovich and his co-workers and B. Weiss for kindly providing preprints. Research was supported in part by a grant from the Association pour la Recherche sur le Cancer (No. 6791).

REFERENCES

Abdollahi, H. and J.W.T. Wimpenny. 1990. Effects of oxygen on the growth of *Desulfovibrio desulfuricans*. *J. Gen. Microbiol.* **136:** 1025.

Amano, A., S. Shizukuishi, H., Tamagawa, K. Iwakura, S. Tsumasawa, and A. Tsunemitsu. 1990. Characterization of superoxide dismutases purified from either anaerobically maintained or aerated *Bacteroides gingivalis*. *J. Bacteriol.* **172:** 1457.

Arai, K. and H. Oguchi. 1990. Proliferation and lipid peroxidation of *Streptococcus mutans* associated with superoxide dismutase. *J. Dent. Res.* **69:** 376.

Aruoma, O.I., B.V. Halliwell, and M. Dizdaroglu. 1989. Iron-dependent modification of bases in DNA by the superoxide radical generating system hypoxanthine/xanthine oxidase. *J. Biol. Chem.* **264:** 13024.

Bagg, A. and J.B. Neilands. 1987a. Ferric uptake regulation protein acts as a repressor, employing iron(II) as a cofactor to bind the operator of an iron transport operon in *Escherichia coli*. *Biochemistry* **26:** 5471.

———. 1987b. Molecular mechanism of regulation of siderophore-mediated iron assimilation. *Microbiol. Rev.* **51:** 509.

Beaman, B.L., C.M. Black, F. Doughty, and L.V. Beaman. 1985. Role of superoxide dismutase and catalase as determinants of pathogenicity of *Nocardia asteroides:* Importance in resistance to microbial activities of human polymorphonuclear neutrophils. *Infect. Immun.* **47:** 135.

Beaman, B.L., S.M. Scates, S.E. Moring, R. Deem, and H.P. Misra. 1983. Purification and properties of a unique superoxide dismutase from *Nocardia asteroides*. *J. Biol. Chem.* **258:** 91.

Beaman, L. and B.L. Beaman. 1990. Monoclonal antibodies demonstrate that superoxide dismutase contributes to protection of *Nocardia asteroides* within the intact host. *Infect. Immun.* **58:** 3122.

Beck, B.L., L.B. Tabatabai, and J.E. Mayfield. 1990. A protein isolated from *Brucella abortus* is a Cu-Zn superoxide dismutase. *Biochemistry* **29:** 372.

Beyer, W.F., Jr. and I. Fridovich. 1991. In vivo competition between iron and manganese for occupancy of the active site region of the manganase superoxide dismutase of *Escherichia coli*. *J. Biol. Chem.* **266:** 303.

Beyer, W., J. Imlay, and I. Fridovich. 1991. Superoxide dismutases. *Prog. Nucleic Acid Res. Mol. Biol.* **40:** 221.

Bloch, C.A. and F.M. Ausubel. 1986. Paraquat-mediated selection for mutations in the manganese-superoxide dismutase gene *sodA*. *J. Bacteriol.* **168:** 795.

Bowler, C., L. Van Kaer, W. Van Camp, M. Van Montagu, D. Inze, and P. Dhaese. 1990. Characterization of the *Bacillus stearothermophilus* manganese superoxide dismutase gene and its ability to complement copper/zinc superoxide dismutase deficiency in *Saccharomyces cerevisiae*. *J. Bacteriol.* **172:** 1539.

Bowler, C., T. Alliotte, M. Van den Bulcke, G. Bauw, J. Vandekerchhove, M. Van Montagu, and D. Inze. 1989. A plant manganese superoxide dismutase is efficiently imported and correctly processed by yeast mitochondria. *Proc. Natl. Acad. Sci.* **86:** 3237.

Braun, V. 1985. The unusual features of the iron transport systems of *Escherichia coli*. *Trends Biochem. Sci.* **10:** 75.

Britton, L. and I. Fridovich. 1977. Intracellular localization of the superoxide dismutases of *Escherichia coli*: A reevaluation. *J. Bacteriol.* **131:** 815.

Brown, O.R. and F. Yein. 1978. Dihydroxyacid dehydratase: The site of hyperbaric oxygen poisoning in branched-chain amino acid biosynthesis. *Biochem. Biophys. Res. Commun.* **85:** 1219.

Cadet, J. and M. Berger. 1985. Radiation-induced decomposition of the purine bases within DNA and related model compounds. *Int. J. Radiat. Biol.* **47:** 127.

Carlioz, A. and D. Touati. 1986. Isolation and superoxide dismutase mutants in *E. coli:* Is superoxide dismutase necessary for aerobic life? *EMBO J.* **5:** 623.

Craig, N. and H. Nash. 1984. *E. coli* integration host factor binds to specific sites in DNA. *Cell* **39:** 707.

Cunningham, R.P. and S.M. Saporito. 1990. Cloning radiation repair genes in *E. coli. UCLA Symp. Mol. Cell. Biol. New Ser.* **136:** 261.

de Lorenzo, V., F. Giovannini, M. Herrero, and J.B. Neilands. 1988. Metal ion regulation of gene expression Fur repressor-operator interaction at the promoter region of the aerobactin system of pColV-K30. *J. Mol. Biol.* **203:** 825.

de Lorenzo, V., S. Wee, M. Herrero, J.B. Neilands. 1987. Operator sequence of the aerobactin operon of plasmid ColV-K30 binding the ferric uptake regulation (*fur*) repressor. *J. Bacteriol.* **169:** 2624.

Dougherty, H.W., S.J. Sandowski, and E.E. Baker. 1978. A new iron-containing superoxide dismutase from *Escherichia coli*. *J. Biol. Chem.* **253:** 5220.

Drury, L.S. and R.S. Buxton. 1985. Identification and sequencing of the *Escherichia coli cet* gene which codes for an inner membrane protein, mutation of which causes tolerance to colicin E2. *Mol. Microbiol.* **2:** 109.

Farr, S.B., R. D'Ari, and D. Touati. 1986. Oxygen-dependent mutagenesis in *Escherichia coli* lacking superoxide dismutase. *Proc. Natl. Acad. Sci.* **83:** 8268.

Farr, S.B., D. Touati, and T. Kogoma. 1988. Effects of oxygen stress on membrane functions in *Escherichia coli:* Role of HPI catalase. *J. Bacteriol.* **170:** 1837.

Fee, J.A., E.C. Niederhoffer, and C. Naranjo. 1988. First description of a variant of *E. coli* lacking superoxide dismutase activity yet able to grow efficiently on minimal oxygenated medium. In *Trace element metabolism in man and animals* (ed. L. Hurley et al.), p. 239. Plenum Press, New York.

Flint, D.H. and M.H. Emptage. 1990. Dihydroxy acid dehydratase: Isolation, characterization as an iron sulphur protein and sensitivity to inactivation by oxygen radicals. In *Biosynthesis of branched-chain amino acids* (D. Chipman et al.), p. 285. Balaban, Philadelphia.

Fontecave, M., A. Gräslund, and P. Reichard. 1987. The function of superoxide dismutase enzymatic formation of the free radical of ribonucleotide reductase. *J. Biol. Chem.* **262:** 12332.

Franzon, V.L., J. Arrondel, and P.J. Sansonetti. 1990. Contribution of superoxide dismutase and catalase activities to *Shigella flexneri* pathogenesis. *Infect. Immun.* **58:** 529.

Fridovich, I. 1986. Superoxide dismutases. *Adv. Enzymol.* **58:** 68.

Gardner, P.R. and I. Fridovich. 1991a. Superoxide sensitivity of the

Escherichia coli 6-phosphogluconate dehydratase. *J. Biol. Chem.* **266:** 1478.

———. 1991b. Superoxide sensitivity of the *Escherichia coli* aconitase. *J. Biol. Chem.* **266:** 19328.

Greenberg, J.T. and B. Demple. 1989. A global response induced in *Escherichia coli* by redox-cycling agents overlaps with that induced by peroxidase stress. *J. Bacteriol.* **171:** 3933.

Greenberg, J.T., P. Monach, J.H. Chou, P.D. Josephy, and B. Demple. 1990. Positive control of a global antioxidant defense regulon activated by superoxide-generating agents in *Escherichia coli*. *Proc. Natl. Acad. Sci.* **87:** 6181.

Gregory, E.M. 1985. Characterization of the O_2-induced manganese-containing superoxide dismutase from *Bacteroides fragilis*. *Arch. Biochem. Biophys.* **238:** 83.

Gregory, E.M. and I. Fridovich. 1973. Induction of superoxide dismutase by molecular oxygen. *J. Bacteriol.* **114:** 543.

Gregory, E.M., W.E.C. Moore, and L.V. Holdeman. 1978. Superoxide dismutase in anaerobes: Survey. *Appl. Environ. Microbiol.* **35:** 988.

Gruber, M.Y., B.R. Glick, and J.E. Thompson. 1990. Cloned manganese superoxide dismutase reduced oxidative stress in *Escherichia coli* and *Anacystis nidulans*. *Proc. Natl. Acad. Sci.* **87:** 2608.

Gutteridge, J.M.C. 1985. Superoxide dismutase inhibits superoxide-driven reaction at two different levels. *FEBS Lett.* **185:** 19.

Haas, A. and W. Goebel. 1990. Cloning and expression in *Escherichia coli* of a gene encoding superoxide dismutase from *Listeria ivanovii*. *Free Radical Res. Commun.* **12:** 371.

Halliwell, B. and J.M.C. Gutteridge. 1984. Oxygen toxicity, oxygen radicals, transition metals and disease. *Biochem. J.* **219:** 1.

Hantke, K. 1982. Negative control of iron uptake systems in *Escherichia coli*. *FEMS Microbiol. Lett.* **15:** 83.

———. 1984. Cloning of the repressor protein gene of iron-regulated systems in *Escherichia coli* K12. *Mol. Gen. Genet.* **197:** 337.

Hassan, H.M. 1989. Microbial superoxide dismutases. *Adv. Genet.* **26:** 65.

Hassan, H.M. and I. Fridovich. 1977. Regulation of the synthesis of superoxide dismutase in *E. coli*: Induction by methyl viologen. *J. Biol. Chem.* **252:** 7667.

———. 1979. Intracellular production of superoxide radical and of hydrogen peroxide by redox active compounds. *Arch. Biochem. Biophys.* **196:** 385.

Hassan, H.M. and C.S. Moody. 1987. Regulation of manganase superoxide dismutase in *E. coli*. Anaerobic induction by nitrate. *J. Biol. Chem.* **262:** 17173.

Hassett, J.D. and M.S. Cohen. 1989. Bacterial adaptation to oxidative stress: Implications for pathogenesis and interaction with phagocytic cells. *FASEB J.* **3:** 2574.

Hoerter, J., A. Eisenstark, and D. Touati. 1989. Mutations by nearultraviolet radiation in *Escherichia coli* strains lacking superoxide dismutase. *Mutat. Res.* **215:** 161.

Hopkin, K.A. and H.M. Steinman. 1992. Are manganese and iron superoxide dismutases of *Escherichia coli* K12 physiologically equivalent? In *Oxidative damage and repair: Clinical, biological, and medical aspects* (ed. K.J.A. Davies), p. 31. Pergamon Press, New York.

Hutchinson, F. 1985. Chemical changes induced in DNA by ionizing radiation. *Prog. Nucleic Acid Res. Mol. Biol.* **32:** 115.

Imlay, J.A. and I. Fridovich. 1991a. Isolation and genetic analysis of a mutation that suppresses the auxotrophies of superoxide dismutase deficient *Escherichia coli* K12. *Mol. Gen. Genet.* **228:** 410.

——. 1991b. Assay of metabolic superoxide production in *Escherichia coli*. *J. Biol. Chem.* **266:** 6957.

——. 1992. Suppression of oxidative envelope damage by pseudoreversion of superoxide dismutase-deficient mutant of *Escherichia coli*. *J. Bacteriol.* **176:** 953.

Imlay, J.A. and S. Linn. 1987. Mutagenesis and stress-responses induced in *Escherichia coli* by hydrogen peroxide. *J. Bacteriol.* **169:** 2967.

Iuchi, S. and E.C.C. Lin. 1988. arcA (dye), a global regulatory gene in *Escherichia coli* mediating repression of enzymes in aerobic pathways. *Proc. Natl. Acad. Sci.* **85:** 1888.

Iuchi, S., D.C. Cameron, and E.C.C. Linn. 1989a. A second global regulator gene (arcB) mediating repression of enzymes in aerobic pathways of *Escherichia coli*. *J. Bacteriol.* **171:** 868.

Iuchi, S., D. Funlond, and E.C.C. Lin. 1989b. Differentiation of arcA, arcB, and cpxA mutant phenotypes of *Escherichia coli* by sex pilus formation and enzyme regulation. *J. Bacteriol.* **171:** 2889.

Iuchi, S., Z. Matsuda, T. Fujwara, and E.C.C. Lin. 1990a. The arcB gene of *Escherichia coli* encodes a sensor-regulator protein for anaerobic repression of the arc modulon. *Mol. Microbiol.* **4:** 715.

Iuchi, S., V. Chepuri, H.A. Fu, R.B. Gennis, and E.C.C. Lin. 1990b. Requirement for terminal cytochromes in generation of the aerobic signal for the arc regulatory system in *Escherichia coli*: Study utilizing deletions and *lac* fusions of cyo and cyd. *J. Bacteriol.* **172:** 6020.

Kow, Y.W. and S.J. Wallace. 1985. Exonuclease III recognizes urea residues in oxidized DNA. *Proc. Natl. Acad. Sci.* **82:** 8354.

Kogoma, T., S.B. Farr, K.M. Joyce, and D.O. Natvig. 1988. Isolation of gene fusions (soi::lacZ) inducible by oxidative stress in *E. coli*. *Proc. Natl. Acad. Sci.* **85:** 4769.

Kuo, C.F., T. Mashino, and I. Fridovich. 1987. α, β-Dihydroxyisovalerate dehydratase: A superoxide-sensitive enzyme. *J. Biol. Chem.* **262:** 4724.

Laudenbach, D.E., C.G. Trick, and N.A. Strauss. 1989. Cloning and characterization of *Anacystis nidulans* R2 superoxide dismutase gene. *Mol. Gen. Genet.* **216:** 455.

Liochev, S.I. and I. Fridovich. 1991. Effects of overproduction of superoxide dismutase on the toxicity of paraquat towards *Escherichia coli. J. Biol. Chem.* **266:** 8767.

Loeb, L.A., T.J. McBride, T.M. Reid, and K.C. Cheng. 1991. Mutagenic spectra of oxygen free radicals. *FASEB J.* (in press).

Lowrie, D.B. 1983. How macrophages kill tubercle bacilli. *J. Med. Microbiol.* **16:** 1.

Martin, M.E., B.R. Byers, M.O.J. Olson, M.L. Salin, J.E.L. Arceneaux, and C. Tolbert. 1986. A *Streptococcus mutans* superoxide dismutase that is active with either manganese or iron as a cofactor. *J. Biol. Chem.* **261:** 9361.

May, B.P. and P.P. Dennis. 1990. Unusual evolution of a superoxide dismutase-like gene from the extremely halophilic archaebacterium *Halobacterium-cutirubrum. J. Bacteriol.* **172:** 3725.

Meier, B., D. Barra, F. Bossa, L. Calabrese, and G. Rotilio. 1982. Synthesis of either Fe- or Mn-superoxide dismutase with an apparently identical protein moiety by a anaerobic bacterium dependent on the metal supplied. *J. Biol. Chem.* **257:** 13977.

Miyake, K. 1986. Effect of nitrate on the level of superoxide dismutase in anaerobically grown *Escherichia coli. J. Gen. Appl. Microbiol.* **32:** 527.

Moody, C.S. and H.M. Hassan. 1984. Anaerobic biosynthesis of the manganese containing superoxide dismutase in *E. coli. J. Biol. Chem.* **259:** 12821.

Naik, S.M. and H.M. Hassan. 1990. Use of site-directed mutagenesis to identify an upstream regulatory sequence of *sodA* gene of *Escherichia coli* K12. *Proc. Natl. Acad. Sci.* **87:** 2618.

Nakayama, K. 1990. The superoxide dismutase-encoding gene of the obligately anaerobic bacterium *Bacteroides gingivalis. Gene* **96:** 149.

Natvig, D.O., K. Imlay, D. Touati, and R.A. Hallewell. 1987. Human copper-zinc superoxide dismutase complements superoxide dismutase deficient *E. coli* mutants. *J. Biol. Chem.* **262:** 14697.

Neilands, J.B. 1990. Parallels in the mode of regulation of iron assimilation in all living species. In *Iron transport and storage* (ed. P. Pomka et al.), p. 41. CRC Press, Boca Raton, Florida.

Ni Bhriain, N., C.J. Dorman, and C.F. Higgins. 1989. An overlap between osmotic and anaerobic stress responses: A potential role for DNA supercoiling in the coordinate regulation of gene expression. *Mol. Microbiol.* **3:** 933.

Niederhoffer, E.C., C.M. Naranjo, K.L. Bradley, and J.A. Fee. 1990. Control of *Escherichia coli* superoxide dismutase (*sodA* and *sodB*) genes by ferric uptake regulation (*fur*) locus. *J. Bacteriol.* **172:** 1930.

Pagel, J.M. and G.W. Hatfield. 1991. Integration host factor-mediated expression of the *ilvGMEDA* operon in *Escherichia coli*. *J. Biol. Chem.* **266:** 1985.

Pennington, C.D. and E.M. Gregory. 1986. Isolation and reconstitution of iron- and manganese-containing superoxide dismutases from *Bacteroides thetaiotaomicron*. *J. Bacteriol.* **166:** 528.

Privalle, C.T. and I. Fridovich. 1988. Induction of superoxide dismutases in *Escherichia coli* under anaerobic conditions. Accumulation of an inactive form of the manganese enzyme. *J. Biol. Chem.* **263:** 4274.

———. 1990. Anaerobic biosynthesis of the manganese-containing superoxide dismutase in *Escherichia coli*. *J. Biol. Chem.* **265:** 21964.

Privalle, C.T. and E.M. Gregory. 1979. Superoxide dismutase and O_2 lethality in *Bacteroides fragilis*. *J. Bacteriol.* **138:** 139.

Privalle, C.T., W.F. Beyer, Jr., and I. Fridovich. 1989. Anaerobic induction of Pro-Mn-superoxide dismutase in *Escherichia coli*. *J. Biol. Chem.* **264:** 2758.

Puget, K. and A.M. Michelson. 1974. Isolation of a new copper-containing superoxide dismutase, bacteriocuprein. *Biochem. Biophys. Res. Commun.* **58:** 830.

Pugh, S.Y.R. and I. Fridovich. 1985. Induction of superoxide dismutase in *Escherichia coli* B by metal chelators. *J. Bacteriol.* **162:** 196.

Pugh, S.Y.R., J.L. DiGiuseppi, and I. Fridovich. 1984. Induction of superoxide dismutase in *Escherichia coli* by manganese and iron. *J. Bacteriol.* **160:** 137.

Rowley, D.L. and R.E. Wolf, Jr. 1991. Molecular characterization of the *Escherichia coli* K-12 *zwf* gene encoding glucose 6-phosphate dehydrogenase. *J. Bacteriol.* **173:** 968.

Sakamoto, H. and D. Touati. 1984. Cloning of iron superoxide dismutase gene (*sodB*) in *Escherichia coli*. *J. Bacteriol.* **159:** 418.

Salin, M.L., M.V. Duke, D. Oesterhelt, and D.-P. Ma. 1988. Cloning and determination of the nucleotide sequence of the Mn-containing superoxide dismutase gene from *Halobacterium halobium*. *Gene* **70:** 153.

Saporito, S.M. and R.P. Cunningham. 1988. Nucleotide sequence of the *nfo* gene of *Escherichia coli* K12. *J. Bacteriol.* **170:** 5141.

Scott, M.D., S.R. Meschnick, and J.W. Eaton. 1987. Superoxide dismutase-rich bacteria: Paradoxical increase in oxidant toxicity. *J. Biol. Chem.* **262:** 3640.

Smith, H. 1977. Microbial surfaces in relation to pathogenicity. *Bacterial Rev.* **41:** 475.

Steinitz, Y., Z. Mazor, and M. Shilo. 1979. A mutant of the cyanobacterium *Plectonema boryanum* resistant to photooxidation. *Plant Sci. Lett.* **16:** 327.

Steinman, H.M. 1982. Copper-zinc superoxide dismutase from *Caulobacter crescentus*. *J. Biol. Chem.* **257:** 10283.

―――. 1985. Bacteriocuprein superoxide dismutases in pseudomonads. *J. Bacteriol.* **162:** 1255.

―――. 1987. Bacteriocuprein superoxide dismutase of *Photobacterium leiognathi:* Isolation and sequence of the gene and evidence for a precursor form. *J. Biol. Chem.* **262:** 1882.

Steinman, H.M. and B. Ely. 1990. Copper-zinc superoxide dismutase of *Caulobacter crescentus.* Gene cloning, sequencing and mapping, and periplasmic location of the enzyme. *J. Bacteriol.* **172:** 2901.

Stock, J.B., A.J. Ninfa, and A.M. Stock. 1989. Protein phosphorylation and regulation of adaptive responses in bacteria. *Microbiol. Rev.* **53:** 450.

Storz, G., L.A. Tartaglia, and B.N. Ames. 1990. Transcriptional regulator of oxidative stress-induced genes: Direct activation by oxidation. *Science* **248:** 189.

Takao, M., A. Oikawa, and A. Yasui. 1990. Characterization of a superoxide dismutase gene from the Archaebacterium *Methanobacterium-thermoautotrophicum. Arch. Biochem. Biophys.* **283:** 210.

Takeda, Y. and H. Avila. 1986. Structure and gene expression of the *E. coli* Mn superoxide dismutase gene. *Nucleic Acids Res.* **11:** 4577.

Tardat, B. and D. Touati. 1991. Two global regulators repress the anaerobic expression of MnSOD in *Escherichia coli*: Fur (ferric uptake regulation) and Arc (aerobic respiration control). *Mol. Microbiol.* **5:** 455.

Thangaraj, H., F.I. Lamb, E.O. Davies, P.J. Jenner, L.H. Jeyakumar, and M.J. Colston. 1990. Identification, sequencing, and expression of *Mycobacterium leprae* superoxide dismutase, a major antigen. *Infect. Immun.* **58:** 1937.

Touati, D. 1983. Cloning and mapping of the manganese superoxide dismutase gene (*sodA*) of *Escherichia coli* K-12. *J. Bacteriol.* **155:** 1078.

―――. 1988a. Transcriptional and post-transcriptional regulation of manganese superoxide dismutase biosynthesis in *Escherichia coli*, studied with operon and protein fusions. *J. Bacteriol.* **170:** 2511.

―――. 1988b. Molecular genetics of superoxide dismutases. *J. Free Radicals Biol. Med.* **5:** 393.

―――. 1989. The molecular genetics of superoxide dismutase in *E. coli.* An approach to understanding the biological role and regulation of SODs in relation to other elements of the defense system against oxygen toxicity. *Free Radical Res. Commun.* **8:** 1.

Touati, D., B. Tardat, and I. Compan. 1992. Regulation of MnSOD in response to environmental stimuli. In *Oxidative damage repair: Clinical, biological, and medical aspects* (ed. K.J.A. Davies), p. 13. Pergamon Press, New York.

Tsavena, F.R. and B. Weiss. 1990. *soxR* a locus governing a superoxide response regulon in *Escherichia coli* K12. *J. Bacteriol.* **172:** 4197.

Van Camp, W., C. Bowler, R. Villarroel, E.W.T. Tsang, M. Van Montagu, and D. Inze. 1990. Characterization of iron superoxide dismutase from plants obtained by genetic complementation in *Escherichia coli. Proc. Natl. Acad. Sci.* **87:** 9903.

Vignais, P.M., A. Terech, C.M. Meyer, and M.F. Henry. 1982. Isolation and characterization of a protein with cyanide-sensitive superoxide dismutase activity from the prokaryote *Paracoccus denitrifans. Biochim. Biophys. Acta* **701:** 305.

Walkup, L.K.B. and T. Kogoma. 1989. *Escherichia coli* proteins inducible by oxidative stress mediated by the superoxide radical. *J. Bacteriol.* **171:** 1476.

Weiss, B. and J. Wu. 1992. Control of a superoxide response regulon of *Escherichia coli.* In *Oxidative damage repair: Clinical, biological and medical aspects* (ed. K.J.A. Davies), p. 188. Pergamon Press, New York.

Welch, D.F., C.P. Sword, S. Brehm, and D. Dusanic. 1979. Relationship between superoxide dismutase and pathogenic mechanisms of *Listeria monocytogenes. Infect. Immun.* **23:** 863.

Wu, J. and B. Weiss. 1991. Two divergently transcribed genes, *soxR* and *soxS*, control a superoxide response regulon of *Escherichia coli. J. Bacteriol.* **173:** 2864.

Yang, C.C. and H.A. Nash. 1989. The interaction of *E. coli* IHF protein with its specific binding sites. *Cell* **57:** 869.

Zhang, Y.Z., R. Lathigra, T. Garbe, D. Catty, and D. Young. 1991. Genetic analysis of superoxide dismutase, the 34 kilodalton antigen of *Mycobacterium tuberculosis. Mol. Microbiol.* **5:** 381.

The Human CuZn Superoxide Dismutase Gene and Down's Syndrome

Y. Groner,[1] O. Elroy-Stein,[1] K.B. Avraham,[1]
M. Schickler,[1] H. Knobler,[1] D. Minc-Golomb,[1]
O. Bar-Peled,[1] R. Yarom,[2] and S. Rotshenker[3]

[1]Department of Molecular Genetics and Virology
The Weizmann Institute of Science, Rehovot, Israel
Departments of [2]Pathology and [3]Anatomy and Embryology
Hebrew University, Hadassah Medical School
Jerusalem, Israel

The human copper zinc superoxide dismutase (h-CuZnSOD) is encoded by a gene residing on chromosome 21. This chromosome when triplicated causes the phenotypic expression of Down's syndrome (DS) (Cooper and Hall 1988), and elevated production of CuZnSOD, due to gene dosage, is commonly found in DS patients (Sinet et al. 1974; Sinet 1982; Groner et al. 1986). DS is the most common human genetic disorder, occurring once in every 600–800 live births. The affected individuals suffer from a wide range of symptoms. Most obvious among these are morphological defects such as hypotonia in the newborn, short stature, and the epicanthic eye folds which give rise to the eye shape characteristic of the syndrome. Patients are mentally retarded, and those who survive past their mid-thirties usually develop Alzheimer's disease (Epstein 1986). The risk of a child being born with trisomy 21 sharply increases as maternal age progresses into the fourth decade of life. Because presently many couples in Western societies postpone parenthood until this age, the incidence of DS is expected to increase. Current techniques for prenatal screening for DS (amniocentesis and chorion villi biopsy) are costly and not without risk; they are therefore applied routinely only to at-risk pregnancies, with the result being that most pregnancies are not screened at all for DS. Moreover, because of continuous improvements in all aspects of clinical treatment, the life expectancy of DS patients has tripled during the past two decades; middle-aged DS patients are no longer a rare occurrence. Thus, despite the medical and technological advances of recent years, the prevalence of DS individuals in society is not likely to be significantly decreased in the near future. Although DS was described more than a century ago and the relationship between trisomy 21 and the Down's phenotype has been known for more than 30 years (Lejeune et al. 1959), very little is known about the way in which the additional chromosome 21 causes the disease. The syndrome is distinguished from most other genetic disorders; the latter are for the most part the result of a defect in a single gene causing a reduction in the activity of a gene product, whereas in DS, the wide range of symptoms is caused by an overexpression of several otherwise normal genes. The current concept is that the presence of extra copies of chromosome 21 genes results in the synthesis of increased amounts of gene products, which creates an imbalance in various biochemical pathways; this in turn causes the physiological defects giving rise to the clinical picture of the syndrome. One of the main research efforts in the molecular genetics of DS is directed toward cloning of genes residing on chromosome 21 (Patterson and Epstein 1990) and relating the consequences of their enhanced expression to the clinical symptoms of the syndrome through the use of model systems (Cooper and Hall 1988; Epstein et al. 1990; Groner and Elson 1991).

MODEL SYSTEMS FOR GENE DOSAGE EFFECTS OF CUZNSOD IN DS

CuZnSOD is a key enzyme in the metabolism of oxygen free radicals (see other chapters in this volume). Overexpression of this gene in DS may upset the steady-state equilibrium of active oxygen species within the cell, resulting in oxidative damage to biologically important molecules. Indeed, Yim et al. (1990) have recently reported that CuZnSOD is able to catalyze the formation of hydroxyl radicals from hydrogen peroxide, and others have shown that elevated levels of SOD actually enhance the cytotoxicity of active oxygen radicals (Finazzi-Agro et al. 1986; Scott et al. 1987; Iwahashi et al. 1988). Excess oxidative activity produced by increased activity of CuZnSOD could in part be responsible for the phenotypic defects found in DS (Sinet 1982; Groner et al. 1985). The possible involvement of CuZnSOD overproduction in the etiology of the syndrome was investigated by us during recent years through the use of two types of model systems.

The reasons we tried to develop a suitable system for studying the molecular events underlying gene dosage effects in DS are twofold. The first stems from difficulties attendant in research on humans. Most of the pathological consequences of trisomy 21 are manifested during fetal development; research on human subjects, especially in utero, is ethically complicated and practically impossible. The second reason has to do with the need to identify and sort out the quintessence from the large number of genes residing on chromosome 21. It is not clear how many genes are involved in determining the characteristic DS phenotype and which one is doing what. Two types of model systems were developed: a cellular system, consisting of cultured cells stably transfected and overexpressing the CuZnSOD gene, and an animal model, employing transgenic mice harboring the h-CuZnSOD gene and producing increased levels of active enzyme.

TRANSFECTED CELLS WITH INCREASED ACTIVITY OF CUZNSOD HAVE ALTERED PROPERTIES

When the CuZnSOD gene is introduced by expression vector into cultured animal cells, the recipients resemble trisomy 21 cells except for one important difference: The imbalance is limited to one particular gene, rather than affecting the whole chromosome. A cellular system of this type permits the study of the biochemical effects of the altered dosage of CuZnSOD in a defined background, irrespective of the overexpression of other chromosome 21 genes. The vector constructed, pg-SOD-SV*neo* (Fig. 1), contained a 12-kb *Eco*RI-*Bam*HI fragment en-

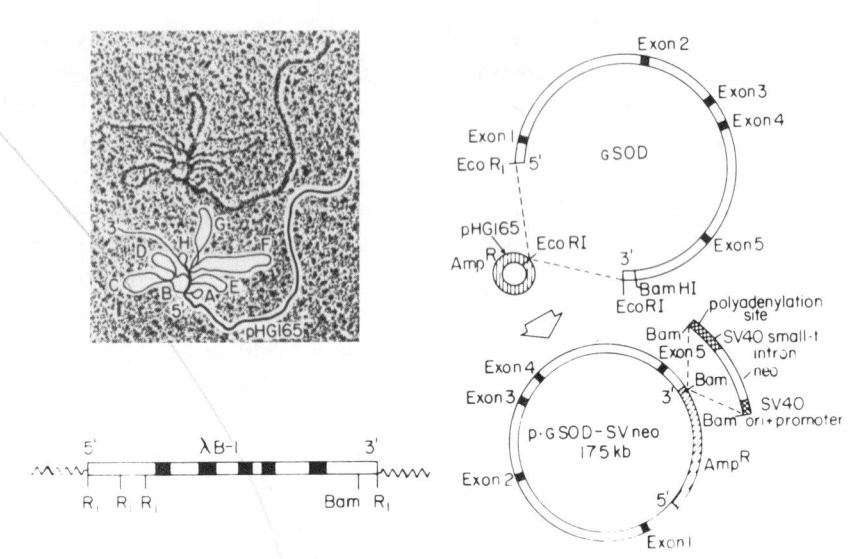

FIGURE 1 Structure of h-CuZnSOD gene and recombinant plasmid. (*Right*) Shuttle vector for expression of the CuZnSOD gene. Hatched segments represent the plasmid pG165 DNA that contains the pBR322 origin of DNA replication, the β-lactamose gene, and a polylinker. Open segment represents the *neo* gene, and the crosshatched segment represents the SV40 transcription regulatory elements. Closed segments indicate the exons of the CuZnSOD gene. (*Left*) Electron micrograph and tracing of heteroduplex molecules between the CuZnSOD cDNA and the genomic fragment present in pg-SOD-SV*neo* that was also used for microinjecting mouse fertilized eggs. (Reprinted, with permission, from Groner et al. 1990.)

compassing the h-CuZnSOD gene including its regulatory sequences (Levanon et al. 1985) and a 2.7-kb *Bam*HI fragment containing the *neo* transcriptional unit. pg-SOD-SV*neo* was introduced into mouse L cells, and stable transformants expressing elevated levels of authentic enzymatically active h-CuZnSOD were isolated. The L-SOD clones exhibited altered properties; they were resistant to the toxic effect of paraquat and had increased lipid peroxidation, an indication for altered oxidative balance (Elroy-Stein et al. 1986). The pg-SOD-SV*neo* vector was also transfected into rat pheochromocytoma PC12 cells, and clones expressing increased amounts of h-CuZnSOD were obtained. Although outwardly maintaining their response to nerve growth factor and their typical appearance of cultured neurons, the cells expressing the extra gene had a greatly reduced capacity to take up certain types of neurotransmitters like dopamine (DA) or norepinephrine (NE), whereas the uptake of other neurotransmitters like ascorbate or choline was normal (Fig. 2).

Following detailed analysis of the phenomenon, it was discovered that in the transformant-CuZnSOD cells, the chromaffin granules (the cellular organelles responsible for accumulating neurotransmitters), have a lesion in the transport mechanism. It was found that the pH gradient (ΔpH) across the membrane, which is the main driving force for neurotransmitter transport, was diminished in the PC12-SOD granules (Fig. 3A). This deficiency could have important consequences for neurons in the central nervous system that use a similar organelle (the synaptic vesicle) for accumulation of neurotransmitters. If a released transmitter substance persists for an abnormally extended period, new signals cannot get through at the proper rate. This observation demonstrated that even at the cellular level, an imbalance in the expression of the CuZnSOD gene has a deleterious effect which, if it occurs in the central nervous system, produces alterations in neuron function; this would impair the transduction of signals and mimic defects found in DS (Elroy-Stein and Groner 1988). The PC12-SOD clones were also found to have impaired biosynthesis of prostaglandin E_2 (PGE_2). Prostaglandins (PGs) are unsaturated lipids derived from membrane-bound arachidonic acid, a known substrate for lipid peroxidation. The biosynthetic pathway of PGs involves free radical-mediated reactions and is greatly influenced by the oxidative balance within the cell. As shown in Figure 3B, the synthesis of PGE_2 was significantly reduced in PC12-SOD clones. The rate-limiting step in the biosynthesis of PGs is the cleavage of arachidonic acid from cellular phospholipids. This step is usually induced by calcium ionophores and cholinergic agonists. Addition of the calcium ionophore A23187 or the cholinergic agonist carbachol to PC12 cells greatly enhanced the biosynthesis of PGE_2 (Fig. 3B), but even in the presence of these inducers, the formation of PGE_2 by PC12-SOD amounted to only 50% of the noninduced PC12 control, indicating that

△--△ - PC12-Con, ▲--▲ - PC12-hSOD.

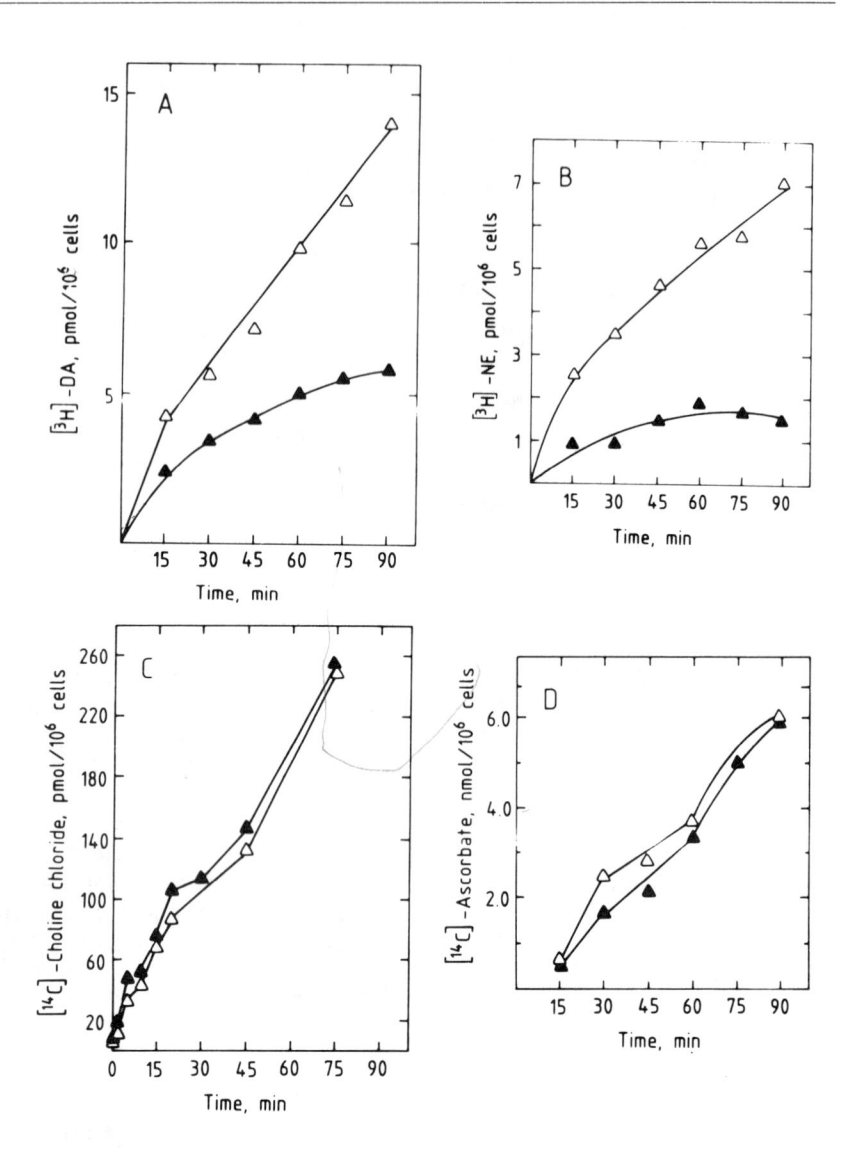

FIGURE 2 Accumulation rate of neurotransmitters by PC12-SOD transformants. Cultures were exposed for the lengths of time indicated to medium containing (A) 2.3 × 10^{-7} M ^{3}H-DA, (B) 10^{-7} M ^{3}H-NE, (C) 1 μM ^{14}C-cholin, and (D) 50 μM ^{14}C-ascorbate. Each point represents the average of two determinations on duplicate cultures. (Open triangles) PC12-Con; (closed triangles) PC12-SOD. (Reprinted, with permission, A and B from Elroy-Stein and Groner [1988, copyright held by Cell Press] and C and D from Groner et al. [1990].)

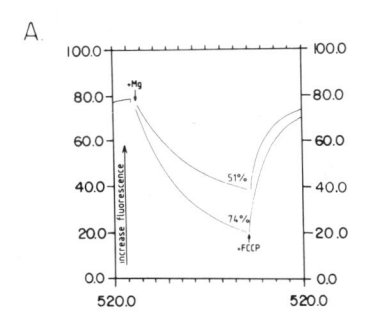

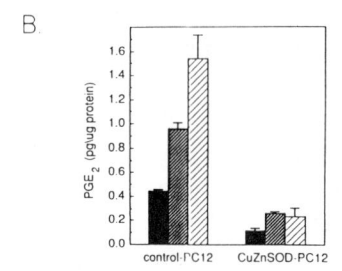

FIGURE 3 CuZnSOD gene dosage effects in PC12-SOD transformants. (A) Proton uptake by chromaffin granules measured by acridine orange fluorescence quenching. Reactions were initiated by addition of MgSO$_4$ and terminated after 10 min by addition of FCCP. The quenching of fluorescence due to acidification amounted to 74% for PC12-Cont and 51% for PC12-SOD. (B) Release of PGE$_2$ from control and PC12-SOD cells. A representative experiment, performed in triplicate. The vertical bars on top of the histograms represent the S.E.M.. The amount of PGE$_2$ is expressed as pg PGE$_2$ released per mg cell protein. Untreated (solid bars); treated with A23187 (heavy hatching) or with carbachol (light hatching). (Reprinted, with permission, from Elroy-Stein and Groner 1988 [copyright held by Cell Press].)

the process of generating free arachidonic acid was not impaired. Similar reduction of PGE$_2$ was measured in fibroblasts obtained from DS patients, who bear, among other anomalies, elevated levels of CuZnSOD. In addition, primary cells isolated from three lines of transgenic CuZnSOD mice showed similar reduction of PGE$_2$ release (see below). The findings strongly suggest that gene dosage of CuZnSOD causes a reduction in the formation of PGE$_2$ and link it to a similar reduction observed in trisomy 21 fibroblasts.

TRANSGENIC CUZNSOD MICE WITH ELEVATED LEVELS OF THE ENZYME DISPLAY PHENOTYPIC FEATURES OF DS

Gene transfer into mice, leading to the creation of transgenic mice, can be achieved by microinjecting a foreign DNA into a fertilized egg. The exogenously introduced gene becomes stably integrated into the mouse chromosome, and the resultant embryo develops into a mouse that carries an extra gene and transfers it to subsequent generations in a Mendelian fashion. Transgenic mice harboring the h-CuZnSOD gene have an advantage over transfected cultured cells in that they resemble more closely the natural situation; the transgene is present in every cell of the animal, and its influence is manifested throughout its entire developmental history. The results obtained with the transfected cells suggest that transgenic mice overexpressing CuZnSOD provide an animal model for investigating whether increased CuZnSOD has adverse phenotypic effects. Transgenic CuZnSOD mice harboring the h-CuZnSOD gene were produced by microinjecting fertilized eggs with a linear 14.5-kb EcoRI-BamHI fragment (Fig. 1) of human genomic DNA containing the entire gene, including its regulatory sequences (Epstein et al. 1987). Several strains of transgenic mice, designated TgHS-41, -51, -69, and -70 and carrying from one to several copies of the h-CuZnSOD gene, were obtained. These animals expressed the transgene in a manner similar to that of humans, with two RNA transcripts of 0.9 kb and 0.7 kb (Sherman et al. 1983, 1984). The level of RNA in the transgenic animals correlated well with the activity of the human enzyme assessed after electrophoresis, which separated the human and mouse enzymes (Fig. 4A). Increased CuZnSOD activity was recorded in nearly all tissues of the TgHS strains. Although the ratio of transgenic h-CuZnSOD to control activity varied from tissue to tissue, it remained constant in successive generations of heterozygous offspring. In all TgHS lines, expression was high in brain and relatively low in liver (Fig. 4A). Comparable with other transgenic systems, there was no correlation between the number of gene copies integrated and the observed level of transgene expression. This situation likely results from a combination of two factors: (1) the influence of the site of integration on the regulation of ex-

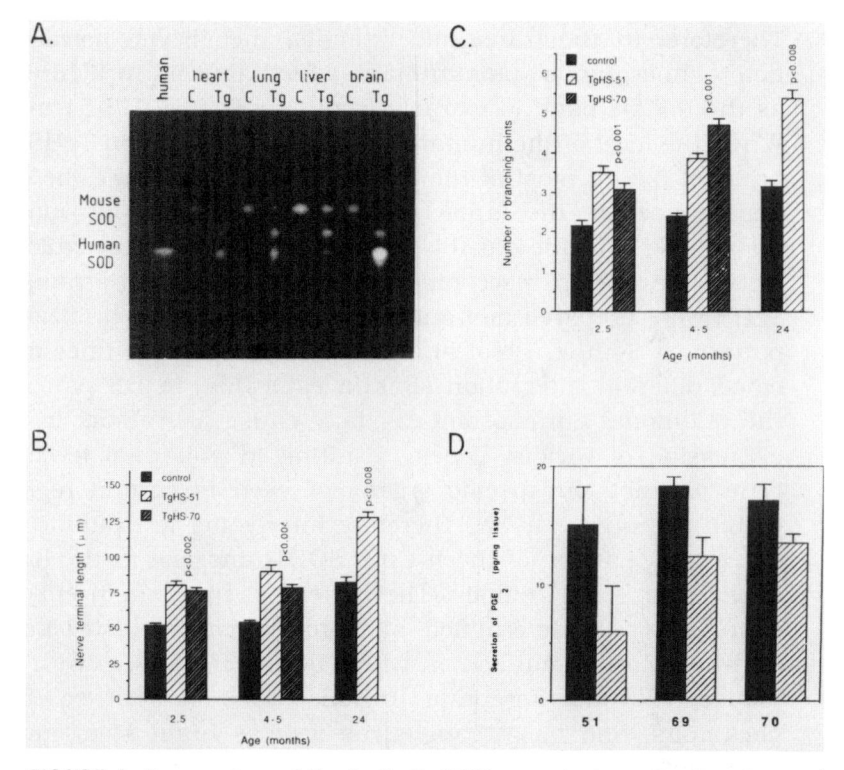

FIGURE 4 Expression of the h-CuZnSOD gene in transgenic mice and the consequent phenotypic effects. (A) Polyacrylamide gel analysis of CuZnSOD enzymatic activity in tissues of the transgenic mice: TgHS 51 (Tg lanes) and littermate (C lanes). (B,C) NMJ morphometry in soleus muscle. (B) Nerve terminal length and (C) number of nerve terminal branching points. At each age, 4–6 mice were analyzed. Results are expressed as mean ± S.E.M. The changes between controls and transgenics are indicated. Control mice (2.5 and 4–5 months old) differed significantly from 24-month-old controls ($p<0.001$). Transgenic mice (2.5 and 4–5 months old) differed significantly from 24-month-old transgenics ($p<0.001$). (D) Release of PGE_2 from kidneys of transgenic CuZnSOD and nontransgenic control mice. Experiments were performed in triplicate with each of the three strains 51, 69, and 70 of TgHS mice. The vertical bars on the top of the histograms represent the S.E.M. The amount of PGE_2 released is expressed as pg PGE_2 per mg of tissue. Nontransgenic (solid bars); transgenic CuZnSOD (hatched bars). (Reprinted, with permission, *A* from Groner et al. [1990] and *B*, *C*, and *D* from Avraham et al. [1991].)

pression and (2) alterations in the DNA sequences of some of the integrated transgenes when multiple copies are present. CuZnSOD is a dimer composed of two identical subunits. Therefore, in the transgenic animals, the enzyme formed a heterodimer (mouse plus human), which appears in Figure 4A as the middle band in extracts from tissues of TgHS-51 mice. When the level of the human enzyme was high, as in TgHS-51 brain, it bound most of the mouse subunits, causing the disappearance of the upper band representing the mouse homodimer. To rule out the possibility that the phenotypic effects were due to insertional mutagenesis, several strains of TgHS were prepared and analyzed. The different hybridization pattern in Southern blot analysis of the transgenic lines indicated different integration sites in each line. In DS patients, the additional chromosome 21 may cause alterations in the expression of various genes, resulting in abnormal levels of gene products in specific organs or even in certain regions within the organ. It was therefore interesting to examine the expression pattern of the h-CuZnSOD transgene in the brain of the transgenic animals. The pattern of the human enzyme expression was determined by immunohistochemistry using anti-h-CuZnSOD antibodies, which do not cross-react with the mouse enzyme. Forebrain regions such as cortex, hippocampus, and basal ganglia, as well as brain stem nuclei (substantia nigra, central gray, locus coeruleus) exhibited higher expression of the transgene as compared to other regions. This expression pattern was similar to that of the mouse enzyme visualized in control littermate mice. These results indicated that overexpression of the h-CuZnSOD was confined to brain regions in which the resident mouse gene was expressed. Outwardly, the transgenic CuZnSOD mice appeared normal, without any obvious deformities. This was not surprising, because there was no reason to expect that elevation of CuZnSOD activity alone would cause the major dysmorphic features of DS. Rather, it was anticipated that overexpression of CuZnSOD would affect more subtle aspects of tissue function and integrity, particularly in those tissues that might be affected by altered metabolism of oxygen free radicals. Bearing in mind the effect observed in the cellular system (PC12-SOD), the neurotransmitter serotonin was quantitated in the blood of the TgHS mice and was found to be significantly lower than the value in nontransgenic littermate mice (Schikler et al. 1989). This observation generated much interest, because reduced concentration of blood serotonin has for some time been a well-known clinical symptom among DS patients. When this deficiency was first noticed in the 1960s, it aroused considerable attention because of the possible relevance of serotonin uptake by blood platelets to neurotransmitter function in the central nervous system, and hence its involvement in the hypotonia and mental retardation of DS (Bazelon et al. 1967, 1968). At that time, attempts were made to raise the levels of blood serotonin in DS infants by administration of its precursor, 5-hydroxytryptophan; muscular tone, motor activity, and sleep abnormalities were reported to improve concomitantly with its administration. However, the development of infantile spasms, a severe seizure syndrome, in 17% of the patients receiving the drug brought these studies to a halt (Coleman 1971). Serotonin is an important neurotransmitter in the central nervous system, both in the embryonic state and in infants. It usually does not appear free in circulating blood because of its efficient uptake by platelets, where it is accumulated and stored in the dense granules. Detailed analysis of platelets isolated from the transgenic CuZnSOD mice revealed that the uptake process in these granules was impaired, and this caused the reduced concentration of blood serotonin in these animals. The dense granules of the platelets are in many respects similar to the chromaffin granules of PC12 cells, which, as described above, were affected by the increased activity of CuZnSOD. It is intriguing that this same lesion appears both in the PC12-SOD cellular system and in the transgenic CuZnSOD mice and that the consequent defect is a well-known deficiency diagnosed in DS.

The transgenic CuZnSOD mice also had abnormalities in the connections between nerve terminals and the muscles, the so-called neuromuscular junctions (NMJ). Such a defect was first noticed in the tongue muscle of DS patients. In many of them, the large protruding tongue is a striking clinical feature. The reasons for the protrusion have been considered as related to hypotonia of the lower lip and localized tongue enlargement in the region of the lingual tonsil. Studies conducted by Yarom and co-workers (1986, 1987) on the tongue muscle of patients with DS revealed pathological changes in the NMJ. The pathology consisted of terminal axon degeneration and changes in the end plates and myofibers. In addition, significant increases in the concentrations of copper and calcium were recorded in the DS tongues. It was suggested that the latter may be due to the abnormally high levels of CuZnSOD in these individuals. When the tongue muscle of the TgHS mice was examined, it was found that the NMJ exhibited significant pathological changes: withdrawal and destruction of some terminal axons and the development of multiple small terminals. The ratio of terminal axon area to postsynaptic membrane decreased, and secondary folds were often complex and hyperplastic (Avraham et al. 1988). These ultrastructural changes closely resembled the lesions found in the tongue muscle of DS, thus correlating gene dosage of CuZnSOD with specific phenotypic effect found in patients with DS. The studies on deterioration of NMJ were extended to the hindlimb muscles of the TgHS mice. Three parameters of NMJ morphology were examined: nerve terminal length, number of nerve terminal branching points, and incidence of sprouting that results in synapse formation. It was found that these parameters increased with advanced age and that the increase occurred earlier in the transgenic CuZnSOD mice (Fig. 4B,C). These results indicated that the transgenic mice were undergoing premature aging with respect to NMJ morphology, most probably due to a gene dosage effect of CuZnSOD. These data constitute an example of how CuZnSOD gene dosage accelerated a naturally occurring deteriorative process (Avraham et al. 1991). Following the observation that PC12-SOD and DS fibroblasts exhibited reduced biosynthesis of PGE_2, cells and organs of transgenic CuZnSOD mice were analyzed for prostaglandin release. Primary cell cultures prepared from embryos of TgHS mice synthesized and released PGE_2 to the external medium at a greatly reduced rate. The decrease was similar to that recorded for human trisomy 21 fibroblasts and PC12-SOD cells. The impaired biosynthesis of prostaglandins was not confined to cells grown in culture, since secretion of PGE_2, the major metabolite of arachidonic acid in the kidney, was significantly lower in transgenic kidneys as compared with nontransgenic littermate mice (Fig. 4D). These findings show that overexpression of the CuZnSOD gene induces a demotion in PGE_2 formation in both the cellular and animal models and establish a connection between alteration of prostaglandin biosynthesis in trisomy 21 cells and gene dosage of CuZnSOD (Minc-Golomb et al. 1991).

DISCUSSION

Considering the data obtained with the two model systems, it seems that increased CuZnSOD activity causes alterations of structural and functional elements that are responsible for neurotransmission efficacy and may therefore contribute to the neurobiological abnormalities found in DS. For example, several previous reports described specific abnormalities in cells of DS subjects, which are reminiscent of the diminished amine uptake. Yates and co-workers (1983) found a substantial reduction in the content of neurotransmitters, particularly norepinephrine in nerve terminals of DS brains. The low blood serotonin discussed above is another example. In addition, the abnormal electrophysiological properties found in both DS (Scott et al. 1983; Nieminen et al. 1988) and mouse trisomy 16 (Orozco et al. 1988; Ault et al. 1989) neurons in culture have been attributed to a defect in the membrane K^+ and Na^+ permeability.

The biochemical mechanism by which elevated activity of CuZnSOD produces the deleterious effects described above is not yet clear. Elevated levels of CuZnSOD could cause a local excess of hydrogen peroxide, which may be further converted to hydroxyl radicals (Yim et al. 1990). For example, the electron transfer process within the PC12 chromaffin granules that involves the enzyme dopamine β-hydroxylase, the electron donor ascorbate, and the transmembrane electron carrier cytochrome b-561, generates substantial amounts of superoxide which, upon dismutation, could increase the steady-state level of hydrogen peroxide in PC12-SOD. It is possible that the reduced ΔpH in the PC12-SOD granules and in TgHS platelets is due to a slippage of the proton translocating

ATPase, caused by oxidation of its essential SH group. The enzyme can also act as a peroxidase catalyzing the peroxidation of fatty acids (Hodgson and Fridovich 1975), or it can simply eliminate superoxide radicals that are actually needed for proper cellular metabolism. Increased CuZnSOD activity in transfected cells resulted in enhanced lipid peroxidation (Elroy-Stein et al. 1986; Elroy-Stein and Groner 1988), conceivably due to an increase in the steady-state concentration of H_2O_2 and production of •OH radicals. The cascade of prostaglandin biosynthesis constitutes a controlled radical process highly sensitive to the oxidative state. Low doses of H_2O_2, for example, produce rapid inhibition of cyclo-oxygenase activity, a key step in the biosynthesis of PGE_2 (Whorton et al. 1985).

Moving to DS, a pathogenetic mechanism attempting to explain and resolve how trisomy 21 causes the abnormalities in morphogenesis and mental function should combine information regarding genes overexpressed in DS with means to relate the increased amount of their gene products to the genesis of specific defects found in the syndrome. By overexpressing individual genes in transgenic mice and transfected cells, it might be possible to dissect the trisomy phenotype gene by gene. The data summarized here illustrate the potential of this approach. Exploration of those model systems will further increase as more candidate genes from chromosome 21 are cloned. Although exacting and technically tedious, it represents the most direct approach to the issue of relating gene dosage of a particular chromosome 21 gene with specific phenotypic features found in the syndrome. Transfer of cloned candidate genes to the mouse germ line, followed by a mating program between strains of transgenic mice, will enable the development of a mouse strain that overexpresses several transgenes. Eventually, a battery of transgenic strains with a full complement of the DS locus triplicated will become available. Such an animal model will not only lead to identification of the genes participating in the etiology of DS, but may also provide a test system for therapeutic or ameliorative procedures. Although the application of gene therapy for DS is still remote, it may include genetic means to eliminate the overexpression of genes that have been identified as causing clinical symptoms.

Such approaches require gene targeting; homologous recombination between DNA sequences residing in chromosome 21 and newly introduced DNA. This technique may one day permit a selective inactivation of the extra gene copies that contribute to the Down's phenotype. Currently, this technology is being applied in introducing mutations that will silence the human CuZnSOD carried by the transgenic mice.

ACKNOWLEDGMENTS

The work described in this chapter was supported by grants from the National Institutes of Health (HD-21229), the Minerva Foundation (Munich, Germany), and the Weizmann Institute Leo and Julia Forchheimer Center of Molecular Genetics.

REFERENCES

Ault, B., P. Caviedes, and S.I. Rapoport. 1989. Neurophysiological abnormalities in cultured dorsal root ganglion neurons from the trisomy 16 mouse fetus, a model for Down syndrome. *Brain Res.* **485:** 167.

Avraham, K.B., H. Sugarman, S. Rotshenker, and Y. Groner. 1991. Down syndrome: Morphological remodeling and premature aging at the neuromuscular junction of transgenic CuZn-superoxide dismutase mice. *J. Neurocytol.* **20:** 208.

Avraham, K.B., M. Schickler, D. Sazpoznikov, R. Yarom, and Y. Groner. 1988. Down's syndrome: Abnormal neuromuscular junction in tongue of transgenic mice with elevated levels of human Cu/Zn-superoxide dismutase. *Cell* **54:** 823.

Bazelon, M., A. Barnet, A. Lodge, and S.A. Shelbourn. 1968. The effects of high doses of 5 hydroxytryptophan on a patient with trisomy 21. *Brain Res.* **11:** 397.

Bazelon, M., R.S. Paine, V.A. Corvie, P. Hunt, J.H. Houck, and D. Mahanand. 1967. Reversal of hypotonia in infants with Down syndrome by administration of 5 hydroxytryptophan. *Lancet* **I:** 1130.

Coleman, M. 1971. Infantile spasms associated with 5-hydroxytryptophan administration in patients with Down syndrome. *Neurology* **21:** 911.

Cooper, D.N. and C. Hall. 1988. Down's syndrome and the molecular biology of chromosome 21. *Prog. Neurobiol.* **30:** 507.

Elroy-Stein, O. and Y. Groner. 1988. Impaired neurotransmitter uptake in rat PC12 cells overexpressing human Cu/Zn-superoxide dismutase. Implication for gene dosage effects in Down's syndrome. *Cell* **52:** 259.

Elroy-Stein, O., Y. Bernstein, and Y. Groner. 1986. Overproduction of human Cu/Zn-superoxide dismutase in transfected cells: Extenuation of paraquat-mediated cytoxicity and enhancement of lipid peroxidation. *EMBO J.* **5:** 615.

Epstein, C.J., ed. 1986. The consequences of chromosomal imbalance. In *Principles, mechanisms and models.* Cambridge University Press, New York.

Epstein, C.J., N.C. Berger, J.E. Carlson, H.P. Chan, and T.-T. Huang. 1990. Models for Down syndrome: Chromosome 21-specific genes in mice. *Prog. Clin. Biol. Res.* **360:** 215.

Epstein, C.J., K.B. Avraham, M. Lovett, S. Smith, O. Elroy-Stein, G. Rotman, C. Bry, and Y. Groner. 1987. Transgenic mice with increased CuZn-superoxide dismutase activity: An animal model of dosage effects in Down's syndrome. *Proc. Natl. Acad. Sci.* **84:** 8044.

Finazzi-Agro, A., A. Diziulio, G. Amicrosante, and C. Grifo. 1986. Photohemolysis of erythrocytes enriched with superoxide dismutase, catalase and glutathione peroxidase. *Photochem. Photobiol.* **43:** 409.

Groner, Y. and A. Elson. 1991. Down syndrome molecular genetics. In *Encyclopedia of human biology* (ed. R. Delbecco), p. 615. Academic Press, New York.

Groner, Y., O. Elroy-Stein, K.B. Avraham, R. Yarom, M. Schickler, H. Knobler, and G. Rotman. 1990. Down syndrome clinical symptoms are manifested in transfected cells and transgenic mice overexpressing the human CuZn-superoxide dismutase gene. *J. Physiol.* **84:** 53.

Groner, Y., O. Elroy-Stein, Y. Bernstein, N. Dafni, D. Levanon, E. Danciger, and A. Neer. 1986. Molecular genetics of Down's syndrome: Overexpression of transfected human CuZn-superoxide dismutase gene and the consequent physiological changes. *Cold Spring Harbor Symp. Quant. Biol.* **51:** 381.

Groner, Y., J. Lieman-Hurwitz, N. Dafni, L. Sherman, D. Levanon, Y. Bernstein, E. Danciger, and O. Elroy-Stein. 1985. Molecular structure and expression of the gene locus on chromosome 21 encoding the Cu/Zn superoxide dismutase and its relevance to Down's syndrome. *Ann. N.Y. Acad. Sci.* **450:** 133.

Hodgson, E.K. and I. Fridovich. 1975. The interaction of bovine superoxide dismutase with hydrogen peroxide: Chemiluminescence and peroxidation. *Biochemistry* **14:** 5299.

Iwahashi, H., T. Ishii, R. Sugata, and R. Kido. 1988. Superoxide dismutase enhances the formation of hydroxyl radicals in the reaction of 3-hydroxyanthranilic acid with molecular oxygen. *Biochem. J.* **251:** 893.

Lejeune, J.M., M. Gautier, and R. Turpin. 1959. Etude des chromosomes somatique de neuf enfants mongoliens. *C.R. Acad. Sci.* **248:** 1721.

Levanon, D., J. Lieman-Hurwitz, N. Dafni, M. Wigderson, L. Sherman,

Y. Bernstein, Z. Laver-Rudich, E. Danciger, O. Stein, and Y. Groner. 1985. Architecture and anatomy of the chromosomal locus in human chromosome 21 encoding the Cu/Zn superoxide dismutase. *EMBO J.* **4:** 77.

Minc-Golomb, D., H. Knobler, and Y. Groner. 1991. Gene dosage of CuZnSOD and Down syndrome: Diminished prostaglandins synthesis in human trisomy 21, transfected cells and transgenic mice. *EMBO J.* **10:** 2119.

Nieminen, K., B.A. Suarez-Isla, and S.I. Rapoport. 1988. Electrical properties of cultured dorsal root ganglion neurons from normal and trisomy 21 human fetal issue. *Brain Res.* **474:** 246.

Orozco, C.B., C.J. Epstein, and S.I. Rapoport. 1988. Voltage activated sodium conductances in cultured normal and trisomy 16 dorsal root ganglion neurons from the fetal mouse. *Dev. Brain Res.* **38:** 265.

Patterson, D. and C.J. Epstein, eds. 1990. Molecular genetics of chromosome 21 and Down syndrome. *Prog. Clin. Biol. Res.* **360**.

Schickler, M., H. Knobler, K.B. Avraham, O. Elroy-Stein, and Y. Groner. 1989. Diminished serotonin uptake in platelets of transgenic mice with increased Cu/Zn-superoxide dismutase activity. *EMBO J.* **8:** 1385.

Scott, B.S., L.E. Becker, and J.L. Petit. 1983. Neurobiology of Down's syndrome. *Prog. Neurobiol.* **21:** 199.

Scott, M.D., S.R. Meshnick, and J.W. Eaton. 1987. Superoxide dismutase-rich bacteria: Paradoxical increase in oxidant toxicity. *J. Biol. Chem.* **262:** 3640.

Sherman, L., N. Dafni, J. Lieman-Hurwitz, and Y. Groner. 1983. Nucleotide sequence and expression of human chromosome 21-encoded superoxide dismutase mRNA. *Proc. Natl. Acad. Sci.* **80:** 5465.

Sherman, L., D. Levanon, J. Lieman-Hurwitz, N. Dafni, and Y. Groner. 1984. Human Cu/Zn superoxide dismutase gene: Molecular characterization of its two mRNA species. *Nucleic Acids Res.* **12:** 9349.

Sinet, P.-M. 1982. Metabolism of oxygen derivatives in Down's syndrome. *Ann. N.Y. Acad. Sci.* **396:** 83.

Sinet, P.-M., D. Allard, J. Leujeune, and H. Jerome. 1974. Augmentation d'activitié de la superoxyde dismutase érithrocytaire dans la trisomie pur le chromosome 21. *C.R. Acad. Sci.* **278:** 3267.

Whorton, A.R., M.E. Montgomery, and R.S. Kent. 1985. Effect of hydrogen peroxide on prostaglandin production and cellular integrity in cultured porcine aortic endothelial cells. *J. Clin. Invest.* **76:** 295.

Yarom, R., U. Sagher, Y. Havivi, I.J. Peled, and M.R. Wexler. 1986. Myofibers in tongues of Down's syndrome. *J. Neurol. Sci.* **73:** 279.

Yarom, R., Y. Sherman, U. Sagher, I.J. Peled, and M.R. Wexler. 1987. Elevated concentrations of elements and abnormalities of neuromuscular junctions in tongue muscles of Down's syndrome. *J. Neurol. Sci.* **79:** 315.

Yates, C.M., J. Simpsin, A. Gordon, A.F.J. Maloney, Y. Allison, I.M. Ritchie, and A. Urguhart. 1983. Catecholamine and cholinergic enzymes in presenile and senile Alzheimer-type dementia and Down's syndrome. *Brain Res.* **280:** 119.

Yim, M.B., P.B. Chack, and E.R. Stadtman. 1990. Copper-zinc superoxide dismutase catalyses hydroxyl radical production from hydrogen peroxide. *Proc. Natl. Acad. Sci.* **87:** 5006.

Index

AAPH. *See* 2,2′-Azobis(2-amidinopropane) hydrochloride
Aconitase, 215–217, 238
Active oxygen. *See* Oxidants
Aging, 1–2, 10, 12, 23
Agrobacterium tumefaciens, 142–144
Albumin. *See* Antioxidant
Anaerobiosis, 109, 250–251
Antioxidant. *See also* Ascorbate peroxidase; Catalase; Glutathione peroxidase; Superoxide dismutase (SOD)
  AAPH scavenging, 38
  albumin, 23, 27, 37
  ascorbic acid (vitamin C), 7, 11, 15, 27–35, 38–39
  bilirubin, 27, 30–35, 39
  β-carotene, 7, 24, 27, 38–39
  copper-induced oxidant scavenging, 37–38
  deficiency, 10
  ferredoxin, 118
  folic acid, 12
  glutathione, 24–25, 27
  lutein, 39
  lycopene, 38–39
  peroxyl radical scavenging, 29–35
  phytofluene, 38
  protein thiols, 27–35
  singlet oxygen scavenging, 39
  α-tocopherol, 7, 24, 27–35, 38–40
  ubiquinols, 24, 27–28, 38, 40
  uric acid, 24, 26–34, 37, 39
Arc (aerobic respiration control), 245, 247–249
Arthritis, 23, 26, 34
Ascorbate. *See* Antioxidant, ascorbic acid
Ascorbate peroxidase, 119, 180–184, 186, 188–189
Ascorbic acid. *See* Antioxidant
Atherosclerosis, 23, 39–40
Autoimmune disease, 34
2,2′-Azobis(2-amidinopropane) hydrochloride (AAPH), 29, 31–33, 36, 38

Bilirubin. *See* Antioxidant
Blood plasma, 24, *31*
Brucella abortus, 233, 236

cAMP, 109, 159
Cancer, 1–2, 10–11, 13, 15. *See also* Mitogenesis; Oxidative stress
Carcinogenesis, 8, 11
β-Carotene. *See* Antioxidant
Catalase
  Bacillus subtilis, 97, 111
  bacterial, 97
  barley mutant, 134
  DNA damage, inhibition, 49
  Escherichia coli, 97–101, 103–104, 107, 109, 169
  extracellular activity, 36
  maize (*Zea mays*), 120–124, 129–130, 132–138, 141–142
  peroxisomal, 185
  reaction, 119
  yeast (*Saccharomyces cerevisiae*), 154–165
Cell transformation. *See* Tumor induction
Cercosporin, 130, 142
Chlorella, 219
2-Chloroethylphosphoric acid. *See* Ethephon
Chloroplasts, 117–118, 173, 218
Cigarette smoke, 23, 34, 36, *48*
Copper, 26, 37, 49, 61

Italicized page numbers refer to figures or tables.

Copper-zinc superoxide dismutase (CuZnSOD)
  evolution, 184-185
  half-life, 76
  hydrogen peroxide inhibition, 178-179
  intracellular localization, 126, 184-185
  maize (*Zea mays*), 125, 129-*131*, 138-141
  overexpression, 265-275
  rice, 185
Crystallography. *See* Iron superoxide dismutase (FeSOD); Manganous superoxide dismutase (MnSOD)
Cyanide-resistant respiration, 122, 219
Cyanobacteria, 184, 186, 188-189, 233-234
Cycloheximide, 73-74, 159
Cytochrome *f*, 176
Cytochrome oxidase, 3

Dehydroascorbate reductase, 181-182
α,β-Dihydroxy acid dehydratase, 215, 237, 238
DNA damage
  assays, 58, 60
  measuring in humans, 13
  by neutrophils, 11
  by nuclease activation, 50-51
  oxidative, 3-6, 15, *48*-52, 55-57
  strand cleavage, 62
  by white blood cells, 2
DNA repair, 2, 3, 5, 13, 103
DNA supercoiling, 249
Down's syndrome, 263-277

Escherichia coli, 52, 82, 294. *See also* Catalase; Iron superoxide dismutase (FeSOD); Manganous superoxide dismutase (MnSOD)
Ethephon, 139

Fatty acid oxidation. *See* β-Oxidation

Fenton reaction, 26, 118, 231
Ferredoxin. *See* Antioxidant
Folate. *See* Antioxidant, folic acid
Folic acid. *See* Antioxidant
Free radical detection, 54, *58*
Free radicals. *See* Oxidants
Fur (ferric uptake regulation), 245-247

Gas chromatography (GC), 60
GC/MS. *See* Gas chromatography; Mass spectrometry
β-Glucuronidase, 143-144
Glutathione. *See* Antioxidant
Glutathione peroxidase, 7, 24-25, 186, 189
Glutathione reductase, 24, 119
Glutathione S-transferase, 24
Glycosylase. *See* DNA repair
Glyoxylate cycle, 118
Glyoxysomes, 121-122

Haber-Weiss reaction, 214, 221, 231
Heat shock, 160, 164-165
Heme, 26, 37, 154
Hemoglobin, 26, 37
High performance liquid chromatography (HPLC), 58
HPLC. *See* High performance liquid chromatography
Hydrogen peroxide, 3, 19, 34-36, 47, 117, 178-180
Hydroperoxidases, 97, 213, 231
8-Hydroxydeoxyguanosine (oh^{8}dG), 5-7
8-Hydroxyguanine (8-OH-Gua), 55-59
Hydroxyl radical
  inhibition of formation, 214
  production, 3, 26, 47, 49, 53
  reactions with DNA products, 55, 57
5-Hydroxymethyluracil, 3-4
Hypochlorous acid, 2, 14
Hypomethylation, 12

INDEX 283

Interleukin-1, 75, 81
Ionizing radiation, 52-53
Iron, 26, 37, 49, 118, 214, 231.
  See also Heme
Iron box, 246
Iron superoxide dismutase (FeSOD)
  crystal structure, 194
  Escherichia coli expression, 213
  hydrogen peroxide inactivation, 233
  inhibitor binding, 200-201
  metal coordination, 197
  reaction scheme, 204-206
  reduction, 200
  structure, 197, 201-204

katF, 103-110

Lipid peroxidation, 4, 23, 26, 31-35, 37, 57
Low-density lipoprotein (LDL), 23, 36-39
Lutein. See Antioxidant
Lycopene. See Antioxidant

Malondialdehyde (MDA), 57
Manganese. See Manganous superoxide dismutase (MnSOD)
Manganous superoxide dismutase (MnSOD)
  antisense RNA, 75
  crystal structure, 194, *196*
  half-life, 76
  induction, 74-79, 219
  intracellular localization, 126-127
  maize (*Zea mays*), 125, 144-146
  metal coordination, 197, *199*
  overexpression, 75-76, 82, 222
  reduction, 200
  regulation in *E. coli*, 219, 241-252
  structure, 197, 201-202
Mass spectrometry (MS), 60
Menadione, 51
Metal-binding proteins, 37

Metastases, 13-15
Mitochondria, 23, 84-87, 122, 126-127
Mitogenesis, 1, 8, 10-11, 13, 15
Monodehydroascorbate reductase, 181-182
Mouse (*Mus musculus*), 12, 265-275
Mutagens, 2, 5, 10, 12
Mutation, 1, 8-9, 15, 52

Near-ultraviolet radiation (NUV), 103, 239
Neurotransmitter, 267, 272, 275
Neutrophils. See White blood cells
Nocardia asteroides, 236

Oncogenes, 8, 11
Osmotic stress, 252
Oxidants, 3, 7-8, 14-15, 118
β-Oxidation, 118
Oxidative damage, 1, 11
Oxidative stress, 23, 33-37, 48, 53-54
Oxidative stress response, 231
OxyR, 100-102, 110, 216, 221-222
Ozone, 48, 131, 138

P700 (chlorophyll), 174
Paraquat, 48, 130, 144, 218-223
Peroxidases, H_2O_2-scavenging, 186-188. See also Ascorbate peroxidase; Glutathione peroxidase
Peroxisomes, 121, 153, 155-156
Peroxyl radical, 29-34
Phagocytic cells. See White blood cells
6-Phosphogluconate dehydratase, 215, 238
Phospholipase, 37
Photobacterium leiognathi, 233
Photoinhibition, 177-178, 233
Photorespiration, 118, 126
Photosynthetic efficiency, 173
Photosystem I, 174-*175*

Phytofluene. *See* Antioxidant
Plastocyanin, 176
Polymorphonuclear leukocytes. *See* White blood cells
Prostaglandins, 267, 276
Protein thiol. *See* Antioxidant

Reoxygenation injury, 23, 34, 76, 216
Reperfusion injury. *See* Reoxygenation injury

Sigma transcription factors, 103, 105
Singlet oxygen, 3, 23, 56, 117
soxR, 216, 221-223, 244
soxS, 222, 244
Spirogyra, 184
Starvation stress, 104, 158-160
Sulfur dioxide (SO_2), 138
Superoxide dismutase (SOD). *See also* Copper-zinc superoxide dismutase (CuZnSOD); Iron superoxide dismutase (FeSOD); Manganous superoxide dismutase (MnSOD)
  cambialistic, 235, 242
  cercosporin response, 142
  deficiency, 231
  excess, 232, 241
  extracellular activity, 36
  metal-binding specificity, 206-208, 234-235
  mutants, 52, 237-240
  prokaryotic, 233
  reaction, 119
  superoxide scavenging rate, 179
  three-dimensional structure, 193

Superoxide radical, 3, 34-36, 117, 174-177, 179, 215-217

Thermus thermophilus, 194
Thylakoid membranes, 175-177, 183
Thymine glycol, 3, 4
TNF-α. *See* Tumor necrosis factor
Tobacco, 82
α-Tocopherol. *See* Antioxidant
Total radical-trapping antioxidant parameter (TRAP), 29-31
Trisomy 21. *See* Down's syndrome
Tumor induction, 10, 53
Tumor necrosis factor (TNF), 69-75, 78-81, 83
Tumor suppressor genes, 8-9

Ubiquinol. *See* Antioxidant
Uric acid. *See* Antioxidant
Urine, 3, 5-6, 57

Vitamin C. *See* Antioxidants, ascorbic acid

White blood cells
  neutrophils, 11, 34, 36, 38
  phagocytic cells, 3, 11, *48*, 235
  polymorphonuclear leukocytes, 2, 14

Xanthine oxidase, 34-36, 117

Yeast (*Saccharomyces cerevisiae*), 144-146, 153. *See also* Catalase